PLANNING FOR TOURISM

Towards a Sustainable Future

CABI TOURISM TEXTS are an essential resource for students of academic tourism, leisure studies, hospitality, entertainment and events management. The series reflects the growth of tourism-related studies at an academic level and responds to the changes and developments in these rapidly evolving industries, providing up-to-date practical guidance, discussion of the latest theories and concepts, and analysis by world experts. The series is intended to guide students through their academic programmes and remain an essential reference throughout their careers in the tourism sector.

Readers will find the books within the CABI TOURISM TEXTS series to have a uniquely wide scope, covering important elements in leisure and tourism, including management-led topics, practical subject matter and development of conceptual themes and debates. Useful textbook features such as case studies, bullet point summaries and helpful diagrams are employed throughout the series to aid study and encourage understanding of the subject.

Students at all levels of study, workers within tourism and leisure industries, researchers, academics, policy makers and others interested in the field of academic and practical tourism will find these books an invaluable and authoritative resource, useful for academic reference and real world tourism applications.

Titles available

The Entertainment Industry: an Introduction
Edited by Stuart Moss

Practical Tourism Research
Stephen L.J. Smith

Leisure, Sport and Tourism, Politics, Policy and Planning, 3rd Edition
A.J. Veal

Events Management
Edited by Peter Robinson, Debra Wale and Geoff Dickson

Food and Wine Tourism: Integrating Food, Travel and Territory
Erica Croce and Giovanni Perri

Strategic Management in Tourism, 2nd Edition
Edited by L. Moutinho

Research Methods for Leisure, Recreation and Tourism
Edited by Ercan Sirakaya-Turk, Muzaffer Usyal, William E. Hammitt and Jerry J. Vaske

Facilities Management and Development for Tourism, Hospitality and Events
Edited by Ahmed Hassanien and Crispin Dale

Events as a Strategic Marketing Tool
Dorothé Gerritsen and Ronald van Olderen

Entertainment Management: Towards Best Practice
Edited by Stuart Moss and Ben Walmsley

Introduction to Tourism Transport
Sven Gross and Louisa Klemmer

Tourism Information Technology, 2nd Edition
Pierre Benckendorff, Pauline Sheldon and Daniel R. Fesenmaier

Planning for Tourism: Towards a Sustainable Future
Edited by Nigel D. Morpeth and Hongliang Yan

PLANNING FOR TOURISM

Towards a Sustainable Future

Edited by

Nigel D. Morpeth
Leeds Beckett University, Leeds, UK

Hongliang Yan
Leeds Beckett University, Leeds, UK

www.cabi.org

CABI is a trading name of CAB International

CABI
Nosworthy Way
Wallingford
Oxfordshire, OX10 8DE
UK

Tel: +44 (0)1491 832111
Fax: +44 (0)1491 833508
E-mail: info@cabi.org
Website: www.cabi.org

CABI
38 Chauncy Street
Suite 1002
Boston, MA 02111
USA

Tel +1 800 552 3083 (toll free)
Tel +1 (0)617 395 4051
E-mail: cabi-nao@cabi.org

A catalogue record for this book is available from the British Library, London, UK.

Library of Congress Cataloging-in-Publication Data

Planning for tourism : towards a sustainable future / edited by Nigel D. Morpeth, Leeds Beckett University, UK and Hongliang Yan, Leeds Beckett University, UK.
pages cm. -- (CABI tourism texts)
Includes bibliographical references and index.
ISBN 978-1-78064-458-5 (alk. paper)
1. Sustainable tourism. 2. Tourism--Planning. I. Morpeth, Nigel D. II. Hongliang, Yan.

G156.5.S87P53 2015
910.68'4--dc23

2014041432

ISBN-13: 978 1 78064 458 5

Commissioning editor: Claire Parfitt
Editorial assistant: Alexandra Lainsbury
Production editor: Lauren Povey

Typeset by SPi, Pondicherry, India.
Printed and bound by Gutenberg Press Ltd, Tarxien, Malta.

Contents

Contributors

Dzintra Atstāja is Professor at the Department of Entrepreneurship and IT at the BA School of Business and Finance, Latvia, Doctor of Social Sciences in Economics and scientific expert in Management Sciences of the Latvian Council of Science. She is one of the founders and is a board member of the Latvian Economic Association, a member of the Green Economics Institute, an expert at the IFO Institute for Economic Research in Munich providing regular Latvian macroeconomic expertise, and a member of the network of SMART civil society organizations and the Society of Professors of Latvia's universities. Dzintra specializes in macroeconomics, green solutions and sustainable development, business control systems and management, civil defence and work safety. She has more than 25 years of experience in business consulting, more than 80 publications and is editor or co-author of a range of books. Mailing address: School of Business and Finance, K. Valdemara str. 161, Riga, LV1013, Latvia. E-mail: dzintra.atstaja@ba.lv

Kathryn Beardmore is currently Director of Park Services for the Yorkshire Dales National Park Authority. Throughout her 25-year career Kathryn has had extensive involvement in managing and promoting recreation and tourism in England's National Parks having worked for three National Park Authorities. Kathryn's work has also involved policy development and implementation at both a local and a national level. Over the past 3 years, Kathryn has been instrumental in determining the future direction for tourism in the Yorkshire Dales by forming the Dales Tourism Partnership and guiding development of the Yorkshire Dales Sustainable Tourism Strategy. Kathryn's first-hand experience of exploring wild places, both here and overseas through climbing and other adventurous activities, ensures she always brings a passionate, but pragmatic approach to her work. Mailing address: Yorkshire Dales National Park Authority, Yoredale, Bainbridge, Leyburn, North Yorkshire, DL8 3EL, UK. E-mail: kathryn.beardmore@yorkshiredales.org.uk

Ivars Brīvers was born in Riga on 18 April 1957 and in 1980 graduated from the Latvian State University, Faculty of Economics, specializing in economic cybernetics. He is Professor of Economics in the BA School of Business and Finance, Riga, having become a Doctor of Economic Sciences in 1994. Topics of research include assessment and analysis of the macro-economic situation of Latvia, economic modelling and analysis. He is one of the founders of the Latvian Economic Association and in 2009 was Chairman of the Board of the Association, and he has been a member of the European Society for the History of Economic Thought since 2006. Since 1997 he has been an expert at the IFO Institute for Economic Research in Munich providing regular Latvian macroeconomic expertise and since 2011 he has been an economics expert on the Latvian Council of Science. In 2003 he became a member of the panel of the Terminology Commission in Economics of the Latvian Academy of Sciences. Mailing address: School of Business and Finance, K. Valdemara str. 161, Riga, LV1013, Latvia. E-mail: ivars.brivers@ba.lv

Ian Brown currently works for Canterbury City Council's Department of Planning as Head of Planning and Regeneration. He has wide-ranging responsibilities which include: development management and enforcement, conservation and countryside services, planning policy, transportation, building control, and land charges and strategic housing. He is project leader on a number of regeneration and joint developer partner projects. After graduating from the School of Architecture, Planning and Landscape at Newcastle University, he has worked in a range of multidisciplinary planning roles in a number of UK local authorities and also has a Masters degree which focused on the ecological and landscape design principles of planning. Mailing address: Head of Planning and Regeneration, Department of Planning, Canterbury City Council, Council Offices, Military Road, Canterbury, Kent, CTI 1YW, UK. E-mail: ian.brown@canterbury.gov.uk

Alan Clarke currently works at the University of Pannonia, in Hungary and has had a significant role in establishing the English versions of the full undergraduate and Masters programmes. He has a wide range of research and consultancy experience in tourism planning, strategy and sustainable development. He has published widely in sustainable tourism, cultural heritage, community festivals and events, religious tourism and semiotic understandings of these developments. His current research focuses on aspects of governance and stakeholder involvement. He has adopted the Balaton and its cultures as a favourite place, warmed by the sun, refreshed by the waters and challenged to know more about the wines of the region. Mailing address: University of Pannonia, Egyetem Utca 10, Veszprém, Hungary 8200. E-mail: alanhungary@hotmail.com

Neus Crous-Costa is an assistant professor and junior researcher at the Faculty of Tourism (University of Girona) and Tourism Officer at the Museu d'Art de Girona (Girona province art museum). She graduated in tourism studies and holds a Masters degree in tourism planning and management (cultural tourism programme). She is involved in several research projects about cooperation, history of tourism, religious tourism products and destination imagery.

Mailing address: Faculty of Tourism, University of Girona, Plaça Ferrater Mora, 1, 17071 Girona, Spain. E-mail: neus.crous@udg.edu

Sofia Teixeira Eurico has been a lecturer at the School of Tourism and Maritime Technology (ESTM), Polytechnic Institute of Leiria, Portugal, for the last 11 years. At present, she assumes the coordination of the Degree in Tourism, coordinates a working group dedicated to the theme of employability for tourism and is the co-editor of the journal EJTHR – *European Journal of Tourism, Hospitality and Recreation*. She obtained her PhD in tourism from the Faculty of Economics, University of Algarve and as an author has participated in several national and international scientific meetings, with some international publications in the field of tourism. She has reviewed scientific papers and participated in international projects. She mainly conducts research in the areas of employability; higher education in tourism; and ethics, social responsibility and sustainable tourism. Mailing address: Escola Superior de Turismo e Tecnologia do Mar, Instituto Politécnico de Leiria, Campus 4 – Santuário Nossa Senhora dos Remédios, 2520-641 Peniche, Portugal. E-mail: sofia.eurico@ipleiria.pt

Nour Farra-Haddad is a senior researcher, a tourism consultant, an author and she teaches at St-Josef University of Beirut (Lebanon) and at the American University of Science and Technology. She has experience in development and management with a strong business consultancy background and in-depth experience in the Middle East tourism industry. She is a specialist in pilgrimages and religious tourism. She is now managing her own travel consultancy company NEOS. Her credentials include a PhD in religious anthropology, a Masters in anthropology and two BSc degrees in archeology and anthropology. She holds a tourist guide diploma from the Ministry of Tourism and was designated as Vice President of the Lebanese Tour Guides Syndicate. She is also a founding member of the Association for the Development of Pilgrimages and Religious Tourism affiliated to the Maronite Patriarchate in Bkerke. Mailing address: NEOS Tourism, 5th Floor, Vanlian Building No. 12, Pasteur Street, Cityrama, Dekwaneh, Beirut, Lebanon. E-mail: nour@neoslb.com

Craig Grimes began working in tourism for disabled people through writing travel articles for various disability magazines in the UK. In 2001 he moved to Barcelona and established AccessibleBarcelona, an online guide and destination management company providing services such as accommodation bookings, adapted airport transfers, guided tours and equipment hire for disabled people. Through AccessibleBarcelona he developed innovative ways of providing tourism information for people with disabilities through the use of descriptive text, photographs, diagrams and video, utilized to show facilities for disabled people at hotels and tourism attractions. This video is not just for wheelchair users but for those from any disability perspective. Craig established Experience Community in 2011, which is a social enterprise to provide video guides, day trips and holidays for disabled people in the UK and it is also involved in making films and providing access advice to local authorities, large non-governmental organizations and National Parks. Craig also teaches undergraduate students at Leeds Becket University with a focus on issues in rural and city break tourism for disabled people.

Mailing address: Experience Community CIC, The Media Centre, 7 Northumberland Street, Huddersfield, HD1 1RL, UK. E-mail: cgrimes@experiencecommunity.co.uk

Filareti Kotsi, PhD, is an associate professor at the College of Communication and Media Sciences at Zayed University, Dubai, United Arab Emirates (UAE). She teaches a variety of courses related to tourism, heritage and culture as well as planning for events and festivals in the undergraduate and graduate specialization of 'Tourism and Cultural Communication'. She has worked in various universities in Europe (Greece, Cyprus and Belgium) and her research interests include religious tourism and pilgrimage, commodification of culture, reflexive anthropology and auto-ethnography, educational tourism, mobility and internationalization of higher education. She is currently working on UAE-related projects such as destination branding strategies. After receiving her BA in English literature from the Aristotle University of Thessaloniki, Greece, Dr Kotsi received a second bachelor degree and her MA in communications from the Catholic University of Louvain, Belgium and her doctorate in communications from the Ecole Normale Supérieure, Lyon, France. Mailing address: Zayed University, PO Box 19282, Dubai Academic City, Al Ruwayah, Dubai, United Arab Emirates. E-mail: filareti.kotsi@zu.ac.ae

Agita Līviņa is Director of the Economic, Social and Humanities Research Institute, Associate Professor at the Faculty of Social Sciences at the Vidzeme University of Applied Sciences, Latvia, and a member of the Latvian Association of Economics. She gained her PhD in regional economics in 2005. She has worked for 13 years in the university, teaching tourism planning, tourism strategic management, dark tourism, management and politics of nature parks and the student groups consultancy module. She reviews doctoral theses in different universities in Latvia. Her research interests are tourism planning, protected areas, sustainable development, dark tourism and regional planning. She received the US Fulbright Fellowship for research in 2012 and Fellowships of the Government of India for the International Training Programme in 2014. She likes trekking, biking, boating and to discover new places together with the family. Mailing address: Institute of Social, Economic and Humanities Research, Vidzeme University of Applied Sciences, Cesu street 4, Valmiera, LV4201, Latvia. E-mail: agita.livina@va.lv

Ian Michael is Professor of Marketing at Zayed University, Dubai, UAE. He completed a Doctor of Business Administration (DBA) degree from Victoria University, Melbourne, Australia. His thesis investigated the area of electronic/Internet marketing, and was titled 'Consumer behavior in a computer mediated environment – implications for marketers'. Ian also has an MBA; the research thesis was in tourism marketing and was titled 'The economic impact of overseas students on tourism in Victoria, Australia'. Ian has around 18 years of industry experience in India and Australia where he held senior positions in the travel and tourism, banking and finance, hospitality, advertising and marketing, pharmaceutical, and transportation and logistics industries. He is a Fellow of the Australian Marketing Institute (FAMI), and a Certified Practicing Marketer (CPM). He has a number of refereed articles published in journals. He has presented at leading conferences around the world in tourism and marketing. His research was instrumental in Victoria's strategic marketing plan for tourism. Mailing address: Zayed University, PO Box 19282, Dubai Academic City, Al Ruwayah, Dubai, United Arab Emirates. E-mail: i.michael@zu.ac.ae

Nigel D. Morpeth (BA, MSc, PhD, PGCHE) currently is a senior lecturer in Tourism Management, School of Events Tourism and Hospitality, at Leeds Beckett University, UK. He started his academic career in 1989, after working for three UK local authorities within the field of community-based leisure provision, and has also taught at the University of Lincolnshire and was Visiting Scholar at Southern Cross University, New South Wales, Australia. He has published widely in books and journals and has jointly edited two tourism books with CABI. His current research projects and interests are in community and sustainability initiatives and the visual arts and tourism planning and development as well as in pilgrimage and tourism. Mailing address: Tourism Department, School of Events, Tourism and Hospitality, Carnegie Faculty, Leeds Beckett University, Leeds, LS6 3QU, UK. UK. E-mail: n.morpeth@leedsmet.ac.uk

Fernanda Maria Fernandes Oliveira has been a lecturer at the Polytechnic Institute of Leiria, Portugal, since 1997/8 (in the School of Tourism and Maritime Technology and in the School of Education and Social Sciences). At present, she is responsible for the coordination of the Degree in Tourism and Hotel Management. She obtained her PhD in tourism from the Faculty of Economics, University of Algarve, and as an author has participated in several national and international scientific meetings, with some international publications in the field of tourism. She has reviewed scientific papers and participated in international projects. She mainly conducts research in the areas of environmental management, tourism planning, environmental economy, social responsibility and sustainable tourism. Mailing address: Escola Superior de Turismo e Tecnologia do Mar, Instituto Politécnico de Leiria, Campus 4 – Santuário Nossa Senhora dos Remédios, 2520-641 Peniche, Portugal. E-mail: mfaolivia@fc.ul.pt

Girish Prayag, PhD, is a lecturer in Marketing in the Department of Management, Marketing, and Entrepreneurship at the University of Canterbury, New Zealand. His research interests are broadly related to destination management, tourist behaviour, emotions and sustainable tourism. Within such topics, his research is focused on emerging markets and island destinations. Being a native of Mauritius, he has published several papers on marketing issues for the island such as destination image, place attachment, visitor satisfaction and sustainability. His work has appeared in leading tourism and marketing journals such as *Tourism Management*, *Journal of Travel Research*, *Journal of Business Research* and *Current Issues in Tourism*. Mailing address: Department of Management, Marketing and Entrepreneurship, University of Canterbury, Private Bag 4800, Christchurch 8140, New Zealand. E-mail: girish.prayag@canterbury.ac.nz

Ágnes Raffay is Head of the Department of Tourism at the University of Pannonia, in Veszprém, Hungary. She has worked for Tourinform in Veszprém and was the inaugural Tourism Referent in the Mayor's Office of the local authority. Her PhD was a study of stakeholder involvement in urban tourism planning and was awarded by the University of Derby in the UK. She has worked around the Balaton for many years, visiting regularly and running training for the DMOs (Destination Management Organizations) in the region. Her research interests and publications focus on destination management, stakeholder involvement, governance and involvement. Mailing address: University of Pannonia, Egyetem Utca 10, Veszprém, Hungary 8200. E-mail: raffay@turizmus.uni-pannon.hu

Ian Tempest worked for more than 30 years for York City Council, UK (now the City of York Council), initially in Local Plans work but for the majority of the time in Economic Development and Tourism, with a particular focus on inward investment promotion and fulfilment, tourism product development, management and monitoring of Yorkshire Forward and Welcome to Yorkshire contracts and tourism strategy development. Ian is a member of the Tourism Society and the York Civic Trust and has a long-standing interest in York's built and social history and in British economic history in general. Mailing address: 56 Celtic Close, Acomb, York, YO26 5QJ, UK. E-mail: ian.tempest1@btinternet.com

Dolors Vidal-Casellas, PhD, is Dean and a senior lecturer at the Tourism Faculty of the University of Girona. She is Founder of the Official Master in Cultural Tourism, a member of the Contemporary Culture and Art Chair and also member of the research groups 'Contemporary Art Theories' and 'Higher Institute of Tourism Studies' (INSETUR). She is director of the collection 'Turisme cultural' (cultural tourism) by Edicions Vitel·la (http://www.edicionsvitella.com). She joined different cooperation programmes with universities from Algeria, Sao Paolo, Piracicaba (Brazil) and San Martín de Porres (Peru). She is also a member of cultural tourism international networks: Ibertur, Atlas and Unitwin 'Culture, tourisme et développement' (UNESCO). Her research interests include cultural tourism, religious tourism, history of tourism, heritage management, destination imagery and tourism planning and policies. Mailing address: Faculty of Tourism, University of Girona, Plaça Ferrater Mora, 1, 17071 Girona, Spain. E-mail: dolors.vidal@udg.edu

Hongliang Yan (BA, MSc, MA, PhD, PGCHE) is a senior lecturer in Tourism Management in the School of Events, Tourism and Hospitality at Leeds Beckett University, UK. He started his academic career in tourism and hospitality management in 2004. His research interests are in tourism, modernity and sustainable development which relates to the themes of tourism planning and policy making, governance, destination management, tourism and identity building. He has published widely on heritage tourism, tourism modernity and identity building, and tourism planning. Mailing address: Tourism Department, School of Events, Tourism and Hospitality, Carnegie Faculty, Leeds Beckett University, Leeds, LS6 3QU, UK. UK. E-mail: h.yan@leedsbeckett.ac.uk

Preface

As editors of *Planning for Tourism: Towards a Sustainable Future*, we acknowledge the previous pioneering work of a range of esteemed authors, who have been instrumental in enabling generations of students of tourism and related studies, to develop an in-depth understanding of what constitutes tourism planning. We have been fortunate as academics to have collectively over 30 years experience of working with tourism students within higher education and to use these marquee texts as resources to underpin pedagogy. These texts have enabled us to share ideas of the concepts of tourism, planning and sustainability, in lively and productive discussions with students. Our hope is that some of those students will now be employed within the destinations where they have the opportunity to implement approaches to tourism planning, which were inspired on the basis of their studies in classroom environments and that principles of sustainability will inform their everyday work in those locations.

We were inspired to develop a new type of book, which incorporates a range of different and innovative approaches but which acknowledges the bedrock of tourism planning epistemology. For example, in order to respond to the temporal challenge of how to conceptualize sustainability, in evolving pedagogy, we have 'gone back' to the past, to be able to understand the future, by utilizing the 1936 work of the author E.H. Gombrich's ground-breaking book: *Eine kurze Weltgeschichte fur junge Leser* (English translation *A Little History of the World*; Gombrich, 1985). This was written in a 6-week period, in which he chronicled the major events which have shaped the world and humankind, over a 6 million-year timeframe. This book served to add perspective to chronicling the brief timescales and to be able to intellectualize about a 'brief' history of contemporary tourism and an even briefer period of time, to make sense of what the grand narratives of sustainability are, and how we might combine them into the field of tourism studies, but crucially everyday life.

In finding a starting point to chronicle the concept of sustainability, in common policy, academic and everyday parlance to communicate in our work with students, we are conscious that the history of sustainability has largely been written from Western perspectives. Within this text, we therefore started our analysis of sustainability using Eastern perspectives of the physical environment, combined with the social world, to provide more holistic perspectives on the analysis of sustainability and explore ways to transcend the anthropocentric limitations of our perceptions of the world.

The book is divided into three distinct parts. In Part I, *Processes and Approaches to Planning and Tourism*, we begin by presenting ideas on classical approaches to tourism planning in Chapter 1, and this is followed by two chapters which offer distinctive and innovative ways to stimulate students to approach the study of planning, tourism and sustainability through 'new eyes'. In doing so, we suggest thinking about the physical environment from different interdisciplinary and social science perspectives and to consider in the first instance viewing localities from critical theoretical spatial perspectives. This was intended to combine intersectional considerations of how social groups engage with the built and natural environments, as part of developing visions for sustainability underpinned by the concepts of social and environmental equity.

Part II, *Practitioner Approaches to Planning and Tourism*, provided the opportunity to combine theoretical insights from the first three chapters with approaches used in practice, and we invited four practitioners who have worked in the fields of tourism and planning to communicate their everyday experiences of planning for tourism. Specifically they provide invaluable insights into planning in a UK National Park, a prominent historic destination and highlight personal insights into what it means to be a planner in a busy urban and World Heritage destination in the UK, in the 21st century. We were keen to consider perspectives on disability and pioneering views as to how to plan localities using the principles of universal design to respond to the needs of diverse groups of people within destinations, so Part II includes a chapter that addresses this issue. It was not our intention to 'privilege' insights of disabled people, as representative as to what Prince (2008) argues are 'excluded others' but rather this chapter exemplifies the capacity to move from theory to praxis, in so far that the pioneering work of disabled people can instigate change and develop innovative approaches to planning for people in localities.

In Part III, *Global Approaches to Planning and Tourism*, we are grateful for the invaluable contribution of a range of academics, from different global destinations, who have written seven case study chapters, which provide an invaluable cross section of planning for tourism perspectives within diverse cultural contexts. The authors displayed great patience and commitment to progress their chapters following an initial brief of ideas. They separately consider planning from national- to local-level perspectives and from emerging to developed destinations, combining a range of diverse approaches to planning for tourism towards a sustainable future. These case study chapters incorporate exciting insights from the Lebanon, Dubai, Peru, Portugal, Hungary, Mauritius and Latvia and include destinations which appear for the first time in a collection of textbook case studies.

We hope that this book's interdisciplinary coverage and innovative design and content will follow the tradition of other tourism texts which we have alluded to, in inspiring diverse groups

of students, in different cultural contexts, to engage in pedagogy. Furthermore, we hope that it will encourage students to become critical thinkers and adopt a problem-solving approach to global and sustainable perspectives on tourism and planning. Critical thinkers who consider the challenge of working towards a sustainable future, in diverse locations, can be enlivened by using the prism of critical theoretical perspectives to explore feminist, race and sexual identity discourses, combined with the concepts of universal design and accessible tourism, as keystones of planning processes within the 21st century. This is an uncertain future in which grand narratives of sustainability, over-depletion of non-renewable resources, population growth, poverty, famine, warfare, climate change and lack of social equity provide substantial challenges for local and global communities to respond to. We hope you will be inspired to play your part.
We would like to express our gratitude to CABI for their commitment to the development of this book and particularly Claire Parfitt and Alex Lainsbury, for their advice and guidance in helping to develop ideas at different stages of the progression of the book.

Nigel D. Morpeth and Hongliang Yan
August 2014, Leeds, UK

BIBLIOGRAPHY

Gombrich, E.H. (1985) *A Little History of the World.* Yale University Press, New Haven, Connecticut.
Prince, M.J. (2008) Inclusive city life: persons with disabilities and the politics of difference. *Disability Studies Quarterly* 28(1), 1–13. Available at: http://dsq-sds.org/article/view/65/65 (accessed 9 June 2014).

Acknowledgements

This book is dedicated to our families and to those of the chapter authors, particularly to our children and the children of the authors, to their children, to their children's children, to their children's children's children..........

List of Figures

List of Tables

PART I

Processes and Approaches to Planning and Tourism

The chapters in Part I review different conceptual aspects of planning, tourism and sustainability, explored from different interdisciplinary perspectives. These perspectives preface and underpin four practitioner-based chapters in Part II, and then in Part III, seven applied case study chapters, written by academics from diverse cultural contexts in different parts of the globe. The three chapters comprise:

- Chapter 1: *Approaches to Planning and Tourism* – Hongliang Yan and Nigel D. Morpeth provide a thorough academic overview of the key conceptual and theoretical aspects of classical approaches to tourism planning, in the first instance discussing the relationship between policy and planning. The characteristics and techniques of different planning traditions are reviewed and a case study focuses on how an integrated planning approach is related to sustainable tourism development.
- Chapter 2: *Sustainability: From Theory to Practice* – Nigel D. Morpeth and Hongliang Yan initially consider different discourses of sustainability from both Eastern and Western perspectives. The chapter then considers distinctions between sustainable development and sustainability and the emergence of these concepts, politically, and in pedagogy and policy. Finally, a case study of the UK considers how principles of sustainability have informed planning and policy agendas.
- Chapter 3: *Critical Approaches to Spatial Analysis* – Nigel D. Morpeth and Hongliang Yan use interdisciplinary approaches to review different perspectives of the natural environment and to consider dualist thinking between the natural and the social environment. This chapter then considers how critical theory can be used to create insights as to how spatial analysis can be applied to planning, for the

needs of different intersectional groups, in diverse cultural localities. A case study on mega-events, and specifically the Brazil 2014 soccer World Cup, contextualizes the conceptual aspects of spatial analysis.

chapter 1

Approaches to Planning and Tourism

Hongliang Yan* and Nigel D. Morpeth

Leeds Beckett University, Leeds, UK

LEARNING OBJECTIVES

- To develop an understanding of the importance of tourism planning and the relationship between tourism planning and policy.
- To recognize the characters of different planning traditions and the planning techniques and methods employed by each tradition.
- To outline the different elements in the integrated planning approach and relate them to sustainable tourism development.

INTRODUCTION

The aim of this chapter is to provide an introduction and discussion on the topic of tourism planning, with three main tasks. First, it explores the theoretical scope of planning and policy studies and the relationship between tourism planning and policy. Secondly, it discusses the traditions of tourism planning approaches in tourism development and highlights the main issues, methods, strengths and limitations of each tradition. Finally, a case study explains how integrated tourism planning and processes might be used as a planning approach which exemplifies the principles of sustainability, which can be applied at national, regional and local levels.

UNDERSTANDING TOURISM PLANNING

Tourism

What exactly is meant by the term tourism? Tourism is a complex field of study and

*h.yan@leedsbeckett.ac.uk

difficult to define as an industry and as Leiper (1993) emphasizes from both an academic and an industrial perspective, it is a concept that is subject to a diversity of meanings and interpretations. Additionally, both he and Mathieson and Wall (1982) suggest that tourism can be described as a 'system'. Historically, it is a phenomenon that is well researched by researchers from varied disciplines, such as geography, sociology, anthropology, economics and business studies and as such is interdisciplinary in its scope. The United Nations World Tourism Organization (UNWTO) provides a commonly accepted definition of tourism, as a social, cultural and economic phenomenon, which entails the movement of people to countries or places outside their usual environment for personal or business and professional purposes (UNWTO, 2013). What is without ambiguity is that tourism is an important global activity, generating employment, revenue and general economic, environmental and socio-cultural impacts. UNWTO emphasizes that it is an important economic force for many countries and regions: in 2013, international tourism receipts reached an estimated US$1159 bn worldwide, up 5% in real terms from US$1078 bn in 2012 (UNWTO, 2014). Arguably, planning policies and processes have an important role in harnessing the benefits of tourism and mitigating externalities which have blighted communities in a range of physical environments globally. In the process there have been intensifying calls for greater synergies between tourism and sustainability.

Tourism and sustainability

As far back as 1975, Turner and Ash exploded the myth that rapid tourism development would inevitably provide economic and societal benefits for destinations, regions and countries. The prominent negative impacts that they highlighted to society (the social–physical environment, exacerbated by an age of mega-mass tourism) provide a seemingly insoluble challenge in achieving the goal of long-term sustainability (Fayos-Sola, 1996). Those challenges have been gradually but inconsistently recognized by governments, community and other stakeholders in tourism and stimulated by global engagement in recognizing the challenge of sustainable development through the Brundtland Report (*Our Common Future*) (WCED, 1987) and United Nations Conference on Environment and Development (UNCED) global conferences to set in train initiatives to work towards sustainability. Brundtland and UNCED, as we will see in detail in the next chapter, highlight the importance of sustainability to enable future world economic development to continue. Representatives of many authors that have applied an analysis of sustainability are Bramwell and Lane (1993, p. 2) who define sustainable tourism as 'a positive approach intended to reduce the tensions and friction created by the complex interactions between the tourism industry, visitors, the environment and the communities which are host to holidaymakers'. Later, Bramwell *et al.*'s (1996) book *Sustainable Tourism Management: Principles and Practice* is considered as a milestone in the development of a theoretical framework for sustainable tourism. It recognized the need for tourism to respect the environment so as not to aid its own disappearance (Bramwell *et al.*, 1996).

Although the concept of sustainable tourism has evolved from a narrowly defined term, to embrace all aspects of tourism over the past few decades, stakeholders may still have different views and inevitable contrasting interpretations of this term. Unsurprisingly what is viewed as the basis of sustainability is often difficult to determine, and as will be discussed in the next chapter, global blueprints must be cognizant of the discussion on sustainable tourism which has to take place in the context of the cultural and physical environments where tourism occurs (McCool, 1994). These contrasting localities form the focus of this book in the narratives of planning and sustainability. The challenge of sustainable tourism is that it is often difficult to implement at different levels as there are many stakeholders in the complex system of tourism. Later in the chapter the concept of sustainability will be applied to integrated forms of planning but initially the concept of tourism planning is explored in depth.

Tourism planning

Planning is usually about developing a vision and setting up objectives for the future. It is an ordered sequence of operations and actions that are designed to realize either a single goal or a set of interrelated goals and objectives (Williams, 2009). Murphy (1985, p. 156) suggests 'planning is concerned with anticipating and regulating change in a system, to promote orderly development so as to increase the social, economic, and environmental benefits of the development process'. It is a series of activities that try to predict the future situation, based on the current understanding of tourism at a destination.

Tourism planning is a complex phenomenon with different types of social, cultural, economic and political implications and as with the concepts of tourism and sustainability, the definition and scope of tourism planning is ambiguous (Tosun and Jenkins, 1998). Tourism authors apply their emphasis and concerns on different aspects of tourism planning processes (Getz, 1987; Inskeep, 1991, 1994; Gunn, 1994; Hall *et al.*, 1997; Timothy, 1998, 1999; Tosun and Jenkins, 1998; Dredge and Jenkins, 2007; Hall, 2008). A popular definition given by Getz (1987) suggests that tourism planning is a process which seeks to optimize the potential contribution of tourism to human welfare and environmental quality.

An important consideration of studying tourism planning is to consider the concept of tourism development, with its impacts on destinations and local communities which must be closely studied. Not least, carefully planned developments are likely to experience the most success in terms of high tourist satisfaction level, positive economic benefits, and minimal negative impacts on the local social, economic, and physical environments (Timothy, 1999). On the contrary, it is also widely accepted that places adopting unplanned or insufficiently planned tourism development, provide host communities with environmental and social problems. To those destinations, they often have low destination competitiveness and high conflict resolution costs (Inskeep, 1994; Yuksel *et al.*, 1999). Ironically, the demand of planning for tourism development is a direct response to unwanted effects brought by tourism development (Hall, 2008).

Therefore, there is an inescapable case for effective tourism planning in tourism destinations, if there is to be a likelihood of providing positive socio-cultural and economic benefits, as a prerequisite of realizing the goal of sustainability. Williams (2009) indicates that tourism planning is a mechanism for delivering a range of more specific outcomes. Furthermore, he provides a list of the positive outcomes of tourism planning (Williams, 2009, p. 157), as below:

- the integration of tourism alongside other economic sectors;
- the direction and control of physical patterns of development;
- the conservation of scarce or important resources;
- the active promotion and marketing of destinations; and
- the creation of harmonious social and cultural relations between tourists and locals.

Early studies of tourism planning have a significant focus on the economic benefits and the physical environmental concerns of tourism development. In the past several decades, there has been a growing awareness and emphasis on the socio-cultural aspects and the local community benefits in the study of tourism planning. For instance, Gunn and Var (2002) indicate that tourism must be planned towards goals of enhanced visitor satisfaction, community integration and greater resource protection.

Tourism policy

In the tourism sphere, policy is generally considered as a public policy which is designed to achieve specific objectives, relevant to tourism development at different levels. Tourism policy is an important area for study because of its practical and theoretical importance (Scott, 2011). As with other public policies, Hall and Jenkins (1995) argue that tourism policy is a political activity which reflects society's values and ideologies, economic conditions reflecting the distribution of power institutions, and decision-making processes. A definition given by Goeldner and Ritchie (2012, p. 326) for this term is that 'tourism policy is a set of regulations, rules, guidelines, directives, and development/promotion objectives and strategies that provide a framework within which the collective and individual decisions directly affecting long-term tourism development and the daily activities within a destination are taken'. Tourism policy is significantly influenced by the economic, social and cultural characteristics of society, as well as by the formal structures of government and other features of the political system (Hall, 2008).

Tourism policies are generally created as a mechanism for managing the development of the tourism sector. They should reflect the political attitude on tourism and the overall development policy of the country, region or locality. It has an important role in the promotion of a successful tourism destination. Goeldner and Ritchie (2012, p. 327) recommend that there are six main functions of tourism policy, as below:

- It defines the terms under which tourism operators must function.
- It sets out activities and behaviours that are acceptable for visitors.
- It provides a common direction and guidance for all tourism stakeholders within a destination.
- It facilitates consensus around specific strategies and objectives for a given destination.
- It provides a framework for public/private discussions on the role and contributions of the tourism sector to the economy and to society in general.

- It allows tourism to interface more effectively with other sectors of the economy.

From a long-term view, tourism policies can provide the destination with a clear vision or direction and it helps support and regulate the tourism sector and facilitate the need of tourism development. It also provides a framework for tourism development and acts as the guidance for tourism planning for a country or region.

Tourism planning and policy

There is a strong link between tourism policy and tourism planning. They share many common characters with each other. In particular, both of them involve anticipating future changes of tourism destinations and aim to affect that change by making decisions on future objectives and specific actions. Planning and policy both involve political debate about what the agenda is, what the issues are, who is involved or affected and the alternative courses of action available to address problems (Dredge and Jenkins, 2007).

Although tourism policy and planning have many similarities, there are some important differences between these two terms in the tourism context. First, tourism policy focuses on the politics of decisions. It signifies the formal adoption of a political position by the government, whereas tourism planning would not normally stop by the adoption of political position. Secondly, compared with tourism policy, tourism planning focuses more on the practical side of implementation and monitoring. Policy making is more concerned the determination of 'what should be done in long-term tourism development, while planning may pay more attention to the "how" to realise the achievement of specific destination goals' (Goeldner and Ritchie, 2012).

APPROACHES TO TOURISM PLANNING

Planning for tourism, at all levels, is crucial to achieve successful tourism development and more importantly, to promote long-term sustainable development of tourist destinations (Mathieson and Wall, 1982; Murphy, 1985; Getz, 1986; Inskeep, 1991; Bramwell and Lane, 2000; Gunn and Var, 2002; Hall, 2008). With the development of tourism, the approaches to tourism planning also experience an evolution from narrow concerns with economic and physical planning, to a more balanced form of planning and development (Timothy, 1999). Getz (1986) in his work on models of tourism planning classifies tourism planning traditions into four broad categories:

- boosterism;
- economic, industry-oriented;
- physical/spatial; and
- community based.

This classification is an accessible and popular way to examine issues in tourism planning and it is worth noting that those tourism planning approaches are not mutually exclusive, nor are they necessarily sequential (Getz, 1986). In the following discussion, a detailed review will be provided on the planning assumptions, models and methods of each tradition.

Boosterism

A boosterist approach has been the dominant tradition towards tourism development and planning since mass tourism began, particularly prevailing throughout the 1960s and early 1970s (Hall *et al.*, 1997). Under this approach the process of tourism planning is generally aligned with exploitation of the natural or cultural resources of a destination to maximize economic returns (Williams, 2009). Tourism planning under a boosterist tradition is more concerned with how to attract more tourists to a specific destination and it assumes a simplistic assumption that tourism development is generally 'good' and beneficial for a destination and its inhabitants. A boosterist approach emphasizes the ability to forecast the demand mainly for the purposes of tourism development and promotion (Hall, 2008).

Based on a boosterist view, a tourism development strategy is very much growth oriented. Under this model, decisions are primarily deferred to the public sector, local communities are not included in decision making and planning processes and the carrying capacity of the region is not given adequate consideration (Page, 1995). Cultural and natural resources are promoted as assets, in order to stimulate market interest and increase economic benefits and barriers to development are reduced (Getz, 1987; Dredge, 1999; Hall, 2008). Both the public and the private sectors put significant efforts into increasing visitation for the destination. The analysis and goal setting in the planning of tourism are considered in a marketing context, which often parallels the desires of tourism-related business.

Indicated by Baidal and Josep (2004), under this approach, mass tourism has been deemed as the means to realize a destination's economic prosperity, with little consideration of the social and environmental impacts brought by tourism development. In a similar vein, Hall (2008) also emphasizes that the analysis of tourism destinations under this approach mainly focuses on the demand side rather than the sustainability of resources and social carrying capacity of a region. Because of that, many tourism academics view tourism development under the boosterist approach as a form of unplanned tourism development (Getz, 1987; Tosun and Jenkins, 1998; Hall, 2008). However, because of its obvious political benefits, Getz (1987) argues that boosterism stereotypically will always be practised by politicians who philosophically or pragmatically believe that economic growth is always to be promoted and by other stakeholders who will gain financially from tourism.

Economic, industry-oriented approach

Tourism is often viewed as an industry in its own right, based on the belief that tourism is equal to other industries in terms of its economic generating capacity. It contributes to foreign exchange earnings, improving terms of trade, balancing payments and making a contribution to investment, consumption, employment, gross domestic product (GDP) and taxation. Tourism-related industries become a compelling means to promote economic gains.

The economic, industry-oriented approach suggests tourism as an economic sector therefore can be compared to other economic sectors. It has a special interest in developing the appropriate measurement of what tourism can contribute, or has contributed, to the economic development of a country or destination. However, as it is difficult to separate tourism businesses and jobs from the rest of the economy, various methods have been adopted to measure the economic contribution made by the tourism sector by governments and academics. For example, the most popular framework is the Tourism Satellite Account (TSA) which was accepted and approved by the UNWTO, the Organisation for Economic Co-operation and Development (OECD) and EUROSTAT (Massieu, 2004). By developing a TSA framework, tourism, which sometimes is not defined as an industry in national accounts, has the opportunity to be measured in parallel to other economic sectors (UNWTO, 2000). TSA provides guidelines on the treatment of tourism within a conceptual framework of economic analysis, which can be viewed as an extension of a National Accounts type of analysis, within the specific tourism domain of interest (Massieu, 2004).

The economic, industry-oriented approach is often associated with economic impact statements and feasibility studies. Governments and industry emphasize market segmentation studies and matching product and markets (Hall, 2008). The planners are seen or act as 'experts' and tourism development is defined in economic terms. The main planning issues for tourism under this approach is to put in action tourism as a vehicle of growth, the maximization of income and employment multipliers, influencing consumer choice and allocating economic values for externalities. Marketing becomes a primary tool of this approach and marketing techniques remain dominant as a tourism 'planning' tool to attract visitors (Tosun and Jenkins, 1998). Planning models under this approach include tourism master plans and economic multipliers, while common planning methods are the analysis of tourism supply and demand, which include market segmentation, cost–benefit analyses and development incentives. As Williams (2009) indicates, the economic approach represents a more regulated form of boosterism, as it still constitutes a planning approach that is shaped by economic imperatives. As economic goals have more priority, an economic framework is often used to examine the impacts of tourism and consequently little attention is paid to negative socio-cultural and environmental impacts.

Physical/spatial approach

The planning for tourism under both boosterism and economic, industry-oriented traditions focused on the market and growth, instead of putting the values of destination environment and the community as a central concern (Inskeep, 1991). As unbalanced and sometimes haphazard tourism development routinely causes social and environmental impacts to many destinations, tourism planners and academic scholars had started to question the net benefits and economic efficiency of tourism development and called for a new approach of tourism planning to replace the existing growth or market-led approach. For example, Godfrey (1996) argues if mass tourism is inevitable due to sheer tourism demand, then it is necessary to find a way to make tourism more sustainable.

A physical/spatial tradition of tourism planning is derived from geography studies and considers tourism as part of regional and destination planning. It is one of the traditional forms of environmental protection, based on rational decision making (Hall, 2008). Under this approach tourism is viewed as a 'user' of natural resources and environmental conservation has priority in the planning of tourism. The regulating of physical development and the appropriate ordering of land use and associated infrastructures are the primary concerns within this approach of tourism planning (Williams, 2009).

Carrying capacity versus limits of acceptable change

With the intention to achieve sustainable development of tourism, many academics consider imposing a concept of limited growth, to control the negative impacts of tourism and resolve development-related conflicts at the destinations, and determine an appropriate scale of economic development.

In attempting to understand the ability of destinations to withstand tourist use, the assessment of carrying capacity is developed as an important sign of tourism impact on space (the social–physical environment). It represents an important component of planning of spatial development in tourism, and is regarded as one of the major mechanisms for establishing standards for sustainable tourism under the physical/spatial approach. The concept of carrying capacity considers that any tourism development engenders some changes and maintaining a level of development. A widely accepted definition for carrying capacity is given by Mathieson and Wall (1982, p. 21), stating: 'the maximum number of people who can use a site without an unacceptable alteration in the physical environment and without an unacceptable decline in the quality of the experience gained by visitors'. It emphasizes both the physical and the social aspects of tourism impacts and that there should not be a single tourism carrying capacity at a destination. Generally speaking, carrying capacity has been represented in five ways: physical, economic, ecological, socio-cultural and perceptual carrying capacity.

Physical carrying capacity is normally regarded as the maximum number of visitors that an area is actually able to support. For a specific tourist attraction, it is the maximum number that can fit on the site at any given time and still allow people to be able to move (Mowforth and Munt, 2003). It is more concerned with absolute space (Williams, 2009) so consequently, the 'limit' is often artificially taken as a numerical volume of tourists. In addition to the assessment on tourist space, it also considers the availability of the supporting infrastructure. Physical carrying capacity includes an assessment on whether a destination has adequate availability of transportation, utility and community facilities and services.

Economic carrying capacity emphasizes the economic dimension of planning and management at tourist destinations. It describes the level of unacceptable change within the local economy of a tourist destination. Crucially, the extent to which a tourist destination can accommodate tourist functions without the loss of local activities (Mathieson and Wall, 1982) and whether the economic impacts brought by tourism development would distort the structure of the local economy. For example, popular tourist destinations normally have a higher inflation rate, compared with non-tourist destinations, which is usually reflected in the price of accommodation, local transportation, food and other services. If the increased revenue generated by tourism development is overtaken by the inflation caused by tourism, the destination often reaches the point of its economic carrying capacity.

Ecological carrying capacity, by its name, is the level of use that an ecological system or physical space of the social environment (the ecology of the city, as a notion, complicates the natural connotations which are applied to this concept) can sustain before long-term damage to the environment is experienced (Williams, 2009). It recognizes that there is an ecological tolerance level of tourist use, which when exceeded, is likely to cause environmental damage and to reduce visitor satisfaction and crucially the quality of life of local residents. It conjures up ideas of a threshold of a habitat's ability to regenerate and the scale of tourism should not exceed this threshold, as there will be irreversible destruction to the destination. The assessment of ecological carrying capacity focuses on the conservation of wildlife and natural vegetation and it also measures acceptable levels of air, land, water and noise pollution at destinations.

Tourism development can help maintain local customs and traditions, without detrimental effects. However, if the growth is not well managed, it can also change values and attitudes of local people, reduce life quality of residents, and the authenticity of local culture. Social carrying capacity tries to provide social indicators for the acceptable level of tourism development from the host community. It is the extent of tourism development that can be absorbed without unacceptable adverse impacts to the activities and lifestyles of the local community. It is usually used to assess the issues such as host–guest relations, cultural identity of the local community, and the absorptive capacity of social problems (e.g. crime) caused by tourism development (complex and invariably subjective assessments).

Closely linked with social carrying capacity, perceptual carrying capacity emphasizes tourist experience and their perception about tourist destinations. It relates to the levels of development and numbers of tourists that are compatible with the perceived image of the destination. It also refers to the compatible type of cultural and environmental experience that the visitors are seeking from a destination. For perceptual carrying capacity, the pertinent question asked is how much tourism is acceptable before there is a decline in visitor satisfaction (Coccossis and Mexa, 2004). If the area exceeds saturation levels, the environment and products that tourists want to experience will be destroyed or degraded, and the satisfaction level will be reduced as the destination is degraded in both quality and popularity.

Carrying capacity, with all its idealized and subjective limitations, is regarded as one useful tool which guides tourism planning for sustainable levels of development. As one of the main techniques that have been widely used under physical spatial tradition of tourism planning, it receives a measure of credibility through its grounding in observations that intuitively make sense (Williams, 2009). However, carrying capacity assessment has its own limitation and received critiques from many authors (Stankey and McCool, 1984; Buckley, 1999; McCool and Lime, 2001; Bramwell, 2006). First, in tourism planning practice, determining carrying capacities is often not a simple or precise job. Although scientific methods developed from other disciplines such as geography, biology and economics can be applied in the assessment, it still depends significantly on the assumptions that are made at many destinations. Secondly, carrying capacity seems to have a simple notion that there is a 'limit' or 'threshold' of tourism use and going beyond that 'limit' the sustainability of the destination is threatened. However, it is often wrongly assumed that carrying capacity is the maximum use before degradation actually occurs. The danger for this wrong assumption is that there might be no immediate remedial responses to the impacts until the destination gets to the predetermined threshold. Carrying capacity can sometimes give the impression that a site is better protected than it actually is (Pedersen, 2002). Thirdly, it is difficult to anticipate the numeric limit or threshold for carrying capacity due to the changing environment and various variables existing in tourism planning. For example, the tourist volume can be dramatically changed at a destination due to seasonality. Similarly, because the tourists' behaviour could vary considerably between groups, it often creates extra difficulties to determine the adequate reference for the calculation of carrying capacity. As suggested by Hammitt and Cole (1987), instead of using a mere numeric threshold to determine impact, other factors are often far more important such as: (i) the resiliency of the site; (ii) the size of user group; (iii) the mode of transportation; (iv) the length of stay; (v) the season of the year; and (vi) behaviour patterns of users. Socio-cultural and perceptual carrying capacity varies between individuals and groups depending on values and circumstances. It is sometimes a very subjective rather than scientific process to make decisions about how much adverse impact is too much. Finally, without the identification and articulation of qualitative community views and priorities on development, the application of carrying capacity methodologies might be viewed as being overly simplistic, prescriptive and a normative and reductionist technique to aid planning processes.

As a response to those limitations, the Limits of Acceptable Change (LAC) was gradually developed as an alternative method to address the technical difficulties and societal nature of tourism planning (Stokes, 1990). LAC is a nine-step process designed to guide planners and the management of the tourist destination (Stankey *et al.*, 1985), as shown below.

1. Identify area issues and concerns.
2. Define and describe opportunity classes.
3. Select indicators of resource and social conditions.
4. Inventory existing resource and social conditions.
5. Specify standards for resource and social indicators for each opportunity class.
6. Identify alternative opportunity class allocations reflecting area issues and concerns and existing resource and social conditions.
7. Identify management actions for each alternative.
8. Evaluate and select a preferred alternative.
9. Implement actions and monitor conditions.

The major difference between LAC and carrying capacity is that LAC emphasizes the quality perspective of the destination and the tourism product it offers. By offering a management framework, the LAC is much more flexible to local situations. It focuses on what kinds of resources and social conditions are desired to be maintained rather than trying to determine an arbitrary number, for example the limit of growth. LAC represents a shift from 'How much use is too much?' to 'How much change is acceptable?' (Stankey and McCool, 1984). Besides the support of science in the subjective decision making, to determine the acceptability of destinations also invariably is tempered by subjective human judgement. The planning and management of the destination is based on the objectives desired by not only the tourism planners, but also tourists, and vitally the community. Therefore, it has the potential to encourage a wider participation in determining the acceptability of destinations. To some degree, this could help reduce the subjectivity of carrying capacity.

As an alternative to carrying capacity, LAC has some potentially important improvements when assessing the 'acceptable' conditions of tourism growth. However, the framework of LAC has its practical limitations. Similar to carrying capacity, there are technical difficulties in agreeing and assessing qualitative aspects of tourism development. The LAC often tends to emphasize the conditions of the physical environment but not the needs of tourist markets, the local economy and community. As indicated by Williams (2009), to successfully apply this framework to tourism planning needs a structured planning system and sufficient expertise and capital to operationalize the monitoring and review stages.

Community-based approach

The focus of the physical/spatial approach is environmental conservation and protection. However, the socio-cultural impacts of tourism, such as the uneven distribution of benefits in tourism, rising tensions between the visitors and the host, and exclusive decision making by the public sector in tourism planning and development, have still not been fully addressed.

Stimulated by global economic restructuring, there have been profound changes in society and a growing support for neo-liberal ideologies to dominant Western government political and policy agendas since the 1970s (Peck, 2001). The greater direct participation by the public and interest groups gradually becomes a more imperative issue in the tourism planning arenas to mitigate the unfettered development programmes of neo-liberal ideologies (Bramwell and Lane, 2000). Many tourism planners and scholars recommend a greater recognition of the need for the participation of relevant stakeholders and alternative forms of tourism development as the response to the undesirable impacts brought by tourism (see also the work of Brindley *et al.* (1996) mentioned in Chapter 2, this volume, on 'London Docklands'). The call for stakeholders' involvement in tourism is not only for their role in tourism development but also in the decision-making processes for tourism planning itself (Elliot, 1997). Through the encouragement of collaboration and joint management of destinations, social equity as a vital element of sustainability is attainable (Selin, 1999). Since many tourism planning decisions significantly affect the welfare of the local community, the involvement of local community members in the decision-making process becomes an important requirement to try and attain. Following the involvement of stakeholders, especially the local community, tourism planning has evolved from an uncomplicated view of tourism into a more sophisticated and decentralized view (Timothy, 1998).

A community-based approach for tourism planning advocates a more locally controlled and balanced development of tourism and has a greater recognition of the social and environmental concerns of the public. As one of the first advocates of community-based tourism planning, Murphy (1985) indicates that if tourism is to become a successful and self-perpetuating industry, it needs to be planned and managed as a renewable resource industry, based on local capacities and community decision making. The main principle of a community-based approach reflects the need to develop more socially acceptable guidelines for tourism expansion (Hall, 2008). More importantly as argued by Mitchell and Reid (2001), a community-based planning approach considers local people and communities as the subject of the development not the object of it. It recognizes that tourism should not only serve tourists but also the local community (Timothy, 1999). For the first time the host community becomes the central focus of tourism development and planning. Under the community-based planning approach, the tourism planners are seen as facilitators rather than 'experts'. This, however, does not mean an exclusion of the public sector in the planning, rather it promotes a more democratic and collaborative planning approach among stakeholders. The main planning issues are therefore: (i) a need to understand the impacts of tourism on a community; (ii) their attitudes towards tourism; and, more importantly, (iii) how they can participate in and lead the process of planning for tourism.

Through the involvement of the local community, a community-based approach of tourism planning has the potential to promote more sociological, economic and environmental benefits to the community. It could help facilitate a more sustainable and effective management of tourism's impacts and, in the meantime, it could also help resolve socio-cultural tensions between tourists and local people (Williams, 2009). To many countries and destinations, a decentralized governance system for the management of the tourism industry represents a significant shift of tourism planning from a 'top-down' approach to a more 'bottom-up' approach. Turning to a more practical perspective, there are several ways to achieve more decentralized planning for tourism. The broad participation of community in tourism planning normally includes the opportunity for consultation and collaboration.

Consultation

A consultative approach in planning emphasizes that local people have a right to be consulted about activities which are likely to have effects on their well-being. The basic aim of public participation in tourism planning is to provide the local community with adequate information on issues under consideration, opportunities or even rights (Lucas, 1978). As a popular form to involve the public, a consultative approach of tourism planning is widely adopted in many destinations. For example, Yuksel *et al.* (1999) discussed the application of public consultation in the study of Pamukkale, Turkey. The main principle of this approach is a quest for community inputs through their consultative involvement in the tourism development processes. In this way, Smith (1978) suggests that the participation of the community could help strengthen tourist and community bonds, which will result in socio-cultural harmony. By informing the public in the planning process, the impacts of tourism felt most keenly at destination areas might be mitigated by communities and residents playing an essential part in destination planning (Simmons, 1994).

Within this approach, the planners are normally public sector professionals who gather different views from relevant stakeholders in the planning process. The participatory techniques adopted for public consultations vary from the use of questionnaires, focus groups, interviews, telephone and online surveys or drop-in centres (Yuksel *et al.*, 1999). The public can express their planning preferences and opinions through those communication methods. The purpose of consultation is mainly, on the one side to keep stakeholders, in particular the local community, informed about the tourism development and the possible changes to the community in the near future. While on the other side, it could help the tourism planners to develop an understanding about public interests and concerns, and hence help the planners to address those issues in the planning processes. With meaningful community and stakeholder input there might also be interventions by national government depending on the 'top-down' nature of the political system (Hall, 2008).

One obvious limitation of the consultative approach is that there is little face-to-face dialogue between tourism-related stakeholders and the planners. Although, as indicated above, various communication methods can be applied in the process of tourism planning, the effectiveness of them is questionable in planning practice at many destinations. First, from the technical aspect, there are limitations on tools that the planners can use to get views from everyone. Secondly, not all community members share the same interests in tourism though they might be affected by its development (Sewell and Phillips, 1979). Because of this, the planning team must identify the issues required from the public and develop a clear planning vision based on a cross section of public views to be incorporated in the process of consultation.

The second limitation is that public consultation does not necessarily offer opportunities for feedback, which to a certain extent makes the consultative approach very much a one-way process. Because of that, it raises the question about the use and interpretation of the information from consultation with the public. The incorporation of these views into the tourism plan and policy is often left to the judgement of public-sector 'tourism planners' (it is important to note that tourism might be 'planned' in tandem with other planning priorities) with little or no direct involvement of other stakeholders in the decision-making process. There is also a danger that the consultation becomes a token for more decentralized planning while there might not be any fundamental change in the planning process. As Hall (2008) indicates, if the direction of decisions has already been prescribed by government, nationally, regionally or at a local level, communities rarely have the opportunity to say no.

Although the limitations of the consultative approach are obvious from the planners' perspective, at least they have demonstrated an intention to listen to the views from other stakeholders and incorporate them into tourism planning. The voice from the public can be heard by the planners. Compared with boosterism, economic or even the physical/spatial traditions, it is a more decentralized approach with limited participation of stakeholders in tourism development and planning.

Collaboration

Compared with the consultative approach, the collaborative approach believes that the decisions about tourism development should not only be left to a few politicians, government officials or tourism entrepreneurs. Various stakeholders should have the opportunity to be involved in the process for tourism planning. Stakeholder collaboration has the potential to lead to dialogue, negotiation and the building of mutually acceptable proposals about how tourism should be developed. It involves regular face-to-face interactions between stakeholders who may be in the public realm, private or voluntary sectors, including pressure and interest groups. The intention is to provide a platform for stakeholders and planners to listen to each others' views on the plan for development. Through the dialogue, stakeholders would have the opportunity to discuss and negotiate about planning issues, and to reach an agreed position or consensus about the plan. During this process it is crucial to have the correct mix of stakeholders in the collaborative approach. The identification and selection of stakeholders often takes account of not only their skills capacity and willingness to contribute, but also their political and economic power (Medeiros de Araujo and Bramwell, 1999, 2002). This implies both mutual learning and shared decision making (Bramwell and Lane, 2000).

There are apparent benefits of collaborative planning. First, the wide participation of stakeholders can greatly enhance the democracy of tourism governance, if a cross section of parties is involved in decisions about future development. The communication between stakeholders would encourage them to express views on tourism planning and help understand their positions. In this way, the broadly based ownership of decisions on tourism planning would bring democratic empowerment and equity, operational advantages, and an enhanced tourism product (Jamal and Getz, 1995; Timothy, 1999). Secondly, the wide participation in tourism planning can pool various resources into tourism development and in doing so, two or more stakeholders could also solve a set of problems which neither can solve individually (Selin and Chavez, 1995). The destination product can then be enhanced through this type of collaborative development. Finally, the collaborative decision-making process can help encourage greater coordination of relevant policies and ensure the implementation of the tourism plan is more effective. Since the plan is based on negotiation and discussion, it may help identify the potential problems existing in the plan and therefore help reduce adversarial conflicts in tourism development. As Gunn (1994) indicates, the incorporation of the local community's opinions and desires in the planning process does not mean the absence of the professional opinions of planners, but makes the final decisions potentially more implementable.

Although the benefits of the collaborative approach are quite obvious, there are challenges and potential difficulties for this approach (Bramwell and Lane, 2000). In particular, substantial problems remain in the operation and implementation of this process. First, the involvement of diverse stakeholders at regular meetings and in negotiations can be complex and very time consuming. With a dynamic market environment or emerging crisis at the destination, prompt responses might be needed in the 'planning' for tourism. Regular meetings could be viewed as slowing down the decision-making process and in particular, stakeholders involved in the discussion may direct the focus of the meeting to the bargaining of other interests. Secondly, a collaborative approach might assume that all stakeholders would have an equal position in the decision-making process. However, in the domain of tourism development there are always more powerful stakeholders who have unequal power in the control of or access to various resources (i.e. information, funding, marketing channels, etc.). As a result, some organizations may not take into account the views of other parties on an equal basis in the negotiation process. Furthermore, to protect their own interests, the more powerful stakeholders may continue to work together ignoring others and creating barriers during the implementation of the tourism plan. Thirdly, although regular public meetings could help facilitate the discussion between stakeholders, it can also be exploited by those individuals and organizations who know how to utilize meeting procedures and dynamics in their favour (Hall, 2008).

Compared with the consultative planning approach, the collaborative approach facilitates a much wider and higher level of stakeholder participation. As a 'bottom-up' approach, stakeholders are directly involved in the decision-making process for tourism development. Additionally, stakeholders have an intention to listen to views expressed and to discuss and negotiate around them in meetings and could also encourage stakeholders to reach some form of agreement about how tourism might be developed.

Case Study: The Process of Integrated Sustainable Tourism Planning – Theoretical Insights

It is important to emphasize that this case study is not designed to position this approach within highlighted cultural contexts but rather to re-emphasize, from a theoretical perspective, how the approaches to tourism planning, previously discussed in this chapter, can be combined with complex social, economic, environmental and political dimensions to form the process of integrated sustainable planning.

With more narrow concerns of the role of marketing and the physical environmental values in tourist destinations, many authors have advocated a more sustainable and integrated planning process for tourism planning to derive benefits for communities (Getz, 1986; Inskeep, 1994; Bramwell and Lane, 2000; Gunn and Var, 2002; Dredge and Jenkins, 2007; Hall, 2008; Edgell and Swanson, 2013). Inskeep (1994) suggests that an underlying concept in tourism planning is that tourism should be viewed as an interrelated system of demand and supply factors. Therefore, it is necessary to have a holistic view of the tourism system and also consider how tourism can be planned in a more integrated way, embodying and promoting sustainable development at different levels. Various models to describe the process and application of integrated planning for tourism have been developed by authors (Inskeep, 1994; Gunn and Var, 2002; Dredge and Jenkins, 2007; Hall, 2008). Although there is a diversity of potential applications for planning, the basic nature of the planning process is remarkably uniform in theory (Williams, 2009). An integrated planning process usually involves a normative and systematic approach to desired planning outcomes, which normally includes eight different stages, highlighted in Fig. 1.1.

(*Continued*)

Case Study. Continued.

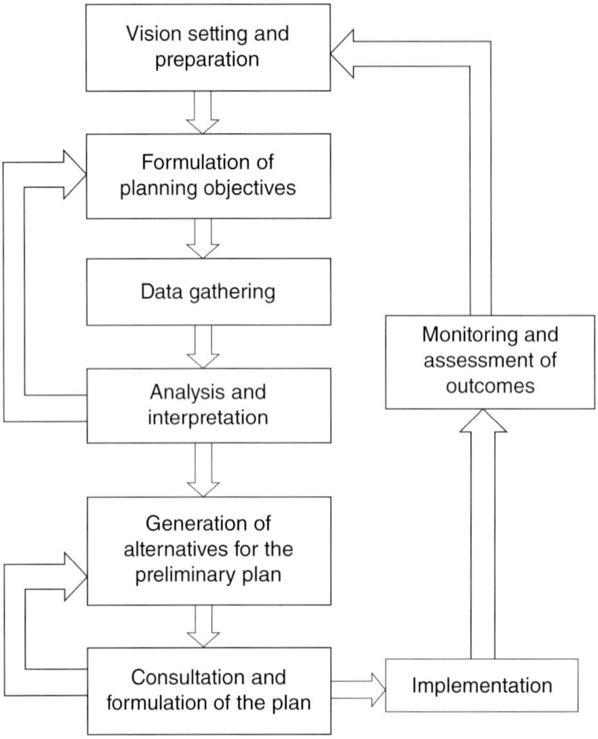

Fig. 1.1. A suggested integrated process for tourism planning.

The elements included in the model sometimes overlap or can be repeated in planning practice and the model suggests an iterative planning and decision-making system which is able to adapt and change to its cultural context. An iterative planning process also allows regular adjustment in the elements of physical development, service delivery and visitor management (Williams, 2009). The model provided is not intended to map out prescriptive steps for tourism planning; rather it reflects some of the latest development of planning theories to be incorporated into considerations of planning for tourism.

Stage One: Vision setting and preparation

The first stage of the planning process normally involves the development of the strategic vision for the plan which, at an early stage of planning, is important to provide an initial picture of the general physical environment and the current development of tourism at the destination. Issues which need to be reviewed include: (i) from the tourism demand side, the existing market characteristics of tourists (both international and domestic); and (ii) from the supply side, accessibility, accommodation and the level of existing infrastructure and facilities. An assessment of the existing level of development provides

(*Continued*)

Case Study. Continued.

a clue to development guidance and prepares for further analysis of the destination in the following stages. In the vision setting of a tourism plan, the planning team needs to determine options and scenarios of tourism development based on relevant economic, environmental, socio-cultural and institutional elements. It is also necessary to set up the horizon year for the plan based on the objectives. To achieve continuity and consistency of tourism development, a timetable for completion should be formulated and a decision on which stakeholders might be involved in the tourism planning process needs to be made as well. As the outcome of the first stage of tourism planning, there should be a statement on the strategic vision developed as guidance for the planning process.

Stage Two: Formulation of planning objectives

Based on a strategic vision developed from the first stage, the second stage considers the more specific objectives for the plan. Objectives are central to developing the strategy as they state the desired results of developing tourism in the country, region or locality (Inskeep, 1994). Compared with the strategic vision setting, the objectives must be set up in a comprehensive and specific way. The formulation of objectives should take consideration of the benefits of tourism development and the potential role that tourism plays in the wider level of the society. Therefore, instead of just focusing on the tourism sector, it also needs to consider how tourism could link with other sectors and contribute to the wider goals of sustainability for the country, region or locality (Bramwell and Lane, 2000). Hence, regional and community goals should also be taken into account when setting up objectives for tourism development. Tourism objectives should not be contradictory to the more general development objectives or objectives already adopted in relevant sectors for a country or destination; rather they should reflect and reinforce them to promote a coherent and consistent pattern of development (Inskeep, 1994). It is worth noting that the process of objectives setting is a political process, as the planning team often has to choose between priorities, especially when there are only limited resources for tourism development. Because of that, at this stage, actors in the process might find contradictory or conflicting objectives between tourism and tourism-related sectors. As integrated planning is based on an iterative process, the objectives can be provisional, which allow them to be changed or refined after the study of destination analysis.

Stage Three: Data gathering

The third stage focuses on the collection of relevant data that could be used as a reference for the destination analysis and planning. According to the preliminary objectives set up in the second stage, data essential to the analysis and development of the planning process should be gathered. Inskeep (1994, p. 13) provides a detailed list for the type of data required in formulating a comprehensive plan for a country or region, as stated below:

- global and regional tourism patterns and trends;
- characteristics of competing tourist destinations;

(*Continued*)

Case Study. Continued.

- tourist arrival trends in the country or region;
- existing and potential tourist attractions and activities;
- existing infrastructure, and tourist facilities and services;
- existing natural resource, land use and land tenure patterns;
- existing economic and employment patterns;
- physical and social development plans;
- environmental characteristics and environmental quality;
- socio-cultural patterns, especially as they relate to the effects of tourism;
- existing private sector organizations in tourism;
- present investment policies and availability of capital for investment in tourism; and
- present tourism legislation and regulations.

Questions regarding what type of data should be collected and how much information should be collected largely depend on the objectives and requirement of the plan. If necessary the public or interest groups should be consulted to help determine any additional information required to assist in the planning process.

Stage Four: Analysis and interpretation

The fourth stage is to analyse the data collected from Stage Three and to provide guidance and a reference to assist the planning for the country or region. At this stage interpretation of information and data gathered will be meaningful to assist the decision-making process. Instead of providing an analysis of each individual element for the country or region, the collected data should be brought together and consideration should be given to relations between the elements. In the analysis, planners will need to develop an understanding of the views from different stakeholders engaged in the process. The assessment of carrying capacity and LAC is often important to provide a reference for the development of compatible infrastructure and tourist facilities. Many methods and models are available to assist the destination analysis in order to obtain a more comprehensive understanding of the situation. For example, for planners who have inadequate expertise and less experience in tourism planning, PESTEL (political, economic, social, technological, environmental, legal) and SWOT (strengths, weaknesses, opportunities and threats) models can be adopted to scan the destination environment and provide internal and external analysis. The results from those two analyses can help identify the potential opportunities and threats or constraints for tourism development. Destination analysis provides focal points for formulating recommendations, and how to take advantage of opportunities and overcome constraints to lead to better development of tourism (Inskeep, 1994). The marketing analysis for the destination is often one of the most important analyses. If tourism has already been developed then it is essential to determine the stage of tourism development at the destination (Butler, 1980). Competing destinations offering similar tourism products should be critically evaluated. Taking a more systematic view on tourism, the analysis of tourism resource

(*Continued*)

Case Study. Continued.

is usually not isolated from the analysis of the tourism demand for the destination. A joint analysis on tourism marketing and tourism resources can often help determine the future position of the destination in the market. For example, Gilbert's (1990) differentiation strategy suggests that destinations should differentiate the products they offer and make efforts to improve their destination image in the market. By linking Gilbert's model with Butler's destination life cycle model, it can help decide the best strategy for destination marketing or even a re-image strategy for the destination. The analysis conducted at this stage provides conclusions and recommendations that could lead to making or conceptualizing a preliminary plan (Goeldner and Ritchie, 2012). If the results from the analysis suggest that the objectives are inappropriate or unachievable for the destination, then based on the review of the results, it is necessary to return to the second stage and formulate a new set of planning objectives.

Stage Five: Generation of alternatives for the preliminary plan

In accordance with the analysis of the situation, the fifth stage of integrated planning involves the generation of alternatives for the preliminary plan. It is very rare to have only one development scenario that could meet development objectives for a destination. Inskeep (1994) argues the best approach to plan formulation is to prepare and evaluate alternative development scenarios. There are a few factors which need to be taken into consideration when outlining and selecting the alternative scenarios and strategies: (i) the potential economic benefits generated by each alternative; (ii) whether there is appropriate community involvement; and (iii) whether each alternative promotes a balanced development at the destination. Based on the evaluation of the alternative future scenarios, the planning team should select the preferred scenario and formulate a preliminary plan. In the formulation process, attention should be paid to the physical solutions, development scale and legal requirement. Specific actions, timescale, financial plans for each objective should also be decided at this stage (Gunn and Var, 2002; Goeldner and Ritchie, 2012). Wider participation should normally be adopted in the generation of the preliminary plan if the resources allow. The involvement of a relevant stakeholder group can promote a mutually beneficial development pattern, by pooling together knowledge, capital and other resources for tourism development.

Stage Six: Consultation and formulation of the plan

After the preliminary plan is formulated, it is necessary to have a wider public consultation about the plan. If wider participation has been achieved in the previous stages, it would be very rare to receive many negative comments or strong objections from stakeholders. Attention should be paid to the opinions and feedback from the groups who are not previously involved in the collaboration and consultation. Their views can normally reflect issues overlooked from the previous stages. According to the results from

(*Continued*)

Case Study. Continued.

the public consultation, if major issues are identified, it is sometimes worth reviewing or regenerating the alternatives from previous stages. If there is no significant contradiction with the preliminary plan, the final plan can be formulated and approved. The detailed plan should include a definition of land use, infrastructure and facilities, timetable and estimates of costs for the development. When determining the land use, a zoning policy should be developed to specify legal types of land use (i.e. location, number, function) and building codes. The demarcation of zones is usually according to the environmental and ecological sensitivity of the specific area. For example, in a national park, it might have: (i) preservation zones which prohibit all use; (ii) scientific research zones for permitted scientific research; and (iii) general use zones for scientific, educational and recreation uses. Besides the physical spatial definition and plans, it is also important to define the responsibilities of stakeholders and the distribution of benefits from tourism. The final plan should be approved and the legal basis of the plan must be established. It could ensure the objectives and planning recommendations are accomplished and more importantly, the implementation would not deviate far from the approved plan.

Stages Seven and Eight: Implementation/monitoring and assessment of outcomes

With the approval of the final plan for tourism development, the techniques for implementation should be specified and applied. Implementation follows through the final plan and promotes operational tourism development. However, it is often wrong to assume that implementation should only be considered at the implementation stage. The implementation should be considered throughout the planning process and by doing that, the final plan and recommendations are realistic to achieve and are implementable (Inskeep, 1994). An integrated planning system should also develop a mechanism of monitoring. The monitoring of tourism planning can provide continuous and consistent feedback on the plan and evaluate the actual results of the plan from implementation. The different stages and programmes of implementation should be constantly examined and checked against the objectives of the plan. Any specific regulations, zoning policies and development standards (e.g. landscape, industrial standards) must be applied and closely monitored. Periodic reports can be produced to provide a record of implementation. The key benefit of monitoring is that it can help detect problems emerging from the implementation of the plan and if necessary, it could facilitate remedial measures and corrective actions to avoid any serious consequences brought to the destination. At the monitoring stage, it is important to work with stakeholders at the destination to determine any inappropriate or impractical programme or actions. Their comments can usually improve the actual results of the plan and help achieve a more balanced and sustainable development at the destination. More importantly, those comments are often useful in determining the objectives for the next planning period. Integrated planning for tourism considers the planning as a structured iterative

(*Continued*)

Case Study. Continued.

and learning process. It accepts that many aspects of strategy emerge continuously and the complexities in 'real' planning situations must and can be fully recognized. Through continuous adjustment, it addresses the available resources at the destination, the requirement of the market and, most importantly, the appropriate distribution of tourism benefits between stakeholders. It promotes a wider participation of stakeholders in the planning process and places value on learning from experience and listening to people. Stakeholders involved in the planning process learn how to be effective in terms of the most appropriate set of goals, objectives, actions, indicators, institutional arrangements and practices (Hall and Page, 1999). A comprehensive planning system based on this approach can help, as an idealized model, achieve balanced and sustainable tourism development.

CONCLUSION

This chapter initially explained and discussed the theoretical landscape of tourism planning studies and the relationship between tourism planning and policy. As one of the fastest growing industries in the world, tourism has become a major industry in the last few decades. However, associated with its economic benefits, the negative environmental and socio-cultural impacts raised concerns from both the industry and the local community at tourism destinations. As a response to that, tourism planning gradually plays an important role nowadays in the management of impacts of tourism and in the promotion of more balanced and sustainable tourism development options for a locality. It is imperative for countries and destinations that wish to employ tourism for regional and national development.

This chapter also provided a detailed review and discussion on the traditions of tourism planning, acknowledging that tourism planning paradigms have evolved from a more simplistic view with an emphasis on economic benefits to a more sophisticated, decentralized and integrated approach with more emphasis on the environmental, socio-cultural and community values. Additionally, Getz's (1986) classification on tourism planning traditions has been reviewed and the various planning methods associated with each tradition discussed.

As suggested in the discussion, the traditions of tourism planning are not exclusive from each other, rather they are interrelated and overlap in the practice of tourism planning. Evolved from the four traditions and aiming to match the tourism demand with tourism supply, integrated planning is based on a strategic, iterative and learning process, with recognition of the role of stakeholders in tourism development. With a review of the characteristics of this approach, this chapter also offered an integrated planning model which could be adopted as a guide for tourism planning to work in an integrated way to achieve the sustainability of the destination. The discussion in this chapter has demonstrated clear links between tourism planning and the principles of sustainability and the next chapter provides an in-depth critique of different aspects of sustainability and sustainable development and linkages with the planning processes.

QUESTIONS

1. What are the impacts of tourism? How could tourism planning address those impacts?

2. Compared with boosterism and economic, industry-oriented traditions of tourism planning, what are the strengths and limitations of the physical/spatial planning approach?

3. By linking with the integrated planning approach, how could community values be addressed in the tourism planning process?

BIBLIOGRAPHY

Baidal, I. and Josep, A. (2004) Tourism planning in Spain: evolution and perspectives. *Annals of Tourism Research* 31(2), 313–333.

Bramwell, B. (2006) Actors, power, and discourses of growth limits. *Annals of Tourism Research* 33(4), 957–978.

Bramwell, B. and Lane, B. (eds) (2000) *Tourism Collaboration and Partnerships: Politics, Practice and Sustainability*. Channel View Publications, Clevedon, UK.

Bramwell, W. and Lane, B. (1993) Sustainable tourism: an evolving global approach. *Journal of Sustainable Tourism* 1(1), 1–5.

Bramwell, W., Henry, I., Jackson, G., Prat, A.G., Richards, G. and van der Straiten, J. (1996) *Sustainable Tourism Management: Principles and Practice*. Tilburg University Press, Tilburg, The Netherlands.

Brindley, T., Rydin, Y. and Stoker, G. (1996) *Remaking Planning: The Politics of Urban Change*. Routledge, London.

Buckley, R. (1999) An ecological perspective on carrying capacity. *Annals of Tourism Research* 26(3), 705–708.

Butler, R. (1980) The concept of a tourism area cycle of evolution: implications for resources. *Canadian Geographer* 24(1), 5–12.

Coccossis, H. and Mexa, A. (2004) *The Challenge of Tourism Carrying Capacity Assessment: Theory and Practice*. Ashgate, Aldershot, UK.

Dredge, D. (1999) Destination place planning and design. *Annals of Tourism Research* 26(4), 772–791.

Dredge, D. and Jenkins, J. (2007) *Tourism Planning and Policy*. Wiley, Brisbane, Australia.

Edgell, D.L. and Swanson, J.R. (2013) *Tourism Policy and Planning: Yesterday, Today and Tomorrow*, 2nd edn. Routledge, New York.

Elliot, J. (1997) *Tourism: Politics and Public Sector Management*. Routledge, London.

Fayos-Sola, E. (1996) Tourism policy: a midsummer night's dream? *Tourism Management* 17(6), 405–412.

Getz, D. (1986) Models of tourism planning: towards integration of theory and practice. *Tourism Management* 7(1), 21–32.

Getz, D. (1987) Tourism planning and research: traditions, models and futures. Paper presented at the Australian Travel Research Workshop, 5–6 November. Bunbury, Western Australia.

Gilbert, D. (1990) Strategic marketing planning for national tourism. *The Tourist Review* 1, 18–27.

Godfrey, K. (1996) Towards sustainability? Tourism in the Republic of Cyprus. In: Harrison, L.C. and Husbands, W. (eds) *Practicing Responsible Tourism: International Case Studies in Tourism Planning, Policy, and Development*. Wiley, New York, pp. 58–79.

Goeldner, C.R. and Ritchie, J.R.B. (2012) *Tourism: Principles, Practices, Philosophies*, 12th edn. Wiley, Hoboken, New Jersey.

Gunn, C.A. (1994) *Tourism Planning: Basics, Concepts, Cases*, 3rd edn. Taylor & Francis, Washington, DC.

Gunn, C.A. and Var, T. (2002) *Tourism Planning: Basics, Concepts, Cases*, 4th edn. Taylor & Francis, New York.
Hall, C.M. (2008) *Tourism Planning: Policies, Processes and Relationships*, 2nd edn. Pearson Prentice-Hall, Harlow, UK.
Hall, C.M. and Jenkins, J.M. (1995) *Tourism and Public Policy*. Routledge, London.
Hall, C.M. and Page, S. (1999) *The Geography of Tourism and Recreation*. Routledge, London.
Hall, C.M., Jenkins, J. and Kearsley, G. (1997) *Tourism Planning and Policy in Australia and New Zealand: Cases, Issues and Practice*. McGraw-Hill, Sydney, Australia.
Hammitt, W.E. and Cole, D.N. (1987) *Wildland Recreation Ecology and Management*. Wiley, New York.
Inskeep, E. (1991) *Tourism Planning: An Integrated and Sustainable Development Approach*. Van Nostrand Reinhold, New York.
Inskeep, E. (1994) *National and Regional Tourism Planning: Methodologies and Case Studies*. World Tourism Organization and Routledge, London.
Jamal, T.B. and Getz, D. (1995) Collaboration theory and community tourism planning. *Annals of Tourism Research* 22, 186–204.
Leiper, N. (1993) Defining tourism and related concepts: tourist, market, industry and tourism system. In: Khan, M.A., Olsen, M.D. and Var, T. (eds) *VNR's Encyclopedia of Hospitality and Tourism*. Wiley, New York, pp. 539–558.
Lucas, A.R. (1978) Fundamental prerequisites for citizen participation. In: Sadler, B. (ed.) *Involvement and Environment: Proceedings of the Canadian Conference on Public Participation*. Environment Council of Alberta, Edmonton, Canada, pp. 47–57.
Massieu, A. (2004) *Benefits Resulting from a Tourism Satellite Account for the Compilation of National Accounts*. Available at: statistics.unwto.org/sites/all/files/docpdf/benefits.pdf (accessed 18 May 2014).
Mathieson, A. and Wall, G. (1982) *Tourism: Economic, Physical and Social Impacts*. Longman, Harlow, UK.
McCool, S.F. (1994) Linking tourism, the environment, and concepts of sustainability: setting the stage. In: McCool, S.F. and Watson, A.E. (eds) *Linking Tourism, the Environment and Sustainability*. United States Department of Agriculture (USDA) General Technical Report INT-GTR-323. USDA, Ogden, Utah, pp. 3–7.
McCool, S.F. and Lime, D.W. (2001) Tourism carrying capacity: tempting fantasy or useful reality? *Journal of Sustainable Tourism* 9(5), 372–388.
Medeiros de Araujo, L. and Bramwell, B. (1999) Stakeholder assessment and collaborative tourism planning: the case of Brazil's Costa Dourada project. *Journal of Sustainable Tourism* 7(3/4), 356–378.
Medeiros de Araujo, L. and Bramwell, B. (2002) Partnership and regional tourism in Brazil. *Annals of Tourism Research* 29(4), 1138–1164.
Mitchell, R.E. and Reid, D.G. (2001) Community integration: island tourism in Peru. *Annals of Tourism Research* 28(1), 113–139.
Mowforth, M. and Munt, I. (2003) *Tourism and Sustainability: Development and New Tourism in the Third World*, 2nd edn. Routledge, London.
Murphy, P.E. (1985) *Tourism: A Community Approach*. Methuen, London.
Page, S. (1995) *Urban Tourism*. Routledge, London.
Peck, J. (2001) Neoliberalizing states: thin policies/hard outcomes. *Progress in Human Geography* 25(3), 445–455.
Pedersen, A. (2002) *Managing Tourism at World Heritage Sites: A Practical Manual for World Heritage Site Managers*. United Nations Educational, Scientific and Cultural Organization (UNESCO) World Heritage Centre, Paris.
Scott, N. (2011) *Tourism Policy: A Strategic Review*. Goodfellow Publishers, Woodeaton, Oxford.
Selin, S. (1999) Developing a typology of sustainable tourism partnerships. *Journal of Sustainable Tourism* 7(3/4), 260–273.
Selin, S. and Chavez, D. (1995) Developing a collaborative model for environmental planning and management. *Environmental Management* 19(2), 189–196.

Sewell, W. and Phillips, S. (1979) Models for the evaluation of public participation programs. *Natural Resources Journal* 19, 337–358.

Simmons, D.G. (1994) Community participation in tourism planning. *Tourism Management* 15(2), 98–108.

Smith, V.L. (ed.) (1978) *Hosts and Guests: The Anthropology of Tourism*. University of Pennsylvania Press, Philadelphia, Pennsylvania.

Stankey, G.H. and McCool, S.F. (1984) Carrying capacity in recreational settings: evolution, appraisal and application. *Leisure Sciences* 6(4), 453–473.

Stankey, G.H., Cole, D.N., Lucas, R.C., Petersen, M.E. and Frissell, S.S. (1985) The limits of acceptable change (LAC) system for wilderness planning. General Technical Report INT-176. United States Department of Agriculture (USDA) Forest Service Intermountain Forest and Range Experiment Station, Ogden, Utah.

Stokes, G.L. (1990) The evolution of wilderness management. *Journal of Forestry* 88(10), 15–20.

Timothy, D.J. (1998) Cooperative tourism planning in a developing destination. *Journal of Sustainable Tourism* 6(1), 52–68.

Timothy, D.J. (1999) Participatory planning: a view of tourism in Indonesia. *Annals of Tourism Research* 26(2), 371–391.

Tosun, C. and Jenkins, C.L. (1998) The evolution of tourism planning in Third World countries: a critique. *Progress in Tourism and Hospitality Research* 4(2), 101–114.

Turner, L. and Ash, J. (1975) *The Golden Hordes: International Tourism and the Pleasure Periphery*. Constable and Robinson, London.

United Nations World Tourism Organization (UNWTO) (2000) Basic Concepts of the Tourism Satellite Account. Available at: statistics.unwto.org/sites/all/files/docpdf/concepts.pdf (accessed 19 May 2014).

United Nations World Tourism Organization (UNWTO) (2013) Understanding Tourism: Basic Glossary. Available at: media.unwto.org/content/understanding-tourism-basic-glossary (accessed 1 December 2013).

United Nations World Tourism Organization (UNWTO) (2014) Tourism Highlights 2014. Available at: mkt.unwto.org/publication/unwto-tourism-highlights-2014-edition (accessed 2 June 2014).

Williams, S. (2009) *Tourism Geography: A New Synthesis*, 2nd edn. Routledge, London.

World Commission on Environment and Development (WCED) (1987) *Our Common Future*. Oxford University Press, Oxford.

Yuksel, F., Bramwell, B. and Yuksel, A. (1999) Stakeholder interviews and tourism planning at Pamukkale, Turkey. *Tourism Management* 20(3), 351–360.

chapter 2

Sustainability: From Theory to Practice

Nigel D. Morpeth* and Hongliang Yan

Leeds Beckett University, Leeds, UK

LEARNING OBJECTIVES

- To consider both Eastern and Western perspectives and different discourses and dimensions of the concept of sustainability.
- To differentiate between the concepts of sustainability and sustainable development and apply these concepts politically and pedagogically.
- To evaluate whether the principles of sustainability have become central to policy and planning agendas globally and in the UK.

INTRODUCTION

Within this chapter we initially explore temporal considerations, in determining how we might assess the extent to which progress towards sustainability has been realized, given the relatively short timescales of the dissemination of discourses of sustainability. We then consider how Eastern perspectives on sustainability can both inform and redress the dominance of narratives of sustainability which have emerged within academic and policy discourse, largely from a Western perspective. In considering the origins and contrasts between sustainable development and sustainability, there is a review of the institutional arrangements and criticisms of these arrangements for advancing the principles of sustainability from the supra-national to the local level. The chapter finally considers the interaction between sustainability and pedagogy, to evaluate how progress towards realizing sustainability can be evaluated both in

*n.morpeth@leedsmet.ac.uk

policy and in educational arenas and inform planning processes.

SUSTAINABILITY AND TEMPORAL AWARENESS

E.H. Gombrich's ground-breaking 1936 book *Eine kurze Weltgeschichte fur junge Leser* (English translation *A Little History of the World*; Gombrich, 1985) created a template for other writers to consider how they might best condense a lengthy commentary on the history of epochs of the world, into one book. It was written in just a 6-week period and chronicled a history of the world, which straddled the Stone Age to the Atomic Age. What was excluded from this condensed narrative of the world, were key dates and so-called facts. However, he was still able to capture what he considered to be significant epochs, within the history of humankind, stretched over a 6 million year timeframe. It was an enormous and unique undertaking in providing such an in-depth overview of the evolution of the world (for a more contemporary compression of the history of time see also Hawking's (1988) *A Brief History of Time: From the Big Bang to Black Holes*).

It is pertinent to consider how a book which chronicles the history of sustainability, as an explicitly articulated concept, would be written. What timeframe would it cover and would it be over a similar timeframe to Gombrich's book? While he did not use the word sustainability, Gombrich identified a world that revolutionized the way in which natural resources were used by humankind, which is at the heart of discourses of sustainability. His final words are redolent of the considerations that underpin contemporary discourses of sustainability and also emphasize his humanist sensibilities when he stated that:

> Whenever an earthquake, a flood or a drought in a far-off place leaves many victims, thousands of people in wealthier countries put their money and their efforts into providing relief. And that, too, used not to happen. Which proves that we still have the right to go on hoping for a better future.
>
> (Gombrich, 1985, p. 284)

While these humanist sentiments are apposite to themes within this chapter, Gombrich's book provides a template for the contemporary reader to explore a horizon line of many millennia to measure the passage of time. This is in contrast to the current age when events are measured by a so-called '24/7' time yardstick (intensified by 24/7 media reporting of global events) that governs the cadence of modern life and our capacity to think beyond the 'immediate' and the present might diminish and impair our visualization of the future, which might be predicated on short time and nano-timescales. As we will see within the progression of this chapter, the building blocks of theorizing about sustainability and the time horizons for planning require the capacity, in part, to conceptualize about long and distant (and unknown and unknowable) timescales. These timescales are infused with complex ideas of intra- and inter-generational and so-called futurity considerations, which have an additional dimension of ethical complexity, when linked to inter-generational equity. O'Riordan (2001) reminds us that considerations of sustainability are about cognisance of histor(ies) and the future.

Such theorizing and visualization might become diminished, if indeed redundant, with an inability to think imaginatively and with prescience beyond the 'moment' and the 'here and now'. As stated in Chapter 3, this volume, the UK coalition government's impatience with the timescales taken to advance planning proposals at a local level, has led them to advocate a reform and acceleration of the planning process and the role of the local state. Indeed the Dockland's development in London, in the UK in the 1980s and 1990s, demonstrated how the local states' stewardship in guiding local visions for development was supplanted by an unelected and unaccountable Development Corporation. The result of this reform, in some people's minds, created an unwanted gentrified 'city within a city', with scant regard to participatory planning, let alone local visions of sustainability (Brindley *et al.*, 1996). Advocates of this style of development and planning argued that it accelerated the process of much-needed regeneration, to an area that was derelict. This emphasizes how there are competing discourses and value judgements on how planning and development scenarios might transform localities.

It is important to note that there have been many starting points in recent decades, and pronouncements both academically and in policy circles, about the significance of sustainability. As a significant concept, it covers a multiplicity of fields of human endeavour related to the everyday lives of people across the globe. These starting points reflect, in part, the role and capacity of the national, regional and local state, in conjunction with local communities, to find a common language and currency of sustainability, to advance actions on working towards sustainability. Indeed, while the significance of the concept of sustainability to the field of tourism has been documented over many decades, arguably the starting points for chronicling progress in realizing sustainable tourism have been written from 'First World' Western perspectives. These North–South observations about how tourism 'works', privilege Western perspectives on the global organization of tourism. With these considerations in mind, we have decided within this book to start with a commentary on sustainability using Eastern perspectives. This counters Western perspectives of following the orthodoxy of pursuing economic growth, while seeking the goal of sustainability (to use a well-known phrase, as a society we want 'jam today and jam tomorrow' – in other words, we want growth *and* sustainability).

EASTERN PERSPECTIVES ON SUSTAINABILITY

Environmental issues and concerns in contemporary society often provide incentives to search for alternative views of the human–environment relationship. Studies of Chinese philosophical traditions are an important contributing factor in considering ecological concerns in the West, as they provide an important conceptual resource in the establishment of a mutually fulfilling and beneficial relationship of humankind and nature (Callicott and Ames, 1989). Chinese traditional philosophies, and in particular Taoism, view the natural world as a fundamentally harmonious dynamic balance, since it is made up of relational opposites which are mutually interdependent and hence not in antagonistic conflict (Jenkins, 1998). As an ancient philosophical tradition, Taoism is normally viewed as a philosophy or school of thought which has significant influence on the culture and societies in Eastern Asia. With the philosophy of Lao Zi and Zhuang Zi being generally recognized as its basic source, the philosophical works of Taoism reflect a wide range of perspectives in ecology. Taoist philosophers believe that nature inclines towards a condition of harmony and balance, not through being forced and constrained but rather through following 'its own way' – Tao (Shi, 2002). The Chinese term Tao, translated as 'way', is regarded as the principle of the universe and the infinite way of the universe, which is without beginning and end and is characterized as 'normalcy, naturalness and selflessness' (Ayers, 1971, p. 134). Taoism views nature as a whole body, rather than fragmented elements and promotes an holistic view when it comes to environmental issues. It emphasizes humankind's symbiotic relationship with natural ecosystems, and the interdependence of mind and nature (Clarke, 1997).

One of the main Taoist principles is 'wu wei', literally translated as 'non-action'. It is regarded as 'acting without artificiality' rather than 'without any action'. From the environmental perspective, it is often considered as action that 'does not go against the grain of nature' (Needham, 1956) or action 'in accordance with nature' (Ip, 1983). It emphasizes the qualities of non-assertion and downplaying the importance of invasive human interventions and suggests human action should be complementary to, rather than contrary to nature. As Jenkins (2002) indicated, 'wu wei' implies that both nature and humanity flourish best without undue coercion, and that techniques and power are best applied non-aggressively, heeding intrinsic environmental properties and natural trends.

Taoist philosophy is often considered as an alternative view on the values of the human–environment relationship in contrast with Western views. For example, David Hall in his work 'On seeking a change of environment' argued that the anthropocentric view of nature in Western culture has caused environmental problems. He indicated that Taoism is a philosophy with an alternative perspective on 'order' and nature and produces a more satisfactory human–environment relationship (Hall, 1989). Compared with the Western view on the human–environment relationship, Taoist philosophy considers the law of nature is the most appropriate norm of human behaviour and a balance between extremes, which thereby creates a dynamic stability (Sabelli, 1998). Similarly, Callicott (1994) also stresses that from an environmental standpoint, Taoism is more compatible with the contemporary explanation of natural systems development, than the traditional Western stance on the environment. Taoist philosophers and practitioners still make their voices heard on environmental issues. In particular, in China, Taoist principles have a unique role in the promotion of environmental responsibility, sustainable practices through education and the environmental conservation at Taoist sacred mountains and sites.

GRAND NARRATIVES OF SUSTAINABILITY

While we can speculate on starting points and 'dates' to analyse sustainability from Western perspectives, it is possible to make broad pronouncements about so-called grand narratives (major themes) of sustainability. Gautier (2009, p. 1) identified that there were three distinct variants of sustainability: (i) environmental sustainability, which is viewed as living within environmental limits, protecting limited natural resources and limiting climate change; (ii) social sustainability, which includes promoting 'healthy' communities and a 'just' society; and (iii) economic sustainability, which builds 'strong' and stable economies. Common concerns about what the elements of sustainability are include considerations that there are inescapable challenges of the so-called 'three pillars' of sustainability that the world community has to address through actions in working towards achieving sustainability. As we see later in Chapter 3, this volume, environmental concerns came to prominence in the 1960s, culminating in 1987, with the Brundtland Report, signalling the start of the popular branding of sustainable development globally. In the context of Gombrich's book which covered a period of 6 million years, the history of sustainability has barely had 30 years of academic and policy coverage, conceptualization and re-conceptualization. Part of the vernacular of this branding is the popular phrase which has emerged in common parlance: 'think globally and act locally'. It encapsulates the aspiration of people in different cultural contexts to consider how their visions for the future capture these global challenges within local actions at a community level.

What has become an important vehicle in the 'branding' of sustainability is the role of the media and almost daily dissemination and narratives of sustainability, highlighting catastrophic observations and predictions of famine, lack of access to clean water, overuse of non-renewable resources, over-population, global conflict and widening inequalities between the rich and poor. One such article, in terms of the reporting of the grand narratives of sustainability, was reported in the UK's *Guardian* broadsheet newspaper. On 31 March 2014, this paper announced that the Intergovernmental Panel on Climate Change (IPCC), the United Nations' (UN's) climate science panel, reporting on climate change, suggested that: 'Every city is at risk. The challenge is too huge.' This paper also reported that: 'The overriding lesson of this report, the scientists said, was that unless governments acted now to reduce greenhouse gas emissions and adopt measures to protect people, nobody would be immune from climate change' (Jenkins, 2014). Starkly, the report emphasized that there had been a rise of 300%, in the number of natural disasters in 2008–2009 compared with 30 years ago. One of the main points of the report was that the poorest people in the world will feel the impact of externalities of global warming caused in the main by Western societies, whose economies historically have been organized towards the unsustainable consumption of non-renewable resources. With regards to this human-centred and anthropocentric approach towards the environment, Porritt highlighted that 'human survival depends on our being able to transcend its anthropocentrism' (1984, p. 206).

An interesting conduit between Western and Eastern perspectives and philosophies about the environment, which in some ways attempts to transcend anthropocentric thinking on sustainability, can be found in the work of James Lovelock and his articulation of the concept of Gaia. Lovelock's (1988) work is in line with the interpretations of deep ecology (strong sustainability) in viewing the biosphere as a 'super-organism', with an ability to recover from ecological and environmental damage, consistent with the rejuvenating qualities of other living organisms. Porritt (1984, p. 206) concurs with Lovelock's Gaia concept of the earth, 'as a self-regulating single system, life-support system, linked to the older universal philosophy of "Mother Earth" protecting the earth' (a phrase viewed as being a patriarchal construct by eco-feminists, see Henderson, 2013). The theory of Gaia is closely allied with the Eastern perspectives previously discussed, and is an ecological interpretation of sustainability, advocating that development should mimic nature and with the adoption of circular flow theory. An example of circular flow theory in this context, is the use in domestic household gardens of circular pipe systems which convert domestic water waste into usable water for irrigation, by the circular motion of unclean water being purified by being moved through a circular pipe-flow system, which filters out the pollutants and impurities in the water. This is an example of small-scale technology that can be replicated in other areas of society, to mitigate the effects of pollution. It is important to note that the theory of Gaia has emerged as a philosophy that has moved from the theoretical realm to be part of planning practice, in the same way that the Taoist philosophy in China was used to inform urban design, demonstrating how urban development can be in harmony with the natural environment. Likewise, the work of the architect Sir Richard Rogers (1997), in China in the 1990s, used the principles of circular flow technology to design buildings that were in harmony with nature, particularly in the ergonomics of design and location. Within a UK context, Gaia theory has moved from planning theory to practice, with organizations such as the Gaia Planning Group adopting the philosophy of Gaia within their professional practice (www.gaiagroup.org).

THEORIZING ABOUT SUSTAINABILITY, ITS ORIGINS AND DISTINCTIVENESS FROM SUSTAINABLE DEVELOPMENT

Sustainability has been characterized as a slippery and nebulous concept, a word with widespread circulation but one that is imprecise in its meaning across different cultural contexts. Nevertheless, the vocabulary of sustainability has gained widespread currency in many policy areas, with claims of the benefits of adopting the principles of sustainability, as part of development strategies and as a key consideration in planning processes. However, without tangible and consistent evidence that these principles can be translated into strategies and policies which are operationalizable, then implementation gaps might appear which will devalue the credibility of the concept of sustainability (Ham and Hill, 1984; Kelder, 1996). As Jackson and Morpeth highlighted 'it is to implementation rather than enunciation and further clarification, that both analysts and the tourism industry must now turn regarding sustainability' (1999, p. 1). Far from wanting to see evidence of the implementation or operationalization of the concept, other academics such as Wheeller (1993) have communicated a scepticism about the rhetorical language of sustainability and, in the case of Sharpley (2009), that the critiques on tourism development have moved 'beyond' sustainability and have assimilated the concept of 'capital(s)' to be used in these critiques.

However, that this book seeks to explore different approaches to planning for tourism, in different cultural contexts, underpinned by the principles of sustainability, emphasizes that far from being redundant, it is viewed as being an important concept for policy makers and planners to seek to implement and operationalize. There is much conjecture about the origins, ideological discourses and interpretations of sustainability and sustainable development. Turner suggests that there is an underlying assumption that sustainability is desirable – 'it is a policy objective we ought to seek to achieve' (1991, p. 209) and Richards identifies that 'sustainable development is a generic policy consideration' (2001, p. 5). Bauler and Hecq note that sustainable development is a continually changing process which poses a challenge to reconcile 'theoretical approaches and practice' (2000, p. 50). Bramwell *et al.* (1996) highlighted that underlying these considerations, are inherent contradictions in melding the concept of sustainability with the concept of development, which is essentially a combination of social egalitarianism and a conservative philosophy. Expanding on Gautier's (2009) so-called 'three pillars' of sustainability, they emphasize that the concept of sustainability should not focus exclusively on the environmental dimension but should integrate economic, cultural, social, policy, managerial and political dimensions, in future planning and development scenarios, in working towards a sustainable future. Their pronouncements in this respect are unusual, in the sense that both academic and policy discourses on sustainability can become 'locked' into the so-called 'three pillars' of sustainability to the exclusion of other dimensions.

TOWARDS A SUSTAINABLE FUTURE?

The phrase 'towards a sustainable future', is not a new one. It was used by the authors Sarre, Smith and Morris in 1991, a year before the first United Nations Conference on the Environment and Development (UNCED) Conference in Rio, which was to set in train a number of conferences on the environment and development, with the themes of sustainability and sustainable development. The words symbolize a quest for global actions on sustainability to protect fragile global resources, which have uncertain and ill-defined actions (and was emphasized in the section on Eastern philosophies; potentially *inactions*). They anticipate that human actions should be modified and constantly recalibrated to avoid the destruction of the environment and the overuse of non-renewable resources. But the conundrum is how to arrive at societal arrangements which achieve this goal, particularly when there is not a consensus in how to work towards these goals. It is important to note that there is interchangeability and nuancing of terms used to reflect these aspirations, for example with the term 'the transition to sustainability' used by O'Riordan in 2001.

Therefore, one approach to seeking to achieve these aspirations was viewed as being through the political and policy arenas of successive world environmental fora and associated reports, popularizing the issue of unsustainable resource use and the need for action. The report of the World Commission on Environment and Development, *Our Common Future* (WCED, 1987) politically mobilized and raised awareness of the concept of sustainable development, on a global scale. It attempted to define the concept of sustainable development for a global audience of politicians, policy makers and citizens. This report was not orientated towards the environment to the exclusion of social equity, and importantly recognized the significance of inter-generational and intra-generational equity, in calling for 'development that meets the needs of the present, without compromising the ability of future generations to meet their own needs' (WCED, 1987, p. 4). Better known as the Brundtland Report, it promoted general economic, social, environmental and political goals for sustainable development, which should 'underlie national and international action on development, preserving vital ecosystems, and natural and cultural diversity' (WCED, 1987, p. 65).

Critics of this definition, identify the simplistic and nebulous qualities embodied within this global clarion call for actions for sustainable development. The report was also criticized for putting forward a global blueprint or 'template' for sustainable development, which might be interpreted as an inappropriate imposition of these development blueprints, on distinctive spatial and cultural localities, which are attempting to implement localized 'sustainable pathways' (Griffiths, 2000). Critics also suggest that these templates are hidebound by discourses of sustainability created within existing 'dominant' economic systems. As such, critics challenged the orthodoxy of traditional approaches to development and economic growth, which eclipse alternative perspectives on how sustainability might be achieved (Shiva, 1992; Galtung, 1995).

Their specific criticism was that the Brundtland Report and successive UNCED conferences promoted a 'dominant economic paradigm', which created a template for growth which serves the interests of developed countries, to the detriment of developing countries (Shiva, 1992, p. 187). Shiva described sustainable development as 'pseudo-sustainability', which was locked into a paradigm of short-termism, based on encouraging economic growth and development (Shiva, 1992, p. 191). Palmer (1992) criticizes the prescriptive development approaches advanced by the Brundtland Report (WCED, 1987) which he interpreted as an imposition of Western culture development patterns and a 'western standard of living on all the world's people, irrespective of their needs and desires' (Palmer, 1992, p. 191). Wiertsema (1995, p. 172) also highlighted the contradictory elements of development, which create non-beneficial changes to the natural environment, arguing, in line with Eastern perspectives discussed earlier, that there should be a focus on the goal of equilibrium rather than growth and an avoidance of blueprints for future development. Like Porritt (1984) the focus should be based on diversity and community-oriented development schemes, in local economies. Wiertsema (1995, p. 231), together with authors such as Fritz (1995), highlighted how technology should mimic natural processes and abandon the 'linear flow' idea (the notion that you transform something from the natural environment, make a product from it, which has a limited lifespan and then it is discarded) which dominates Western production values. Fritz cited Scandinavian eco-villages with 'grass-roots living' as exemplifying a sustainable lifestyle, where villages are designed on the basis of low energy, water usage and recyclable technology, which is ecologically sensitive. This concept was expanded to incorporate eco-municipality projects based on the idea of 'circular flow thinking' and the renewability of resources. These municipal entities also adopted Porritt's (1984) ideal of political decision making on a local level which created their own local visions of sustainability.

Therefore, authors such as Shiva (1992) and Palmer (1992) suggested that in contrast to the contradictory concept of sustainable development, sustainability should be about seeking out sustainable natural processes, cycles and rhythms and in the view of Porritt (1984) the 'principle of diversity', in order to achieve a sustainable society. This would also involve devolved economic and political activity to a local scale, which integrates human as well as physical resources to find new patterns of resource consumption.

THE UN, UNCED AND THE MAINSTREAMING OF SUSTAINABILITY

In 1972, the 'Conference on the Human Environment' held in Stockholm recognized a growing sense of a global ecological crisis, which called for the UN through its United Nations Environmental Programme to monitor global pollution. This programme was designed to monitor a breakdown of 'life support systems' overloaded by human impact on the environment and consider how more prudent use of energy and non-renewable resources could be achieved. However, it was not until the UNCED conference in 1992 in Rio and subsequent conferences through to Rio 2012 (Rio+20) that these considerations were to be the focus of subsequent global environmental fora.

Unsurprisingly, while there was some limited global intra-governmental consensus on what initiatives for mainstreaming sustainability might be, one initiative which emerged from the 1992 UNCED Rio Conference, was Chapter 28 and Agenda 21. This placed a responsibility on 186 signatory nations, that it should be at a micro-level or local level that local authorities across the globe would have the key role in working with local stakeholders and community groups, in establishing local visions for sustainability. This might once again be interpreted as a global blueprint for operationalizing the principles of sustainability, or as a form of branding of shallow sustainability (O'Riordan, 1981). Nevertheless, the Rio conference through this initiative did establish a coherency in suggesting a set of principles, which incorporated complex cultural, environmental, social, economic and political dimensions of sustainability. As will be highlighted later in the chapter, this global commitment to Local Agenda 21 (LA21) had some success, in local municipality authorities, working with local communities to create initiatives for sustainability at a local level.

While this initiative could be viewed as another example of a prescriptive blueprint privileging a Western neo-liberal economic bias, Urban Environmental Management noted:

> that Local Agenda 21 is a local-government-led community-wide, and participatory effort to establish a comprehensive action strategy for environmental protection, economic prosperity and community well-being in the local jurisdiction or area. This requires the integration of planning and development across economic, social and environmental spheres.
>
> (Urban Environmental Management, undated)

In particular the key element of community participation is emphasized and that: 'as the level of governance closest to the people they play a vital role in educating, mobilizing and responding to the challenges of sustainability'. In 2012, the Rio+20 UNCED conference re-stated this commitment to Agenda 21 and produced a report, 'The Future We Want', highlighting 283 aspirations, which range from eliminating world poverty (emphasizing one in five people in the world live in extreme poverty) to greater empowerment for women and communities; stating in general that people should have greater access to decision-making processes on sustainability (UNCED, 2012). Tellingly this report admitted that there had been too little progress in working towards sustainability since the first Rio UNCED conference in 1992. This document provided an update on the progress made in global progress towards sustainability and it demonstrates the perennial search for the answers to the complex questions as to what this concept means and how this concept will persuade humankind to engage in social equitability in the use and allocation of resources. This requires finding the means to alter the unsustainable use of resources and unalterable damage to the environment, ideally through the engagement of communities in participatory and democratic processes.

In 2013, the UN once again reiterated its commitment to sustainable development through the UN General Assembly:

> Emphasizing the need for an improved and more effective institutional framework for sustainable development, the Assembly decided, in the resolution submitted by its President that the High-level Political Forum should provide a dynamic platform for regular dialogue for stocktaking and agenda-setting to advance that process. The New Forum, consistent with its universal intergovernmental character, should also provide political leadership, guidance and recommendations for sustainable development, and review progress in the implementation of related commitments, and enhance integration of the sustainable development's three dimensions (environmental, social and economic) in a holistic and cross-sectorial manner.
>
> (UNCED, 2013)

The conundrum on how to achieve this continues.

THE EUROPEAN COMMUNITY AND SUSTAINABILITY

The European Community has a key role to play in stimulating pan-European initiatives, not least through the European Union (EU) Sustainable Development Strategy (updated 2012). This highlights seven priority areas of sustainability which include: (i) climate change; (ii) transport; (iii) modes of production and consumption; (iv) sustainable management of natural resources; (v) public health; (vi) social exclusion; and (vii) poverty and the fight against global poverty. The challenge for the EU is to encourage European states to instigate initiatives at a local (municipal level) which respond to these priority areas. This is perhaps a task that is made more difficult by no direct references to Agenda 21 in the original 2001 launch of the strategy. Furthermore, it is difficult, 2 years after the 2012 update of the strategy, to find direct references to the current status of the strategy within the work of the EU (EU, 2012).

More widely in continental Europe, in the intervening period between 2001 and 2012, other European organizations have had a role to play in monitoring the progress of European states in working towards sustainability. For example the Federal Office for Spatial Development produced a report titled

'The National Promotion of Local Agenda 21 in Europe', and ironically suggested that as the 'EU is concerned it should also be pointed out that its strategy for sustainable development adopted in 2001 does not refer to Local Agenda 21 as such' (Federal Office for Spatial Development, 2005, p. 17). However, this report does emphasize that in its communications on sustainable development, the EU strategy does encourage LA21 processes at a local level. They go on to say that 'the European Spatial Development Perspective (ESDP) invites member states to take part in Local Agenda 21 (LA21) programmes' (Federal Office for Spatial Development, 2005, p. 17). It was anticipated that these programmes would emerge through legislation, indirect legislation and through programmes that have no specific legal basis. The Federal Office for Spatial Development's report highlighted how they had collaborated with the European Local Action 21 Roundtable, which was involved in a European network which emerged from the Fourth European Conference of Sustainable Cities and Towns in 2004. This report highlighted that in a study of 25 European countries, 22 had National Sustainable Development Strategies (NSDS) and 17 of these countries had LA21 initiatives. Only two countries had introduced LA21 strategies through legislation: Denmark and the UK. Apart from legislative directions, the report suggested that there were other factors which would inform the extent of engagement in LA21 initiatives. These factors might include the size of municipalities, the administrative systems of those states and the extent of political engagement (Federal Office for Spatial Development, 2005, p. 20). The report concluded that national strategies were important to help to ensure coordination of initiatives in regions and localities. The study indicated that it was not possible to calculate the financial and human resources that had been invested by national governments in LA21 processes nationally and therefore calculate success or otherwise. However, what it did emphasize was how important it is that there should be effective communication of LA21 processes and that there should be a variety of different 'players' in LA21 processes at a local level. What was surprising about this report and the 'guidance' at a European level, is that the role of communities in *directing* these processes and establishing local visions for sustainability remained unchronicled and largely invisible. The contemporary existence of the Federal Office for Spatial Development is not known, let alone that there has not been a repeat of their 2005 study and an understanding of how LA21 had evolved within different nation states. While the International Council for Local Environmental Initiatives (ICLEI; an international organization responsible for local government and sustainability) retains a monitoring role of the progress towards sustainability in local authorities on a pan-European level, LA21 does not feature in its monitoring of sustainability in Europe. Currently it has an 'informed cities initiative for sustainability' (ICLEI, 2014) which does have a free diagnostic tool for local authorities to assess their own progress towards sustainability.

OTHER VOICES IN MAINSTREAMING SUSTAINABILITY IN SOCIETY

It is not just governmental organizations that work towards operationalizing the principles of sustainability within civil society, with

O'Riordan (2001) emphasizing that there are multi-levels of governance which are working towards sustainability, including public sector and private sector organizations such as multinational organizations impacting on communities at a local level. The architect Sir Richard Rogers, in the text *Cities for a Small Planet*, stated that: 'Our aim is to achieve a new dynamic equilibrium between society, cities and nature. Participation, education and innovation are the driving forces of the sustainable society' (1997, p. 175). In the opening pages of his book he provided the imagery of the earth seen from a satellite, highlighting both its beauty and its fragility, in peril from pollution, deforestation, industrialization, erosion of the environment and sprawling cities. He speculated that sustainability could be the 'philosophy of the age'. No doubt in the 17 years since publication of his book, satellite visual information will provide a starker set of images of the fragility of the earth.

This challenge of working towards sustainability, in light of the cognisance of the fragility of the earth, has been taken up across the globe, not least at high levels, for example in the words of former Vice-President Al Gore (2007) it remains an 'Inconvenient Truth' which, globally, challenges nation states, through appropriate leadership, to inspire all sections of society into action. Within a UK context, the UK government's commissioned report in 2006, the Stern Review, *The Economics of Climate Change*, had a similar 'big business' audience in mind, with sanguine pronouncements on the impact of climate change on not only developing nations but on all societies on the planet. Stern's report identified that an estimated 5°C increase in global temperatures from pre-industrial levels until 2050 would result in 200 million people being displaced by flooding or drought, would seriously affect food production and create a 15–40% reduction in species. Both Gore and Stern's work were particularly aimed at the business community, on both sides of the Atlantic, but were also designed to spur the global business audience into actions on sustainability.

MEASURING PROGRESS: THE CHALLENGE OF WORKING TOWARDS SUSTAINABILITY

On the question of working towards sustainability, Gower posed the question 'What do we owe future generations?' and in doing so, raised the ethical dilemma of ensuring 'that our savings will be passed on by the intervening generations' (1992, p. 9). As Sefang and Smith suggested: 'Everybody, it appears, is committed to sustainable development. But not everyone is seeking sustainable development in the same way' (2007, p. 584). Having cognisance of how far these aspirations might be achieved is not a precise science, despite a view that without quantifiable sustainability indicators, progress towards sustainability remains unclear.

In relation to Agenda 21 and frameworks for LA21 stimulating progress towards sustainability, Brown (1997) notes that Agenda 21 symbolized a political commitment (rather than a legally binding arrangement) to inform political, environmental, social and economic structures of society with the principles of sustainability. In doing so, Agenda 21 was designed to 'modify' the activities of international organizations, governments, industry and communities to sustain political, environmental, eco- and economic systems (Brown, 1997, p. 75). Brown emphasizes that Agenda 21 should be an evolving document which reflects the dynamic nature of the concept of sustainability and which finds ways of implementing the goals of sustainability (1997, p. 76–77). This raises the issue of the role of LA21 as a policy process which might create policy agendas for sustainability at the local level, which stimulate community scenarios for sustainable development. In a UK context in relation to LA21, Sefang and Smith 'recognized the contribution made by small scale local activities, and has a particular emphasis on *delivery* of sustainable development at all levels' (2007, p. 585). The inference is of 'activists and organizations generating novel bottom-up solutions for sustainable development; solutions that respond to the local situation and the interests and values of the communities involved' (Sefang and Smith, 2007, p. 585). What Sefang and Smith (2007) recognized was that grass-roots actions could be extended beyond these 'organic' activities and be part of a national government 'encouraging' social actions, which would be consistent with government policy agendas.

The work of Morse and Bell (2008), *Sustainability Indicators Measuring the Immeasurable?*, is fundamental in considering the limitations of sustainability indicators. They argue that so much of life is immeasurable and that they describe sustainability indicators as 'a bad application of good science'. Generally they argue that there is a temptation to measure things that can be measured, rather than things that should be measured and that there should be a greater degree of reflection, humility and honesty about the quest to 'move towards sustainability'. In their view in the context of the subjective social construction of society, they describe the sustainable indicators approach as overly reductionist and objective. They advocate a more holistic and subjective approach which is both participative and realistic, known as Systemic Sustainable Analysis (SSA). This they define as 'the participatory deconstruction and negotiation of what sustainability means to a group of people, along with the method of assessment of indicators to assess that vision of sustainability' (Morse and Bell, 2008, p. 147). They argue that with this methodology, any system can be evaluated subjectively, that a range of stakeholders should be heard in any analysis of sustainability and divergent views can form part of negotiation. Furthermore, the creation of indicators as part of this analysis can be both quantitative and qualitative and that assessment of progress can be within a 'band of equilibrium', in which progress towards sustainability can be measured. They emphasize that sustainability indicators have no value in themselves, unless they are to the benefit of communities who have created a vision for their locality. They used the example of the UK local authority area of Norwich, in the east of the UK, and specifically 'the Norwich Agenda 21 Steering Group' which used the framework of LA21 to enable communities (in conjunction with the local authority) to identify and measure what the community values are, as part of visions for sustainability (which covered areas such as governance and equity, transport, services, the environment, the economy, housing and the built environment). Ironically, it is pertinent to the case study (discussed next) that this initiative no longer exists, due to the national government imposing on the local authority a set of sustainability indicators to be used in measurement, which made community visions for sustainability redundant.

Case Study: The UK Approach to Sustainability and Planning Policy

Returning to the start of this chapter and the work of Gombrich, who creatively articulated the history of the epochs of the world within one book, the history of the attempts to mobilize the concept of sustainable development, globally and in a UK context, demonstrates the short timescale that has elapsed, in attempts to operationalize the principles of sustainability at a global, state and local level.

Like many nations across the globe, the UK government created a strategy for sustainable development after the 1992 Rio Conference. In 1993, a national strategy for sustainable development stimulated the operationalization of the principles of sustainability in different policy areas of government. This was done in conjunction with 'bottom-up' solutions for sustainable development through the framework of LA21 initiatives. In this respect, Sefang and Smith noted how 'grassroots action for sustainable development takes different forms, from furniture recycling social enterprises to organic gardening cooperatives, low-impact housing developments, farmers' markets and community composting schemes' (2007, p. 585). By 2000, the UK government introduced the Local Government Act, which stipulated that all UK local authorities should produce a Community Strategy (or Plan), which responded to the social, environmental and economic conditions within the local authority area. This was a departure from the LA21 framework, which enabled local authorities and communities to respond to the challenge of sustainability, on the basis of the local visions emerging rather than a national template for sustainability being imposed. Community Plans and Local Strategic Partnerships eclipsed LA21 as the prime catalyst in the sustainability process, with a lack of evidence of community engagement in sustainability initiatives. As such, this top-down, managerialist approach to sustainability largely dismantled a decade of LA21 initiatives' infrastructure. This has led to an uncertain future for the visibility and viability of sustainability both as a focus for policy and for local actions for sustainability. For example the Norwich Agenda 21 Steering Group described above was replaced by the more managerially led Local Strategic Partnership, which led to the dissolution of the 'branding' of sustainability at a local level as a spur for actions on sustainability.

At a national level, in 2011, the UK coalition government created the 'Visions for Sustainable Development', refreshing the 2005 Strategy for Sustainable Development and highlighted that:

> The coalition Government is committed to sustainable development. This means making the necessary decisions to realise our vision of stimulating economic growth and tackling the deficit, maximising well-being and protecting our environment without negatively impacting the ability of future generations to do the same.
>
> (Defra, 2011)

This strategy might be viewed as an act of debasing the inescapable global imperatives of sustainability, in headlining deficit reduction as the central element of sustainability considerations. The coalition government might argue that prudent use of resources

(Continued)

Case Study. Continued.

(particularly financial), coupled with the 'efficient' procurement of resources, guided by principles of sustainability, used in the public sector, is also a mainstay of the coalition government's environmental considerations of sustainability. While this document recognizes the so-called 'three pillars' of sustainability, it suggests that it is privileging the economic 'pillar' (weak sustainability approach) over the interconnections between society and the natural environment. Coupled with the creation of this strategy, in March 2011, the Sustainable Development Commission (SDC), with an external, independent scrutiny role which offered 'appraisal, advice and advocacy' related to the performance of central government in working towards sustainability, was abolished, leaving no independent body scrutinizing government policy and actions on sustainability. While the coalition government policy 'landscape' for sustainability has many departmental actors, the coordination of sustainability at a national level is led by the Environment Secretary, heading the Department for Environment, Food and Rural Affairs (Defra). The work of Defra is supported by two key pieces of legislation: (i) the 1995 Home Energy Conservation Act (updated in 2012); and (ii) the Climate Change Act, 2008 (and supported by the 2011 Carbon Plan and 2011 Energy Act) with the aim of creating a low carbon economy. More specifically, they provide guidance in business planning, with all government departments having to demonstrate that they are embedding the principles of sustainable development into policies. 'Progress' is monitored in departmental annual reports, using Sustainable Development Indicators (SDIs), which are intended to be a broad statistical measure of economic, social and environmental factors.

The coordinating role of Defra is intended to challenge government departments, in the first instance, to embed the principles of sustainability into policy making and also into practice. The creation of the National Planning Policy Framework, established in 2012, had the intention of integrating the three dimensions (environmental, social and economic) of sustainability. Additionally, in June 2012, local authorities launched the Climate Local Initiative, which offered a framework for UK local authorities to respond to the challenges of climate change. The Department of Energy and Climate Change's Green Deal (2013) replicated this initiative at a national government level. Both at a national and at a local level, climate change and an environmental approach to sustainability is now the key focus, but underpinned by the economic considerations of budget deficit reduction.

National government policy guidelines on sustainability and planning

National government provides guidelines for the operationalization of the principles of sustainability through Planning Policy Guidance (PPGs) notes. However, there is a focus on the environmental aspect of sustainability which perhaps is a reflection of the narrow focus of the UK planning system. At a local authority level there has been an uncertain evolution of policies for sustainability in terms of planning, despite the Department of Communities and Local Government, in 2012, suggesting that the National Planning Policy Framework is infused with principles of sustainability. How far this guidance is emerging at a local level as suggested above is open to question.

THE PEDAGOGY OF SUSTAINABILITY

What then have subsequent generations done in promoting learning and pedagogy related to sustainability and has the epistemology and pedagogy of sustainability evolved?

O'Riordan (2001, p. xx) suggests that in terms of sustainability 'there is no best practice or benchmarking. There is only the memory of learning and the prognosis of learning'. The UN Decade of Education for Sustainable Development (2005–2014) applies a different philosophy of learning related to sustainability (UN, 2005), in contrast to O'Riordan's view. It argues that there are examples of best practice and the creation of sustainability indicators, which can be used as the basis in training and education for organizations globally, to adopt principles which will lead to a sustainable future. Certainly, private industry would argue that the mainstreaming of sustainability through Corporate Social Responsibility (CSR) and sustainable supply-chain initiatives demonstrates the capacity for the principles of sustainability to be mainstreamed into accepted industrialized processes and as an exemplar of corporate learning towards sustainability. Critics might argue that these initiatives might be interpreted as being examples of weak sustainability, in that they are predicated on the motivation to improve economic performance, through more frugal use of resources. The approach taken by national government within the UK to governmental departmental business planning is redolent of an economic prioritization of sustainability.

The case study of the UK and the operationalizing of the principles of sustainability at both a national and a local level, reveals the curtailing of policy initiatives and the rebranding of these initiatives that suggests a lack of policy learning and a move towards shallow forms of sustainability. Policy making for sustainability has been fragmented and is susceptible to the short termism of electoral cycles. It is difficult to recognize policy continuity and consensus about the sustaining of policy cycles based on policy learning within a UK context over a period of just over two decades, particularly in the abandonment of LA21 as a community action for sustainability.

It was Davis *et al.* who emphasized that policy making 'is a procedure without end, in which policies are never finished, only improved or replaced' (Davis *et al.*, 1993, p. 181). Policy-making institutions have to consider how they learn from past policies and to choose 'best' options, based on 'choices between values and competition between resources, all linked by institutional arrangements and ambitions' (Davis *et al.*, 1993, p. 182). This process of policy learning and succession might require the capacity for reflection. The acknowledgement of policy improvements might also be as a result of policy failure. The added dimension of the complexity of policy development underpinned by the principles of sustainability is that the evaluation of its successful operationalization requires an evaluation of community actions.

A good example of policy evaluation and learning that focuses on community actions can be found in the work of Kasemir *et al.* (2003) in their edited text *Public Participation in Sustainability Science: A Handbook*. They made reference to an EU Research and Development programme-funded 'Visions Project' that explored civil society representation and citizen participation in sustainable development policy making, in cities such as Venice, in 1996. This project combined 'physical science' and social science tools to create qualitative research on visions about a sustainable future. In the case of Venice, the visions were the result of a range of residents' views. This study focused on creating four qualitative sustainability scenario visions for Venice in 2050, through a series of resident, industry and political panels and focus groups. This enabled these researchers to explore ways of making issues of sustainable development accessible to a wide range of the population. Significantly, one group of residents who were incorporated into the project were children. Authors such as Galtung (1995) argue that children should be central and be represented in discussions which occur at UNCED conferences, as they are crucial actors to future considerations of the planet (the world they will inhabit as adults).

Following the pioneering work of Bramwell *et al.* (1996) in their consideration of sustainable development management and pedagogy on a European level, they were instrumental in advocating the need for sustainability to be at the heart of the development of curricula which informed tourism education. In a more contemporary sense, Tirone *et al.* highlighted the philosophy of Dalhousie University, who in 2008, created Canada's first College of Sustainability, whose aim is redolent of the central theme of this book, in stating that there is a 'challenge and responsibility of building students' skills and knowledge so they contribute to the creation of more sustainable societies' (Tirone *et al.*, 2013, p. 126). This educational institution provides an interdisciplinary undergraduate programme entitled 'Environment, Sustainability and Society' (ESS) which enables students to combine any degree course with issues of sustainability but also for academic and community partners to address challenges related to social and environmental sustainability.

Tirone *et al.* (2013) contrasted the teaching at Dalhousie University with traditional teaching and learning methods in other institutions that promote linear thinking, which prepares students for 'growth-orientated societies'. They highlighted that Dalhousie University provided a critical thinking, interdisciplinary/multi-voiced approach focused on problem-based learning, stating that the

> non-traditional approach to sustainability education provides a forum for sharing ideas and developing ideas and developing approaches to action that address how society can fulfil the needs and desires of people today in ways that will ensure the ability of future generations to meet their own needs and desires.
>
> (Tirone *et al.*, 2013, p. 127)

A mainstay of the philosophy of pedagogy and sustainability in this institution is related to providing a vision of society which is working towards sustainability, to encourage consumers to engage in 'imaginative consumption'.

CONCLUSIONS

The narratives of sustainability and sustainable development discussed within this chapter used Gombrich's (1985) 'template' for the compression of time across epochs, to offer a broader perspective on human impact on the changing environment of the earth. It was intended that this compression of time would bring into sharp relief the ephemerality of policy cycles, within potentially short-term political timelines, which hinder policy development and learning for sustainability. Bramwell *et al.* (1996) highlighted that not only is sustainable development a contradiction in terms but that the so-called 'three pillars' of sustainability should not exclude political, cultural and managerial considerations.

That major initiatives for sustainable development should emerge at a supra-national level has led to critics viewing attempts to mobilize political and policy actions as being a blueprint for development and economic growth, fashioned from Western perspectives. In advancing Eastern perspectives on sustainability, it is hoped that the dominant Western paradigms on sustainability and sustainable development could be reflected in parallel to the Eastern philosophy of Taoism and 'wu wei' and the philosophy of Gaia, which was explored in tandem with approaches to deep ecology and strong sustainability. The work of Wiertsema (1995), Richards (2001) and Bell and Morse (2008) offer parallel considerations in the quest to move towards realizing the goal of sustainability, suggesting that 'equilibrium' rather than growth should be the goal of sustainability. This might be achieved through the adoption of smaller-scale technologies, in the planning and development of the physical environment but vitally through community participatory arrangements that ensure that community visions for sustainability inform planning processes. The emergence of Agenda 21 and its application through LA21, but in a UK context its premature demise, emphasizes that policy learning for sustainability is susceptible to the machinations of short-term considerations. Therefore planning for a sustainable future has to realize the limitations of anthropocentric and weak approaches to sustainability, which requires patience (and in the view of Bell and Morse (2008) humility and honesty) in the complex challenge to find participative processes to emerge, if community visions of sustainability are to have the opportunity to flourish and inform planning processes.

QUESTIONS

1. Within a national context other than the UK, how has the national state worked with the local state and communities to plan for sustainability?

2. Find a planning initiative which has used sustainable indicators and evaluate how these indicators have been developed and how far they have achieved their purpose.

3. Find two academic papers on Venice and consider what role planning has played in addressing sustainability challenges for the city.

BIBLIOGRAPHY

Ayers, W. (1971) *Chang Chih-tung and Education Reform in China*. Harvard University Press, Cambridge, Massachusetts.

Bauler, T. and Hecq, W. (2000) Some theoretical considerations in response to the claim after 'Information for decision-making'. Proceedings of the 2000 International Sustainable Development Research Conference, 13–14 April, University of Leeds, Leeds, UK.

Bell, S and Morse, S. (2008) *Sustainability Indicators: Measuring the Immeasuarble*, 2nd edn. Earthscan, London.

Bramwell, B., Henry, I., Jackson, G., Goytia, A., Richards, G. and van der Straaten, J. (1996) *Sustainable Tourism Management: Principles and Practice*. Tilburg University Press, Tilburg, The Netherlands.

Brindley, T., Rydin, Y. and Stoker, G. (1996) *Remaking Planning: The Politics of Urban Change*. Routledge, London.

Brown, V. (ed.) (1997) *Managing for Local Sustainability: Policy, Problem-solving, Practice and Place*. Paragon, Canberra.

Callicott, J.B. (1994) *Earth's Insights: A Survey of Ecological Ethics from the Mediterranean Basin to the Australian Outback*. University of California Press, Berkeley, California.

Callicott, J.B. and Ames, R.T. (eds) (1989) *Nature in Asian Traditions of Thought: Essays in Environmental Philosophy*. State University of New York Press, New York.

Clarke, J.J. (1997) *Oriental Enlightenment: The Encounter between Asian and Western Thought*. Routledge, London.

Davis, G., Wanna, J., Warhurst, J. and Weller, P. (eds) (1993) *Public Policy in Australia*. Allen and Unwin, St Leonards, New South Wales, Australia.

The Department of Energy and Climate Change (2013) The Green Deal Framework. DECC, London.

Department for Environment, Food and Rural Affairs (Defra) (2011) Mainstreaming Sustainable Development. Available at: http://sd.defra.gov.uk/documents/mainstreaming-sustainable-development.pdf (accessed 20 April 2014).

European Union (EU) (2012) Sustainable Development. Available at: http://ec.europa.eu/environment/eussd (accessed 16 February 2014).

Federal Office for Spatial Development (ARE) (2005) The National Promotion of Local Agenda 21 in Europe. Available at: http://are.admin.chdokumentation/publication (accessed 16 February 2014).

Fritz, M. (1995) The greening of Sweden. In: Goldsmith, E. (ed.) *The Future of Progress*. Green Books, Totnes, UK, pp. 230–234.

Galtung, J. (1995) Waiting for UNCED: waiting for Godot. In: Rajan, V. (ed.) *Rebuilding Communities: Experiences and Experiments in Europe*. Green Books, Totnes, UK, pp. 25–34.

Gautier, A. (2009) *Green Up! Five Ways to Work with Your Council on the Environment and Sustainability*. Community Development Foundation, London.

Gombrich, E.H. (1985) *A Little History of the World*. Yale University Press, New Haven, Connecticut.

Gore, A. (2007) *An Inconvenient Truth: The Crisis of Global Warming*. Rosdale Press, Emmaus, Pennsylvania.

Gower, B. (1992) What do we owe future generations? In: Cooper, D. and Palmer, J. (eds) *The Environment in Question: Ethics and Global Issues*. Routledge, London, pp. 1–12.

Griffiths, R. (2000) Sustainable development: should we read the obituary? Paper presented at the International Sustainable Development Research Conference, 13–14 April, University of Leeds, Leeds, UK.

Hall, D.L. (1989) On seeking a change of environment. In: Callicott, J.B. and Ames, R.T. (eds) *Nature in Asian Traditions of Thought: Essays in Environmental Philosophy*. State University of New York Press, New York, pp. 99–112.

Ham, C. and Hill, M. (1984) *The Policy Process in the Modern Capitalist State*. Wheatsheaf Books, Brighton, UK.

Hawking, S. (1988) *A Brief History of Time: From the Big Bang to Black Holes*. Bantam, New York.

Henderson, K. (2013) Ecofeminism(s) and just leisure in the 21st century. In: Schwab, K. and Dustin, D. (eds) *Just Leisure: Things That We Believe In*. Sagamore Publishing, Urbana, Illinois, pp. 35–42.

International Council for Local Environmental Initiatives (ICLEI) (2014) Informed Cities: Sustainability Issues. Available at: http://informed-cities.iclei-europe.org (accessed 11 April 2014).

Ip, P.K. (1983) Taoism and the foundations of environmental ethics. *Environmental Ethics* 5, 335–343.

Jackson, G. and Morpeth, N. (1999) Local Agenda 21 and community participation in tourism policy and planning: future or fallacy. *Current Issues in Tourism* 2(1), 39–46.

Jenkins, S. (2014) The IPCC Report Takes Us from Alarmism to Adaptation. Available at: http://theguardian.com/commentisfree/2014/march31/ipcc-report-adaptation-climatechange (accessed 1 April 2014).

Jenkins, T.N. (1998) Economics and the environment: a case of ethical neglect. *Ecological Economics* 26(2), 151–163.

Jenkins, T.N. (2002) Chinese traditional thought and practice: lessons for an ecological economics worldview. *Ecological Economics* 40(1), 39–52.

Kasemir, B., Guimerraes, A. and Pereira, A. (2003) *Public Participation in Sustainability Science: A Handbook*. Cambridge University Press, Cambridge.

Kelder, E. (1996) Policy integration and the 'implementation gap'. Proceedings of the 1995 European Environment Conference, 11–12 September, University of Nottingham, Nottingham, UK.

Lovelock, J.E. (1988) *The Ages of Gaia: A Biography of Our Living Earth*. Oxford University Press, Oxford.

Morse, S. and Bell, S. (2008) *Sustainability Indicators: Measuring the Immeasurable?*, 2nd edn. Earthscan Publications, London.

Needham, J. (1956) *Science and Civilization in China*, Vol 2. Cambridge University Press, Cambridge.

O'Riordan, T. (1981) *Environmentalism*. Pion, London.

O'Riordan, T. (2001) *Globalism, Localism and Identity: New Perspectives on the Transition to Sustainability*. Earthscan Publications, London.

Palmer, J.A. (1992) Toward a sustainable future. In: Cooper, D.E. and Palmer, J.A. (eds) *The Environment in Question: Ethics and Global Issues*. Routledge, London, pp. 181–186.

Porritt, J. (1984) *Seeing Green: The Politics of Ecology Explained*. Basil Blackwell, Oxford.

Richards, G. (ed.) (2001) *Cultural Attractions and European Tourism*. CAB International, Wallingford, UK.

Rogers, R. (1997) *Cities for a Small Planet*. Faber and Faber, London.

Sabelli, H. (1998) The union of opposites: from Taoism to process theory. *Systems Research and Behavioral Science* 15, 429–441.

Sarre, P., Smith, P. and Morris, E. (1991) *One World for One Earth: Saving the Environment*. Earthscan Publications in association with the Open University, London.

Sefang, G. and Smith, A. (2007) Grassroots innovations for sustainable development: towards a new research and policy agenda. *Environmental Politics* 16(4), 584–603.

Sharpley, R. (2009) *Tourism Development and the Environment: Beyond Sustainability*. Routledge, London.

Shi, T. (2002) Ecological economics in China: origins, dilemmas and prospects. *Ecological Economics* 41(1), 5–20.

Shiva, V. (1992) Recovering the real meaning of sustainability. In: Cooper, D. and Palmer, J. (eds) *The Environment in Question: Ethics and Global Issues*. Routledge, London, pp. 185–191.

Stern Review (2006) *The Economics of Climate Change*. Available at: http://wwf.se/source.php/1169157.pdf (accessed 12 April 2014).

Tirone, S., Campbell, C. and Gallant, K. (2013) First college of sustainability: teaching about social and environmental justice and sustainability. In: Schwab, K. and Dustin, D. (eds) *Just Leisure: Things That We Believe In*. Sagamore Publishing, Urbana, Illinois, pp. 126–134.

Turner, R. (1991) Environment, economics and ethics. In: Pearce, D. (ed.) *Blueprint 2: Greening the World Economy*. Earthscan, London, pp. 211–222.

United Nations (UN) (2005) United Nations Decade of Education for Sustainable Development (2005–2014). Available at: http://unesco.org/education/desd.pdf (accessed 20 May 2014).

United Nations Conference on Environment and Development (UNCED) (2012) The Future We Want. Available at: http://uncsd.2012org/futurewewant.html (accessed 31 January 2014).

United Nations Conference on Environment and Development (UNCED) (2013) Political Forum to Replace Commission on Sustainable Development. Available at: http://Un.org/News/Press/docs/2013/ga11393.doc.html (accessed 30 November 2013).

Urban Environmental Management (undated) Local Agenda 21. Available at: http://gdrc.org/uem/la21/la21.html (accessed 21 February 2014).

Wheeller, B. (1993) Sustaining the Ego. *Journal of Sustainable Tourism* 1(2), 121–128.

Wiertsema, W. (1995) Paths to sustainability. In: Goldsmith, E. (ed.) *The Future of Progress: Reflections on Environment and Development*. Green Books, Totnes, UK, pp. 172–178.

World Commission on Environment and Development (WCED) (1987) *Our Common Future*. Oxford University Press, Oxford.

chapter 3

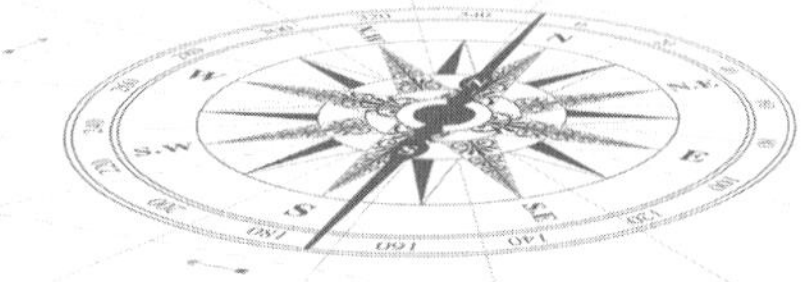

Critical Approaches to Spatial Analysis

Nigel D. Morpeth* and Hongliang Yan

Leeds Beckett University, Leeds, UK

LEARNING OBJECTIVES

- To be able to apply a range of interdisciplinary perspectives on the concept of the natural environment.
- To explore the concept of critical theory to relate to intersectional linkages between different groups in society.
- To use the concept of space and place as important conceptual elements in evaluations of different scenarios for planning for tourism.

INTRODUCTION

In this chapter we explore the concepts of the natural environment, space and the 'social' as elements within spatial analysis, and in doing so highlight the interconnectedness of cultural, economic, political, social and environmental processes. The evaluation of the physical environment, as distinct from the 'social', can lead to a false duality of separating these concepts, so denying the prospect of achieving the interconnectedness of social and environmental justice, as inseparable elements of sustainability. Additionally, an analysis is made of the intersectional needs of different marginalized social groups within localities. In exploring these considerations within the concepts of both leisure and tourism and 'everyday' activities, this acts as a prelude to consider the privileging of urban space for leisure and tourism activity within the context of a World Cup soccer mega-event

*n.morpeth@leedsmet.ac.uk

(large scale) in 2014, in Brazil. This case study contrasts the lack of an adequate planned social infrastructure for communities living in the shadow of exemplary sporting stadia, visited by tourists, in Rio de Janeiro and other urban centres in Brazil. It serves to demonstrate that planning for tourism in isolation of resident needs creates further division in a society lacking in an adequate social infrastructure.

TOURISM: INCENTIVES TO BROADEN THE ANALYSIS BEYOND DISCIPLINARY BOUNDARIES

How we approach and 'enter' destinations for analysis might be guided by a number of considerations and perspectives. We might be constrained (unknowingly) by our sole desire to use the prism of our field of study to make sense of how the destination or localities operate. This approach might appear to have its advantages, where the precision of the 'tunnel vision' of a field of study could be viewed as advancing epistemological insights without having to acknowledge the complexity of interdisciplinary and intersectional considerations, which might seem messy (irrelevant) and overly complex. Furthermore, for those students of tourism who have worked in the tourism industry (perhaps in a sandwich year) and then return to study, there might be a confidence that they understand how tourism 'works', particularly in the destination they have worked in. In short, they have a practical understanding of the key actors (social agents) involved in the organization of tourism in that locality and an insider view of what it takes to 'do' tourism.

There might be little receptiveness then to broaden analysis, beyond the narrow confines of tourism industry actors and the degree of this receptiveness might be predicated on whether the course of study is informed by a tourism studies or tourism management paradigm. Furthermore, applying considerations of how to plan for tourism could be viewed purely as a technical challenge, to put in place an infrastructure, a superstructure and make accessible and market a destination for the enjoyment of tourists. However, if tourist destinations are to be analysed beyond purely 'performance' spaces for tourism, in isolation of the temporal and spatial impacts that impact on host communities, a range of interdisciplinary applications are sought, to stimulate critical thinking about the spatial analysis of localities and destinations. A first consideration then is to demarcate theory and praxis to engage with the popularized notion of the 'real world'.

PLANNING FOR TOURISM IN THE 'REAL WORLD'

The phrase related to understanding how a certain area of society works in the 'real world' is a curious one and suggests that the study of aspects of society might struggle to move from theory to practice or praxis (action informed by the development of critical consciousness – see Friere (1970) *Pedagogy of the Oppressed*) and truly know how things work. Finding the balance between theory and praxis is a challenge, not least to be able to engage with the 'real world' of planning. For example Massey, citing the work of a fellow social scientist Harvey (1989), warns of 'mental maps (of social scientists) not keeping pace with current realities' of places (Massey, 1993, p. 72) and on this theme of striking a balance between the theoretical and the empirical, Morse and Bell (2008, p. 114) differentiate between systems thinking or theorizing and the 'real world'.

So how then do these considerations relate to the study of tourism and specifically the study of planning for tourism? Increasingly educational institutions and tourism educators are marketing courses to potential students with the promise of studying tourism in the 'real world'. But this still begs the question, how do we locate the 'real world' beyond multiple worlds of apparent illusion and imagination?

These considerations are not new ones, for example, as a prescient comment about the need for planning, Donaldson in *Economics of the Real World* (1973) emphasized that there are different approaches, interpretations and perspectives to understanding the realities of how economies and societies 'work'. How, in the words of the artist and cultural theorist Conrad Atkinson (1981), we should find ways of making sense of locating the invisible workings of the system. In 1972, the cultural theorist John Berger in *Ways of Seeing*, once again through the metaphor of art, invited the reader to reconsider how we view, construct and deconstruct the seemingly familiar world through 'new eyes'. Written within this era of critical thinkers, Goodman's book (1972) *After the Planners* was redolent of work which questioned the way the 'system' works (or does not), in regard to the status quo of planning and development, in 1970s' US society. This system in his view enabled the planner, as an 'expert', to engage in 'environmental determinism', to the exclusion of the views of communities who had to live in those technically and expertly 'planned' localities.

Therefore, it is worth considering the application of interdisciplinary academic traditions which have enabled us to apply ways of constructing and deconstructing the realities of societies, globally. In *The Social Construction of Reality*, the same cultural theorists Berger and Luckmann (1966) highlighted how we are able to 'read' society, not 'objectively' but as a subjective reality. They made the distinction between so-called 'everyday' reality (a general body of knowledge) with more specialized 'semantic fields' of knowledge, with planners being an example of one such group with specialist knowledge. The notion of intersectionality enables us to encounter and relate to the different worlds of diverse groups, whose 'everyday' experiences are intertwined with other groups within society. In applying these sociological ideas to the study of planning and tourism, we can combine these ideas with the work of C. Wright Mill (1959) whose book *The Sociological Imagination*, while preceding the work of Berger and Luckmann, espoused a similar philosophical vantage point in encouraging us to be able to transpose our thoughts beyond the 'everyday' and see things anew from different perspectives. From these interlinked social perspectives, there is a parallel with the work of A.J. Ayers (1936) whose philosophical thoughts about the nature of reality, in his work *Language, Truth and Logic* and his phrase 'the self and the common world' (1936, p. 159) invited the reader to consider the limitations of our cognisance and perceptions of the world, which is limited only to certainties about our own existence, with each individual creating their own filtered perceptions of the world.

In this respect and in a more contemporary sense, on the theme of the limitations of humankind's perceptions of and connections with the natural environment, Fox (1995) in *Towards a Transpersonal Ecology: Developing*

New Foundations for Environmentalism, suggested that we are all inevitably hidebound by our anthropocentrism, hindered by what he termed perspectival fallacy. This meant that ultimately seeing our place in the 'bigger scheme of things' is filtered through who we are as individuals (and the labelled stereotypical groups that we represent) and that trying, for example, to hold feminist perspectives as a man, etc. is difficult. Furthermore, he identified that in the search for solutions to environmental problems, the application of technological solutions inevitably sees short-term and pragmatic responses, rather than recognizing an interdependence between humankind and the environment, with long-term considerations (not least the notion of sustainability). It is potentially through the application of critical and cultural theory that we are able to gain further insights into interconnections between the social and the environment, in an attempt to transcend our anthropocentric limitations.

CRITICAL AND CULTURAL THEORY

Spencer and Jostad (2013) highlighted that critical theory emerged from philosophies of the Institute for Social Research, in Germany, an institute influenced by philosophers Kant, Hegel and Marx. Spencer and Jostad characterize critical theory as a radical social theory which was intended to 'produce transformational change in the world' and create 'new discourses of thinking and action that focused on making changes to the social structure of the world and to the ways that individuals construct consciousness' (2013, p. 76). They emphasize that researchers use an 'ethical lens' to explore structures of power (hegemony) in society, to identify if certain groups in society are privileged over others. In the analysis of social equity, an inherent element of deep sustainability, this critical and 'ethical lens' is helpful in making this judgement.

Cavallaro (2001, p. ix) highlights that critical theory might be best used in tandem with cultural theory, to use interdisciplinary approaches to ask searching questions about different aspects of society, through heterogeneous interpretations and discourses of history and individual and collective cultural identity. He suggested that combining critical and cultural theory can enrich our understanding of spatial analysis beyond a purely functionalist analysis of society (Cavallaro, 2001, p. xi). Particularly apposite to the discussion within this chapter on the natural environment and space, is that they (the concepts of the natural environment and space) are socially constructed and culturally created.

The study of tourism and the application of spatial analysis to destinations, applying interdisciplinary and heterogeneous interpretations, potentially enable a more in-depth treatise of the dynamics of a location. For example the work of John Urry (1995) in *Consuming Places* views localities through different theoretical approaches taken from interdisciplinary sociological and geographical perspectives. These perspectives enable students from different disciplines to understand the economic, cultural and social components, as the 'everyday' elements which form the basis of localities. For students of tourism studies, Urry's (1992) *The Tourist Gaze*, perhaps prima facie, provides a more mainstream tourism treatise. But as with *Consuming Places*, Urry (1995) offers us a broader understanding of the everyday elements of localities, through the application of a rich combination of interdisciplinary approaches. Specifically, in *Consuming Places*, a range of theories on spatial and temporal studies, which combine both critical and cultural theory, are vital to understanding the notions of consumption and commodification of localities, beyond a prescriptive and normative approach as to how tourists might inevitably 'consume' that destination. Likewise, Holloway's (1989) concept of 'cloned destination' suggests that the homogenization of destinations, resulting in their 'sameness', means destinations are losing their cultural distinctiveness. Therefore to interrogate this notion requires an incisive application of spatial analysis to understand the complex interrelationships between residents and tourists within localities. (See also Singh's (2014) edited text *Critical Debates in Tourism*, which provides a series of treatises which challenge the status quo of tourism and an opportunity to view tourism through 'new eyes'.)

PERCEPTIONS OF THE NATURAL ENVIRONMENT: THE PITFALLS OF DUALISM

This chapter has already highlighted that to be able to apply an in-depth approach to the analysis of localities requires the capacity to use a range of different interdisciplinary elements, to 'read' the dynamics of a locality or destination. One dimension of the elements for analysis is to understand the origins of the concept of the natural environment and how this relates to the concepts of technocentricism and ecocentrism, which underpin the concept of sustainability initially considered in Chapter 2, this volume.

As a prelude to more in-depth consideration of aspects of spatial analysis investigated later in this chapter, it is important to note that there are different philosophies, which have emerged over time, regarding the relationship between humankind and the natural environment. Silvertown and Sarre (1990) reviewed different historical human responses to the environment, starting with the medieval doctrine of Christian 'Stewardship', which promoted the notion of peaceful coexistence between humans and other 'creatures' contrasted with the philosopher, Francis Bacon's doctrine of Imperialism, which viewed humankind as having scientific 'rights' and control over nature. They highlighted that the philosopher John Stuart Mill advanced the notion of hedonism, which advocated a utilitarian view of how humankind should use the environment. These philosophies were viewed as antecedents to these historical applications of environmentalism and the emergence of eco-philosophy, as a politicized concept in the 1960s. The UK publication of *The Ecologist* pamphlet saw the wider communication of deep ecology as a radical form of environmentalism (Pepper, 1984). It was O'Riordan (1981) who had earlier positioned environmentalists into two 'camps': ecocentrics and technocentrics. While ecocentricism has many different nuances of its applications (Fox, 1990), it generally mistrusts large-scale technology, centralism and materialism, instead embracing small-scale development, decentralization and potentially governmental regulation of resource usage.

Conversely, O'Riordan (1981) suggests that technocentricism, or anthropocentricism, responds to environmental problems through 'technical fixes', with Fox (1990) identifying that anthropocentrism advocates an imperialist view of nature, aligning 'its' human centredness with the doctrine of Utilitarianism.

Reflecting these different strands of technocentrism or anthropocentrism, O'Riordan (1981) identifies that Cornucopians (the doctrine that technical fixes can be applied to problem solving) not only view environmental problems as soluble, but that unlimited growth can be incorporated within this framework. Silvertown and Sarre contrast this human centredness with radical environmentalists known as deep ecologists, who want to 'correct this anthropocentric bias' (1990, p. 259). The late Norwegian eco-philosopher, Arne Naess, epitomized the lifestyle of a deep ecologist, which was predicated on the use of recycled goods. This approach to ecocentricism was contrasted with 'shallow ecology', as environmentalism that was closely allied to utilitarianism, which accepts that detrimental impacts on the environment can be justified, if it is to the benefit of humankind (and so-called continued improvement of the human condition).

Hunt and Sherman (1975, p. 185) made the observation that capitalism required economic growth or it would encounter 'problems of depression, unemployment and stagnation'. However, global concerns about the durability of economic growth and associated environmental degradation were voiced by a number of commentators, with Carson (1962) and Erhlich (1989) questioning the survival of the earth through the exploitation of the environment in favour of short-term economic goals. Goldsmith and Allen's (1972) *Ecologist – A Blueprint for Survival* (which was a prelude to the first world environment conference in Stockholm in 1972) warned that unfettered economic growth, if unchecked, would result in irreversible disruption of 'life-support systems'. These ideas were in line with the 'Club of Rome' study – *Limits to Growth* (Meadows *et al.*, 1972) which innovatively, for the time, produced a computer-generated model to advance future-orientated development scenarios, reflecting the correlation of sustainable development and the problem of pursuing the dominant paradigm of economic growth. Galbraith (1967) and Packard (1960) presented a cogent analysis of a materialist consumer-oriented society, 'obsessed' by what they dubbed 'growthmania' and as Packard highlights: 'a clamour for growth' (1960, p. 30). Mishan (1967) recognized the need for social limits of growth, with Donaldson highlighting that 'if economic growth is used to measure development, we must always ask – growth of what?' (1973, p. 129). Additionally, Schumacher (1975) challenged the wisdom of the existing economic status quo, arguing for different growth and development scenarios, coining the phrase 'small is beautiful'.

In line with these ideas, Pepper (1984, p. 5) felt that the continuation of the dominant Western paradigm of economic growth would lead to depletion of natural resources, at a cost of gaining material prosperity. O'Riordan (1993, p. 62) felt that: 'The dominant ideology of world society is Cornucopian, with a strong belief in the ability of humans to dominate nature, through science and technology and their application via the market place and with the necessary support of the state'. An example of this is that the former UK Prime Minister, Tony Blair, in response to known environmental problems caused by air travel, urged the UK population 'to keep flying' because technologies would be created to mitigate the climate change impact of CO_2 emissions. This is consistent with the ideology of economic growth, known as a dominant paradigm of growth world view and arguably is consistent with an interpretation of weak sustainability (described in Chapter 2, this volume).

Pepper (1993, p. 1) recognized that a new green politics had emerged at the end of the 1980s and that there were different 'shades of green' politics, with ecocentrics viewed as 'Red-Greens' and technocentrics as 'Light-Greens'. He emphasized that 'nature's rights' (biological egalitarianism) are meaningless without human rights (social) and that 'we should proceed to ecology from social justice and not the other way round' (Pepper, 1993, p. 1). He warned of dualist thinking of the separation of society and nature. This meant that sustainability focusing on the physical environment, to the exclusion of social equity, can be viewed as weak or shallow sustainability, rather than deep sustainability. At the heart of these different scenarios of environmental resource usages, according to Turner (1991), is an assessment of the economic cost attached to them. Within the first scenario in Table 3.1, is a resource-exploitative, growth-orientated world view, which demonstrates the continuation of the imperialist view of Francis Bacon. This is that an extrinsic use of natural resources is for the benefit of humankind. At the other end of the spectrum, the extreme resource-preservationist approach, is in tune with the natural environment's intrinsic qualities expounded by 1960s' environmentalists and eco-philosophers. They argued that the natural environment should have rights, equitable to humankind, with the emergence of views such as 'rocks have rights'. This philosophical stance of deep ecology, or deep sustainability, has underpinned the organization of eco-settlements, described in Chapter 2, this volume, and which advocates the adoption of small-scale technology. The Taoist philosophy of 'wu wei' and also the Gaia philosophy, described in Chapter 2, are useful concepts to apply to contemporary planning and policy practice and allow the option to implement and operationalize deep sustainability principles, from both Eastern and Western perspectives.

Table 3.1. Resource usage options for planners and policy makers. (Adapted from Turner, 1991, p. 214.)

Resource usage option	Description
Resource-exploitative, growth-orientated world view	Anthropocentric resource management approach. Cost–benefit approach
Resource-conservationist, managed growth world view	'Brundtland (1987) approach' ensuring that future generations do not have a stock of natural assets which are less than the current generation. Shallow ecology
Resource-preservationist, severely constrained growth world view	Intrinsic value of the environment above economic and instrumental policy goals
Extreme resource-preservationist, zero growth world view	Deep ecology/ecocentric resource management approach

SPATIAL ANALYSIS: THE INTERRELATIONSHIP OF THE SOCIAL AND THE SPATIAL

Whynne-Hammond (1985, p. 22) posited that: 'spatial analysis is a method of analysis that helps the human geographer find patterns and relationships among the apparent complexities of the earth's surface'. Unsurprisingly, Whynne-Hammond focused on the analysis of geometric space from a traditional geographical perspective. In contrast Cavallaro (2001, p. 170) highlights the challenge of defining the concept of space stating: 'The complexity of space both as a concept and a physical reality is testified by the fact that there is no single and universally accepted definition of the term.' Cavallaro (2001) urges us to conceptualize and problematize about space, by understanding its cultural, social and psychological elements (but emphasizing that space is also about cognisance and perception).

The critical human geographer adds to the complexity of spatial analysis, by highlighting the importance of the 'social' in understanding the dynamics of geographical locations. The previously discussed potential for dualist thinking, in the social and the physical aspects of the environment, is useful in guarding against creating a duality of the social and the spatial. There is an established body of knowledge of how to understand and make sense of the dynamics of spatial analysis of places, with authors countenancing against viewing space as an empty vessel or backdrop to activity, rather a space should be defined by the accretion of social activities that have taken place in it. In this respect, Soja suggests that space is partly defined by 'the day to day practices of everyday activities which are the result of human agency' (1985, p. 90). As such space can be viewed as a social construction, with Soja (1985) reminding us that this social construction has cognizant and perceptual elements. Theorists from the 'Chicago School' recognized the interlinked nature social and spatial structures, with Saunders (1985, p. 68) recognizing 'that space is central to social theory and the study of different socio-spatial arrangements'. He and Soja (1985) concur with the work of LeFebvre (1976) that spatiality is the result of the relations of individuals and groups and that space is materially produced by the dominant system of production and consumption. Their analysis focused on a Western perspective of capitalist production and consumption and a globalized world, characterized by social conflict and the marginalization of individuals and groups, under systems dominated by the paradigm of economic growth.

THE CONCEPTUALIZATION OF SPACE AND PLACE: FROM THE GLOBAL TO THE LOCAL

While Massey speculates that: 'Places are unique, different from each other, they have singular characteristics, their own traditions, local cultures and festivals, accents and uses of language' (1993, p. 46). She goes on to speculate as to what happens to the notion of place in an age of globalization and the diminution of a sense of place, with the loss of identity in the characteristics described above. One aspect of spatial analysis is to make sense of the relationship and interplay between the global and the local. As we have seen in Chapter 2, this volume, being able to recognize global influences and impacts on local destinations is an important consideration in understanding how destinations might be planned effectively in an age of globalization.

There are a range of academics, in their exploration of human geography, who recognize that the acceleration in the pace of globalization has restructured places and cultures (Massey and Jess, 1993, p. 1). They argue that with this restructuring and social change there has been a need to redefine and re-conceptualize the nature of local places and a re-imagining of the geography of places. This process of imagining for Massey and Jess is an important aspect in creating the identity of places and how this 'stimulates' the behaviour of people in them (Massey and Jess, 1993, p. 3).

In talking about 'space and place and the global in the local', Massey separately (1993, p. 53) raised the question of the social construction of the concept of urban place. She identified that historically, a place had a 'fixedness' of local identity, forged by tradition and a spatial context, built on social relations and the unifying qualities of community. Within these places were qualities of being 'settled, enclosed and internally coherent' which helps define the identity of a place (Massey, 1993, p. 58). Massey's observation was made against the backdrop of what she observed to be new eras of globalization, resulting in placeless times. Places characterized by 'time–space compression and convergence' (Massey, 1993, p. 53).

The implication for the planning of these urban environments is that places have to be considered in new ways and with the knowledge that they have tangible material elements of physical construction but also these environments are products of our imagination. An example of this is the work of Raban (1985) positing the possibility that urban environments are not solely the creation of hard architecture but are equally about imagined 'personalities', dubbed as the 'soft city'. The city allows the visitor to put their own personal encounters in streetscapes, in adding social agency (the everyday acts of different citizens in society) as an important cultural component of cities. In this respect, tourism is viewed as an industry of illusion and destinations as places of our imagination. (See the work of Walter Benjamin (1999) who provides insights into 'the everyday', by walking the streets of a city with 'new eyes', to consider discovering its 'personality'. See also Boyes (2010) *The Imagined Village: Culture, Ideology and the English Folk Revival*, in which Boyes emphasizes that it is not only the urban environment that has received academic attention as a reappraised place; the rural village environment has been the focus of reappraisal of the changing nature of what constitutes rural culture and ideology.)

Contested space and places

Massey and Jess (1993) suggest that not only are planners engaged in physical planning, which guides the material construction of places, but they must also 'capture' and incorporate the geographical imagination of communities and their sense of place (created through our own sense of identity). As we saw in Chapter 2, this was at the heart of the notion of communities creating their visions for sustainability. They remind us that different groups have rival claims about imposing their own personal identity, on defining a place, which might exclude other people and their interpretation and representation of the concept of place. They argue that this contested issue of place, characterized by 'social and geographical unevenness' (Massey and Jess, 1993, p. 4) is at the crux of determining the future of places and the vital role of planners.

Writing separately, Massey (1993, p. 69) reveals how places are linked together in unequal ways and fashioned by 'geographies of power', and the hegemonic influence of some groups over others. Massey highlights the work of Castells (1989) who wrote of the emergence of 'the informational city', stating that 'the fundamental fact is that social meaning evaporates from places, and therefore from society, and becomes diluted and diffused in the reconstruction of the logic of a space of flows whose profile and origin, and ultimate purposes are unknown' (Castells, 1989, p. 349, cited in Massey, 1993, p. 54). O'Riordan (2001, p. xx) suggests that it is possible for a community to be funded from multinational sources which includes the implementation of a global environment agreement, which is 'regulated by a national organization responsive to global and local needs'. This quote is particularly apposite to tourism as an industry with its 'global–local' reach.

TOURISM AS A SUBSET OF LEISURE

Broadening studies out beyond comfortable disciplinary boundaries no doubt might create uncertainty as to what lies 'beyond' familiar conceptual boundaries into fuzzy disciplinary territory. Students of tourism might go through a course of study without locating the interconnectedness of leisure and tourism, in the evaluation of the dynamic of localities and destinations. In this respect Rojek's (2010) *The Labour of Leisure: The Culture of Free Time* provides a treatise on leisure, in an age of economic scarcity, and suggests that leisure and tourism activities, of individuals and groups, are located in the context of this scarcity. This scarcity has implications for social and environmental justice, as part of our consumption of leisure and tourism as part of the 'everyday' lives of people in destinations.

The concept of social and environmental justice is at the heart of Schwab and Dustin's (2013) edited book *Just Leisure: Things That We Believe In*. This book contains chapters from a range of authors, providing reflexive perspectives from the symposium, *Speaking Up and Speaking Out: Working for Social and Environmental Justice through Park's Recreation, Tourism and Leisure*, held on 17–19 May 2012 at the University of Utah. In adopting a different critique to the utility of leisure (combined with tourism) to consider its role in social and environmental justice, this text is particularly apposite to the discussion within this chapter of spatial analysis. The authors highlight that within the social construction of localities, there are a range of intersectional interests to be considered. These interests form social relations within localities with consideration of the role of social class, sexual identity, racial identity, disabled people and women, as agents who create an accumulation of social activities, which define spatial and temporal elements of both rural and urban spaces. This book highlights those particular intersectional considerations which are challenges for planners to consider in planning for the needs of different social agents and specifically:

- The environment is not simply 'out there' but is a cultural creation and there might be a cultural ideal of what nature should look like within different societies (Samdahl, 2013).
- As a basic human right, leisure is an inalienable right and disabled people should not be excluded from these considerations (Samdahl, 2013).
- Despite growing multiracial populations, 'racial fluidity' and sharing of public spaces, there are still racial boundaries within geographical locations (Roberts, 2013).
- In relation to ecofeminism, it makes visible patriarchal domination in how the environment should be defined beyond male dominance (Henderson, 2013).
- The importance of understanding diversity (e.g. 'creating' gay space) as a physical manifestation of community identity (Barnett and Johnson, 2013).
- The neoliberal transformation of public spaces such as parks into private spaces, particularly limits access to groups such as homeless people (Rose, 2013).
- They advocate experiential learning and critical thinking should encourage students to leave the classroom, and view their surroundings with new eyes and hunt for 'hegemonic artefacts' (evidence of the dominance of other groups over others) (Bruenig and Dear, 2013).

How then are these intersectional considerations embodied within planning processes? Within the next section, the earlier observations of critical theory are applied to broad considerations of the materialist qualities of space.

PLANNING AND INTERSECTIONALITY

Harvey (1996) suggests that there is a danger in prescribing a role for the planner, as creating a normative ordering of the built environment, which can be viewed as a form of environmental determinism and as 'an end in itself'. Furthermore, he suggests that ideally the planner is concerned with creating a 'proper' location, with 'an appropriate mix of activities in space, that make up the totality of the physical structures – the houses, roads, factories, offices, water and sewage disposal facilities, hospitals, schools and the like – constitute the built environment' (Harvey, 1996, p. 176). The inference of Harvey's observations is that these elements are technical requirements, which might not be contingent on recognizing the needs of certain groups.

Likewise, Campbell and Fainstein (1996, p. 1) in their edited text *Readings in Planning Theory* wanted to explore the structure and debates of planning theory from a critical perspective. Their central question was to consider: What is the role of the planner in urban centres operating within the 'constraints of a capitalist political economy and a democratic and political system'? Their focus was on the anatomy of the political economy of the USA and the UK, as representative of capitalist political economic systems. For Campbell and Fainstein (1996, p. 1) there were four main considerations in attempting to expand discourses on planning theory within these cultural contexts, which included:

- Planning theory is informed by interdisciplinary social science discourses and as such it is difficult to demarcate it entirely.
- Planning as a profession is not totally distinct from other related professions. They cite examples of architects, real estate developers and local politicians, suggesting from time to time that they also plan.
- Planning is defined by some as related to 'objects' in terms of 'land use patterns' of both the built and the natural environments and others as 'a process of decision making'.
- Planning is not informed by one methodology but a series of choices in methodologies.

They went on to emphasize that while planners might disregard the theoretical grounding in their undergraduate formative educational experiences, in favour of the pragmatic choices of planning in the 'real world', nevertheless theory enables one to see the 'condition of this pragmatism' (Campbell and Fainstein, 1996, p. 2). Harvey (1996, p. 117) considers that planning can be seen as the ordering of the built environment, as some form of environmental determinism or as a process which focuses on social relations between different groups. Harvey argues that within societies, there is pressure on the role of the planner to reproduce the social order based on the elements of land, labour and capital, the central elements of these capitalist (economic systems) democratic societies. For Harvey, there is a complexity in these composite elements, which come together to form the built environment but in different spatial (social) contexts. In understanding these diverse interests, the planner has to negotiate and understand the competitive elements that emerge between resource uses among different social classes and social groups (with imbalances in power relationships). Harvey argues that the planner is searching for some 'idealized *optimum optimorum* for the city or for regional structure' (1996, p. 182). In considering the different social entities and planning, it is important to consider the different applications of critical theories, which provide insights into a diverse range of social agents within localities.

With regard to race as a concept, Delgado and Stefancic (2012, p. 3) identify that 'critical race theory (CRT) movement is a collection of activists and scholars interested in studying and transforming the relationship among race, racism and power'. In terms of considering the role of critical race theory emerging in planning studies, it is recommended to read the work of Huw Thomas, who provides pertinent theoretical insights on planning in a racialized society, in the UK (see Thomas, 2000, *Race and Planning: The UK Experience*).

Likewise, Fainstein (1996, p. 456) highlights that 'feminist scholarship has provoked a rethinking of fundamental concepts and epistemologies in every discipline including planning'. He contends that the basis of this rethinking was about 'different ways of knowing and modes of social intervention' in response to the belief that men design cities to cater fundamentally for the needs of men

(Fainstein, 1996, p. 456). Furthermore, as planners, this subjugation by men continues with men dominating planning agendas for the urban environment. Fainstein's (1996) supposition about the role of women in planning was extrapolated from the wider role of women within society and the economy. Representative of a forum which is intended to share theoretical and practitioner-based experiences related to women and planning is the London Women and Planning Forum. It is based at Queen Mary University of London and it:

- is an information network for planning officers, architects, academics, students, community and voluntary organizations and other organizations wishing to discuss the urban environment and gender and equity issues;
- promotes feminist thinking in planning organizations and in planning processes; and
- improves the dialogue between academics and practitioners, about the role of women in planning and planning processes (London Women and Planning Forum, 2014).

Prior to the London Olympics this forum organized a seminar which considered: (i) the ergonomics of the redesigned environments, within the five London boroughs, for women; and (ii) the access to sports facilities available for women and their participation in sport in these environments.

In terms of class and the built environment, Harvey (1996, p. 175) reminds us that the concept of class is contingent on the cultural context of particular societies and that there is a complexity to understanding the relations between class and the built urban environment. For example within a UK context, he highlights 'residual feudal institutions such as the church' alongside large 'monetary capitalist' land and property owners. Within this complex myriad of social relations, he argues that there is a role for the planner to take account of social imbalances, to be mitigated by planning processes. (See Brindley *et al.* (1996) for an account of the displacement of working-class communities in the 'planning' and redesign of the urban environment of Docklands, London, in the 1980s and 1990s, cited in Chapter 2, this volume.)

The work of Howard Hughes (2006) on tourism and sexual identity provides insights into perspectives and experiences of marginalized groups in society and highlights expressions of identity and contestation within 'everyday' spaces and for leisure and tourism (Hughes (2006) *Pink Tourism: Holidays of Gay Men and Lesbians*).

In Chapter 6 (Part II, this volume) Craig Grimes provides an in-depth insight to planning and disabled people. His chapter provides a theoretical, practitioner and personal account of issues related to physical design and the everyday experiences of disabled people. Within different cultural contexts, he uses various observations related to planning and disabled people, which he highlights is the concept of universal design and how it might be used positively to plan for all sections of society. In doing so, he emphasizes that universal design can redress planning deficiencies, which have not effectively considered the needs of disabled people in 'traditionally' planned environments.

Case Study: Mega-events (Large Scale) – Planning for Residents or Tourists?

Herbert's (2014) article 'Welcome to the real world for Brazil's poor', published in the UK *i Newspaper* on 27 June 2014, brought into sharp relief the challenge of planning destinations for leisure and tourism events which privilege event considerations over social needs. The contrast is made between the massive investment and improvements in sporting infrastructure for mega-events and the lifestyle of local host communities, who continue to live in poverty with a lack of investment in a social infrastructure.

In considering the role of mega-events within society, it is possible to view them as the high points of the celebration of capitalism, in that they are, par excellence, an opportunity for global capital to inhabit a destination for a short period of time, under the gaze of global media coverage intensified by the so-called 24/7 culture. The requirement for Olympic/Paralympic Games/World Cup soccer is for a destination that enables events to be staged in line with technical requirements but also enables publicity which focuses on sporting prowess against a backdrop of brand market leaders, of the ilk of Coca Cola, McDonald's, etc. The portrayal of mega-event venues, dominated by backdrops of conspicuous global consumption, masks the everyday realities of destinations with populations whose lifestyles are in sharp contrast with the lifestyles that are communicated through brand awareness discourses.

Herbert's article exemplifies the contrast of temporary leisure and tourism environments, juxtaposed with residential locations, which display the realities of poverty and hardship. Herbert's article focuses on this contrast, within the context of the 2014, soccer World Cup in Brazil and talks of what happens when the visitors go away and leave a destination to its everyday realities of poorly planned physical and social environments:

> leaving Moro da Cruz as it is the rest of the year: a random collection of corrugated shacks and half-houses to the east of the city, some with roofs, some without, flung up at random. A community of people who are struggling to shake off the term which its people hate – 'favelados' ('slum dwellers') – and who wonder if anyone might come to collect the stinking refuse which has spilt out on to the street or to fix the sewerage pipe, to take away the stench.
>
> (Herbert, 2014, p. 54)

This article is typical of a number of articles that appeared in the media, which highlighted the opportunity cost of planning for leisure and tourism as priorities over social infrastructure projects. Similarly, Jonathan Watts writing in the UK newspaper *The Guardian*, in his article 'Anti-World Cup protests across Brazil', wrote that 'Anti-World Cup protests in 12 cities have marked the worst day in another awful week for Brazil as the government struggled with strikes, crime and public unease less than a month before the tournament takes off' (Watts, 2014). Watts highlighted that protesters claimed that 200 new schools

(*Continued*)

Case Study. Continued.

could be built for the cost of refurbishing the Maracana football stadium. In a country where such polarities in wealth and poverty live cheek-by-jowl, staging such mega-events as the World Cup and, in the future, the Olympics in 2016 serve to bring into sharp relief the inadequate planning for a social infrastructure. Such concerns in planning priority disparities, highlighted by the staging of mega-events, is not exclusive to Brazil but has beset mega-events over a 40-year period, in which there has been a beautification and redevelopment process of slum dwellers, in order to improve the physical environment for tourist visitation (see www.tourismconcern.org.uk for their work on social disparities and the role of tourism).

In the analysis of mega-events spaces, Soja (1985) warns against viewing space as a backdrop or an empty vessel, which can then be filled in an environmentally deterministic way, which ignores the socially diverse populations who already live there. In this respect, Cavallaro (2001, p. 172–173) reminds us that our subjective responses to space 'combines elements of realism and fantastic distortion'. In other words, we fantasize, mythologize and imagine the qualities of locations. Rutheiser (1996) at the Atlanta Olympics coined the phrase, 'the reimagining' of Atlanta; he catalogued how uncomfortable racialized histories were replaced by narratives of social harmony (an important consideration for Games organizers and multinational organization sponsors). What emerged in this respect with regard to the soccer World Cup in Brazil in 2014, were dominant media images that were broadcast, showing the Copacabana beach, complete with high-rise hotel development, used as a televisual tourist backdrop to symbolize the exotic mythologized tourist territory (a stage set for the girl of Ipanema) rather than nearby spaces, populated by multicultural residents. Higgins-Desbiolles *et al.* (2013, p. 92) highlight searching questions about the rights of host communities more generally but which could be applied to an analysis of destinations hosting mega-events, stating that these communities should have:

- the right to self determination;
- the right to participation;
- the right to refusal; and
- real choice.

They cite the work of Mowforth and Munt (2003) which explores the powerlessness and displacement of indigenous communities and the capacity of communities to 'cultivate' the places where they live and work. In particular, they draw special attention to Mowforth and Munt's (2003) critique on tourism and suggest that it is apposite to the challenges faced by planners in different cultural contexts to consider that:

- The places that tourists see as destinations are also places where communities live, work and play and are integral to those communities' self-determination.

(Continued)

Case Study. Continued.

- Tourism creates conditions of justice and injustice among tourists, tourism operators and residents. Relations of power operate between those who seek to purchase an imagined experience of some place or places and those whose lives are firmly rooted in the place or places that tourists seek to consume.
- 'Imagined' places may be appealing to tourists and lucrative for tourism operators, but at the same time this may be deeply damaging to the people whose lives have been reduced to an entertainment or spectacle for rich outsiders.

CONCLUSIONS

This chapter demonstrates that so-called planning theory presents the 'student' of planning with inevitable challenges, in that the questions of what planning is, and who planners are, are informed by political, economic and overarching ideological considerations, which define the cultural context in which planning takes place. In making these observations we demonstrated that it is possible to consider planning from different critical perspectives and that critical and cultural theory can enable insights into the planning challenges of creating social environments, which include 'everyday' social activities and which incorporate leisure and tourism activities.

These critical perspectives embedded within multidisciplinary traditions of the social sciences offer even more in-depth insights into how we view planning for tourism. Moving beyond the disciplinary boundary and perspectives as to what constitutes a body of knowledge and the field of study of tourism, offers a more holistic approach to understanding the relationship between planning and tourism. The exploration of the concepts of space and place were used as a prelude to a broader analysis of the concept of spatial analysis which observed the interconnectedness and interdependence between intersectional groups within society. It was argued that 'the social' and 'the environment' are inseparable concepts that guide consideration of spatial analysis which underpins judgements on planning for a sustainable future and achieving social and environmental justice. Furthermore, as Lucy (1996) highlights, the process of planning should formulate a 'planning vision' which best serves the public interest. As Lucy (1996) suggests what constitutes the public interest is constantly changed by societal events and the social actions of citizens. The case study of the mega-events contrasts the ability to create technically impressive arenas for sport, which attract ephemeral tourist audiences but which are positioned in cultural contexts of inadequate social infrastructure, where certain parts of host communities live in poverty. In using this case study it is hoped that the theoretical aspects of spatial analysis can be applied to the 'real world', a world in which considerations of ecocentricism and anthropocentricism influence our understanding of how we view the concept of sustainability. This is an important consideration if we are to realize the goal of social and environmental justice as an holistic aspiration and not as separate dualistic elements.

QUESTIONS

1. How would you design 'out of classroom' activities to explore the spatial characteristics and patterns of activities, within urban spaces, within your locality?

2. In researching different social groups described in this chapter, how might planning processes be improved to address their needs?

3. Choose a mega-event and locate media and academic material which highlights the disparities between planning for a social infrastructure and a leisure and tourism infrastructure. Should mega-events be staged in these locations without an adequate social infrastructure?

BIBLIOGRAPHY

Atkinson, C. (1981) *Picturing the System*. Pluto Press, London.

Ayers, A.J. (1936) *Language, Truth and Logic*. Penguin Books, Harmondsworth, UK.

Barnett, J. and Johnston, C.W. (2013) Our town's a drag: drag Queens and Queer space in Athens, Georgia. In: Schwab, K. and Dustin, D. (eds) *Just Leisure: Things That We Believe In*. Sagamore Publishing, Urbana, Illinois, pp. 49–57.

Benjamin, W. (1999) *The Arcades Project*. Harvard University Press, Cambridge, Massachusetts.

Berger, J. (1972) *Ways of Seeing*. Penguin Books, Harmondsworth, UK.

Berger, P. and Luckmann, T. (1966) *The Social Construction of Reality: A Treatise in the Sociology of Knowledge*. Penguin Books, Harmondsworth, UK.

Boyes, G. (2010) *The Imagined Village: Culture, Ideology and the English Folk Revival*. No Masters Cooperative Limited, Leeds, UK.

Brindley, T., Ryan, Y. and Stoker, G. (1996) *Remaking Planning: The Politics of Urban Change*. Routledge, London.

Bruenig, M. and Dear, S. (2013) Experiential education, social and environmental justice through parks, recreation, tourism and leisure. In: Schwab, K. and Dustin, D. (eds) *Just Leisure: Things That We Believe In*. Sagamore Publishing, Urbana, Illinois, pp. 117–125.

Campbell, S. and Fainstein, S. (eds) (1996) *Readings in Planning Theory*. Blackwell, Cambridge, Massachusetts.

Carson, R. (1962) *Silent Spring*. Houghton Mills, Boston, Massachusetts.

Castells, M. (1989) *The Informational City*. Basil Blackwell, London.

Cavallaro, D. (2001) *Critical and Cultural Theory*. The Athlone Press, London.

Delgado, R. and Stefancic, J. (2012) *Critical Race Theory: An Introduction*. New York University Press, New York.

Donaldson, P. (1973) *Economics of the Real World*. Penguin Books, Harmondsworth, UK.

Erhlich, P. (1989) The limits of substitution: meta-resource depletion and a new economic paradigm. *Ecological Economics* 1, 9–16.

Fainstein, S. (1996) Planning in a different voice. In: Campbell, S. and Fainstein, S. (eds) *Readings in Planning Theory*. Blackwell, Cambridge, Massachusetts, pp. 456–450.

Fox, W. (1990) *Beyond a Transpersonal Ecology: Developing New Foundations for Environmentalism*. Green Books, Totnes, UK.

Fox, W. (1995) *Towards a Transpersonal Ecology: Developing New Foundations for Environmentalism*. Resurgence Books, Dartington, UK.

Friere, F. (1970) *Pedagogy of the Oppressed*. Seabury Press, New York.

Galbraith, J.K. (1967) *The New Industrial State*. Penguin Books, Harmondsworth, UK.

Goldsmith, E. and Allen, R. (1972) *Ecologist – A Blueprint for Survival*. Ecosystems Ltd, London.

Goodman, R. (1972) *After the Planners*. Penguin Books, Harmondsworth, UK.

Harvey, D. (1989) *The Condition of Postmodernity: An Enquiry into the Origins of Cultural Change*. Blackwell, Oxford.

Harvey, D. (1996) On planning the ideology of planning. In: Campbell, S. and Fainstein, S. (eds) *Readings in Planning Theory*. Blackwell, Cambridge, Massachusetts, pp. 176–197.

Henderson, K. (2013) Ecofeminism(s) and just leisure in the 21st century. In: Schwab, K. and Dustin, D. (eds) *Just Leisure: Things That We Believe In*. Sagamore Publishing, Urbana, Illinois, pp. 35–42.

Herbert, I. (2014) Welcome to the real world for Brazil's poor. *i Newspaper* (London), 27 June, p. 56.

Higgins-Desbiolles, F., Powys Whyte, K. and Tedmanson, D. (2013) Tourism and environmental justice. In: Schwab, K. and Dustin, D. (eds) *Just Leisure: Things That We Believe In*. Sagamore Publishing, Urbana, Illinois, pp. 91–100.

Holloway, J.C. (1989) *The Business of Tourism*. Pitman, London.

Hughes, H. (2006) *Pink Tourism: Holidays of Gay Men and Lesbians*. CAB International, Wallingford, UK.

Hunt, E. and Sherman, J. (1975) *Economics: An Introduction to Traditional and Radical Views*. Harpers Row, New York.

LeFebvre, H. (1976) *The Survival of Capitalism*. Alison and Busby, London.

London Women and Planning Forum (2014) Available at: http://geog.qmul.ac.uk/lwpf (accessed 4 June 2014).

Lucy, W. (1996) APA's ethical principles include simplistic planning theories. In: Campbell, S. and Fainstein, S. (eds) *Readings in Planning Theory*. Blackwell, Cambridge, Massachusetts, pp. 479–484.

Massey, D. (1993) The conceptualization of place. In: Massey, D. and Jess, P. (eds) *A Place in the World? Places, Cultures and Globalization*. Open University Press, Oxford, pp. 45–86.

Massey, D. and Jess, P. (eds) (1993) *A Place in the World? Places, Cultures and Globalization*. Open University Press, Oxford.

Meadows, D.H., Meadows, D.L., Randers, J. and Behrens, W.W. (1972) *Limits to Growth*. Universal Books, New York.

Mishan, E.J. (1967) *The Costs of Economic Growth*. Weildenfield and Nicolson, London.

Morse, S. and Bell, S. (2008) *Sustainability Indicators: Measuring the Immeasurable*, 2nd edn. Earthscan Publications, London.

Mowforth, M. and Munt, I. (2003) *Tourism and Sustainability: Development and New Tourism in the Third World*, 2nd edn. Routledge, London.

O'Riordan, T. (1981) *Environmentalism*. Pion, London.

O'Riordan, T. (1993) The politics of sustainability. In: Turner, R.K. (ed.) *Sustainable Environmental Development*. Belhaven Press, London, pp. 37–69.

O'Riordan, T. (ed.) (2001) *Globalism, Localism and Identity: Fresh Perspectives on the Transition to Sustainability*. Earthscan Publications, London.

Packard, V. (1960) *The Waste Makers*. Pelican, Aylesbury, UK.

Pepper, D. (1984) *The Roots of Modern Environmentalism*. Croom Helm, London.

Pepper, D. (1993) *Eco-socialism: From Deep Ecology to Social Justice*. Routledge, London.

Raban, J. (1985) *Soft City*. Collins Harvill, London.

Roberts, N. (2013) Meeting at the crossroads: progress for multiracial people or delicate balance amid old divides? In: Schwab, K. and Dustin, D. (eds) *Just Leisure: Things That We Believe In*. Sagamore Publishing, Urbana, Illinois, pp. 27–34.

Rojek, C. (2010) *The Labour of Leisure: The Culture of Free Time*. Sage Publications, Thousand Oaks, California.

Rose, J. (2013) Contesting homelessness: public nature, political ecology, and socio-environmental justice. In: Schwab, K. and Dustin, D. (eds) *Just Leisure: Things That We Believe In*. Sagamore Publishing, Urbana, Illinois, pp. 58–66.

Rutheiser, C. (1996) *Imagineering Atlanta: The Politics of Place in the City of Dreams*. Verso, London.

Samdahl, D. (2013) At whose expense? How our commitment to conservation has propagated social injustice. In: Schwab, K. and Dustin, D. (eds) *Just Leisure: Things That We Believe In*. Sagamore Publishing, Urbana, Illinois, pp. 10–16.

Saunders, P. (1985) Space, the city and urban sociology. In: Gregory, D. and Urry, J. (eds) *Social Relations and Spatial Structures*. Macmillan, Basingstoke, UK, pp. 67–89.
Schumacher, E.F. (1975) *Small is Beautiful: Economics as if People Mattered*. Harper and Row, New York.
Schwab, K. and Dustin, D. (eds) (2013) *Just Leisure: Things That We Believe In*. Sagamore Publishing, Urbana, Illinois.
Silvertown, J. and Sarre, P. (eds) (1990) *Environment and Society*. Houghton and Broughton, London.
Singh, T.E. (ed.) (2014) *Critical Debates in Tourism*. Chanel View Publications, Bristol, UK.
Soja, E.W. (1985) The spatiality of social life: towards a transformative re-theorisation. In: Gregory, D. and Urry, J. (eds) *Social Relations and Spatial Structures*. Macmillan, Basingstoke, UK, pp. 90–127.
Spencer, C. and Jostad, J. (2013) Facebook's status in the lives of generation y: exploring power structures in an online leisure space. In: Schwab, K. and Dustin, D. (eds) *Just Leisure: Things That We Believe In*. Sagamore Publishing, Urbana, Illinois, pp. 75–90.
Thomas, H. (2000) *Race and Planning: The UK Experience*. Routledge, London.
Turner, R. (1991) Environment, economics and ethics. In: Pearce, D. (ed.) *Blueprint 2: Greening the World Economy*. Earthscan, London.
Urry, J. (1992) *The Tourist Gaze: Leisure and Travel in Contemporary Societies*. Sage Publications, London.
Urry, J. (1995) *Consuming Places*. Psychology Press, Abingdon, UK.
Watts, J. (2014) Anti-World Cup protests across Brazil. Available at: http://d/2014/may/16anti-world-cupprotests-across-brazil (accessed 1 June 2014).
Whynne-Hammond, C. (1985) *Elements of Human Geography*, 2nd edn. Allen and Unwin, London.
Wright Mill, C. (1959) *The Sociological Imagination*. Oxford University Press, Oxford.

PART II

Practitioner Approaches in Planning and Tourism

There are four chapters in Part II which are designed to demonstrate different planning approaches and practitioner insights. Three chapters are written by public sector professionals, working in both urban and rural local authority areas in the UK. This part of the book also includes a chapter providing insights from a tourism entrepreneur and academic, on aspects of planning for disabled people, highlighting specific details about the concept and applications of universal design.

- Chapter 4: *How to Plan: the Role of the Planner and Planning in the 21st Century* – Nigel D. Morpeth in conjunction with Ian Brown, Head of Planning and Regeneration at Canterbury City Council in Kent in South East England, explore aspects of Ian's career as a planner and in his current role, in planning for the needs of residents and tourists within an historic urban environment. A case study on tourism within the city highlights the planning challenges of transport and the World Heritage Site of the cathedral.
- Chapter 5: *Planning for Tourists and Residents in an Historic Destination – York UK* – Ian Tempest, former Tourism Manager for the City of York Council, UK, provides a 50-year overview of how tourism has been organized and developed in an historic city, through public–private partnerships. He observes how town planning has guided land uses in the city and, in a case study on transport, highlights how sustainability initiatives have influenced resident and tourist transport uses.
- Chapter 6: *Destination Planning for Disabled People* – Craig Grimes, Experience Community CIC (Community Interest Company), writes from the perspective of an entrepreneur and academic and wheelchair user, and reviews how historically there is an 'invisibility' of disabled people in tourism literature and contrasts this with the developing concepts of accessible tourism and universal design. Craig then applies these concepts to a case study of Barcelona and highlights his pioneering work in the city, not least

in planning and providing bespoke holidays for disabled people.

- Chapter 7: *Spatial Planning – Managing Tourism and Recreation in a Lived-in 'Protected' Area* – Kathryn Beardmore, Director of Park Services, Yorkshire Dales National Park Authority, UK, highlights that an analysis of UK National Parks differs from those in other parts of the world because they are 'lived in' and the majority of land is in private ownership. Kathryn highlights partnership working and a collaborative approach to developing an integrated management plan, which is underpinned by the principles of sustainability, to balance the needs of conservation, recreation and local communities.

chapter 4

How to Plan: The Role of the Planner and Planning in the 21st Century

Nigel D. Morpeth[1]* in conjunction with Ian Brown[2]

[1]*Leeds Beckett University, Leeds, UK;* [2]*Canterbury City Council, Canterbury, UK*

LEARNING OBJECTIVES

- To provide insights on the contextual information on the origins of the planning system within the UK, post-1947, and the Town and Country Planning Act which has relevance to other cultural contexts.
- To explore the evolution of the career of a planner, from their planning education through different professional planning roles, to understand their perception of planning processes and the evolution of planning within a UK context (but with scope to apply these observations to other cultural contexts).
- To understand, within the specific context of the City of Canterbury in the UK, the role of the planner in balancing the planning needs of residents, tourists and pilgrims.

INTRODUCTION

The aim of this chapter is to provide an in-depth profile of a planner who has worked in public

*n.morpeth@leedsmet.ac.uk

sector planning within the UK, for over 30 years. This profile provides insights into the role of the planner and what the changing challenges are, over a period of significant socio-economic and political change within UK society. The chapter initially provides an academic commentary on the public-sector planning context, over that period of three decades, to give an understanding of what the UK planning system is, how it has evolved over that period of time and what the key perceptions of a planner are, in responding to planning challenges over a long period of time. This chapter presents a case study on the challenges of balancing the planning needs of tourism in conjunction with the planning needs of residents and pilgrims.

The main purpose for articulating the profile of a planner is to provide the reader with a human dimension, by personalizing the role of the planner in the planning process, and to enable the reader to understand and to interpret the seemingly abstract and misconstrued 'dry' qualities of planning policy and strategy processes. Planners can be viewed negatively by the public with the interviewed planning officer stating that:

> 'I think that planners and planning, probably together with local government generally, has been the subject of criticism and in many cases quite negative criticism from government, the press and local communities blaming local government for a lot of the issues that are happening within our communities and perhaps not being sympathetic to what planners or what local government is trying to achieve. I think it is probably fair to say that a lot of people would not necessarily understand that the planning system is very much behind what delivers those major proposals (such as the Olympics and the Channel Tunnel project) or just the day-to-day management of making sure that impacts are assessed properly and hopefully moderated. I don't think the public is generally aware of that role.'

As we saw in Chapter 1 'Approaches to Planning and Tourism', there is a rich seam of theory which provides an understanding of how different approaches to planning might work (within the field of tourism) but to coin the phrase from Chapter 3, there is a challenge to interpret planning processes in the 'real world'. In the context of this chapter, the 'real world' of planning provides a contrasting commentary to that of the theoretical world of planning and there are 'real world' insights from the interviewed planning officer. The responses from the planning officer reveal the different dimensions of planners' work within the planning system, with strategic and operational aspects of their work highlighted. Being a planner has particular qualities in approaches to policy, decision making and day-to-day choices that have to be made. It also provides the opportunity to match up theoretical observations of planning processes, highlighted in Chapter 3, which anticipate the realities of being a planner within the public realm in the 21st century. It is hoped that in reading this chapter you will be able to: (i) consider the dynamics and characteristics of planning systems and processes that you might want to apply to other cultural contexts; and (ii) understand how planners balance the needs of residents and tourists within other localities globally.

THE PUBLIC SECTOR CONTEXT OF PLANNING IN THE UK

Prior to providing a profile of a planner as the main focus of this chapter, it is useful to outline the broad context in which this planner works and in which they have based their career over a 30-year period, which is characterized as a public-sector planning system, within the UK. This planning system underpins the work of 468 local authorities (municipal areas) in the UK, which are not uniform in size and incorporate both urban and rural local authority areas. There are different types of local authority areas, from larger urban or metropolitan councils, county councils (for each of the UK's approximately 120 counties – excluding Northern Ireland) to both district and city councils (which are component areas of those counties). For a brief period of time in the late 1990s and early 2000s, the UK flirted with regional government, as an additional level of governance which is no longer in operation. Local authorities have statutory functions, with planning having remained a constant statutory function of local government. It is important to note that tourism is not a statutory function of local government; local authorities have the discretion of not funding tourism, either in its organization or in its promotion, but historically local authorities have chosen to do so.

The interviewed planner has a senior role as a planner in Canterbury City Council, in Kent in the south-east of the UK. It covers the historic and World Heritage Site of Canterbury and its cathedral as well as the coastal towns of Herne Bay and Margate and a largely rural area with a population of 140,000 people (with an additional 30,000 living in the area during term time). A Local Plan is developed in consultation with the population of the local authority area. The current Local Plan was adopted in 2006 and with the consultation stage complete, the Local Plan will now cover the time period up to 2031. A Local Development Scheme sets out the programme of work outlined in planning and strategy documents, which are scrutinized by a sustainability appraisal, which is a national government set of sustainability standards (Sustainable Development Indicators, SDIs). Consistent with other local authority areas within the UK, it creates a Statement of Community Involvement (SCI) which is the set of standards for community involvement in engaging with the planning process. The interviewed planner works within the Planning and Regeneration Division, which contains a Planning Policy team which reviews the Local Plan (Canterbury City Council, 2014a).

The role of the planner has had to adapt to and be responsive to an ever-changing political and policy environment, in which there has been constant reform over many decades. In the UK in 2014, there was a constant question as to whether central government would like to see the dismantling of local government, with so-called 'localism' giving communities powers to run what traditionally have been local authority services (but with central government hoping that this might be done as acts of voluntary engagement). This is an example of a changing role of local authority governance.

Governance

Within a UK context, the concept of governance has led to a number of different organizations and agencies, in addition to local government, being involved in the administration and management of local authority functions, which traditionally would have been purely in the domain of local authority departments. Local government has been encouraged to be more entrepreneurial, particularly in an environment of budget deficit (a public sector trend seen across many countries in continental Europe in 2014). In the context of this entrepreneurial aspect to the work of the interviewed planning officer, he stated that:

> 'We've been working on thoughts for new ways of bringing investment into the district through a new initiative called the Local Enterprise Partnership (LEP) for the South East. We are part of that and the Government is asking for bids to be put forward, so I've been on a group that's been looking at big strategic ideas for East Kent as a whole.'

In terms of what governance means, Boddy and Parkinson (2004, p. 6) highlighted that: 'The idea of governance emphasizes the part played by a wider set of bodies, organizations or associations beyond the formal structure of entity of local authority populations'. This is in theory enhanced by bringing together a range of cross-sectoral organizations and entities as part of service provision. In the realm of planning, planners work across sectors, including working proactively and closely with the private and commercial sectors in the complex arena of the public realm, which is the arena of public policy making. The public realm is made up of a conflict of interests and the local authority has to make an assessment of these interests, partly through public participation. In McCabe's (2010) view the current governance arrangements are a hybridization of the private, public and voluntary sectors, in terms of blurring the boundaries of what previously were state and statutory structures and functions. McCabe also states that the funding of community initiatives (described above as localism) should not be done as a state target to remove tiers of (local) government.

The interviewed planner highlighted how it is important to be able work across sectoral boundaries, stating that:

> 'It is clear that one of the key skills of working as a planner is to have the capacity to communicate with a wide range of cross-sectional groups, and to incorporate a myriad of views in the development of planning policies. It's always a case of what we want to try and achieve, and whether it's in line with what they're trying to achieve. Clearly if what we're trying to do is at odds with one of the other parties then conversations aren't going to be very easy, but we usually find that we have a lot of common ground. But for the county, they have to deal with 12 districts in the county. For the LEP there are, however, many authorities and businesses, and clearly for the Government it is hundreds of councils across the country.'

Public sector planning

Defining planning within this context, Rydin (1993, p. 1) identifies that 'it is decision making to achieve a given goal', which is guided by policy processes that planning systems operate in. The characteristics of planning are:

- It is a future-orientated activity.
- It seeks to devise strategies which lead to desired end states.
- It can be part of private sector activity (and needs to be receptive to private sector processes), but it is primarily a public sector activity.
- While it might involve social and economic planning, it is about planning of the physical environment – both the built and the natural; it is land-use planning which is at the heart of UK planning (Rydin, 1993, p. 1–3).

Rydin identifies that the overall planning 'umbrella' within the UK is determined by the role of Parliament in passing legislation and that the key legislation since the Second World War was the 1947 Town and Country Planning Act (a planning template that has been used in colonial domains and has been adopted and adapted in countries across the globe). This and subsequent legislation required government departments, in conjunction with local-government planning departments, to provide a 'comprehensive regime for the control of development, supplemented by plans setting out intentions in relation to the broad pattern of development' (Rydin, 1993, p. 2). Since the inception of this legislation, Rydin observes that this public-sector planning system was consolidated by the clear and supportive role of local government (and by central governments). In her view this culminated in the professionalization of the career of local government officers, strengthening the role of the planner.

Rydin identified that by the late 1970s, to almost two decades later, the role of both local government and philosophies of planning had been transformed by central government's perception of the function of local government. This was in part through planning incorporating more actors in the process of local governance. In short, in the 1980s there was a curbing of the powers of local government and a favouring of the extension of the number of actors from wider local governance arrangements, described above. Under successive Conservative government administrations from 1979 to 1991, there was a rethinking of what the functions of local government should be, with a greater role for the private sector to take over and run the functions of the delivery of services that had exclusively been the domain of local government. The competitive tendering of services to the highest bidder, either from within the local authority or to outside private companies, was viewed as a way of making local government more financially effective. This continues to be the focus of local authority reform, which in 2014 is anticipated as an ongoing project of budget reduction and a redefining of the role of local government, guided by central government.

It was not only Rydin who highlighted these changes in local governance and the implications for the planning system. Brindley *et al.* (1996) noted how in the UK in the 1980s it was feared that the obituary of planning might be written, through the actions of a Conservative political administration which introduced: 'The relaxation of many controls, the introduction of enterprise zones and simplified planning zones, the transfer of planning powers to urban development corporations, the greater stress on market criteria in development control decisions, all have been lethal attacks on planning itself' (1996, p. 1). While these fears proved to be unfounded, nevertheless the role of post-war consensus planning was superseded by the emergence of market-led planning and this led to concerns about social groups which the market does not cater for being socially and spatially excluded.

Therefore, UK society post-1980 provides an important cultural and economic backdrop to the planning challenges faced at the time and which were challenges that this interviewed planner responded to in his various roles as a planner. The planning challenges from the 1980s to 2014 have focused on urban policy and regeneration in deindustrialized areas and a nation which has moved from an industrial economy to a service and information economy. This is the societal background which the highlighted planner has worked in over the last 30 years. Profiling a planner who has been engaged in the process of planning, within the cultural context of the UK, provides in-depth insights and a longitudinal perspective on planning.

PLANNING AND PEDAGOGY: A UK CONTEXT

This section of the chapter provides an overview of the career development of a planner, from their formative years as a planning student, revealing their motivations for becoming a planner and the importance of the pedagogy of planning and the evolution of synergies between theory and practice.

Q. When and why did you first decide to study planning?

A. 'I decided to study planning after looking at a number of careers linked to urban and physical geography. I considered becoming a surveyor and becoming a chartered surveyor, and I was also particularly interested in architecture and landscape architecture, but decided on planning because of the breadth of issues it covered. I was particularly interested in how settlements work and particularly how they develop over time.'

This reflects the view of Campbell and Fainstein (1996) cited in Chapter 3, this volume, who highlighted the blurred boundaries of what constitutes planning and other professions and also the work of the Royal Town Planning Institute (RTPI; below). Therefore, demarcating what planning is and who a planner is, is not without its challenges. Newcastle University in the north-east of the UK is one of the UK universities with a long history of planning education. This planner highlighted that:

> 'there were a number of key institutions that provided opportunities to engage in pedagogy of planning at university: Herriott Watt, Glasgow (the Art School), Newcastle (where I went), Manchester (a good reputation – a good planning school), Oxford Brookes and in London, University College and Birkbeck and LSE [London School of Economics and Political Science] (postgraduate).'

Q. Why did you decide to study at Newcastle University – what were the key aspects of the course which attracted you to study there?

A. 'I looked at a number of courses in the late 1970s and a number provided a route through to become a member of the Royal Town Planning Institute, which is essential really if you are thinking about a career in planning. I looked at the different courses available and visited a number of universities across Britain, including landscape architecture courses too. What attracted me to Newcastle was the city itself, and the quality of the course and the university and lecturers were committed and enthusiastic about the topic, and a wide range of disciplines were taught as part of the course. This included architecture itself, through to sociology and politics, affecting urban and rural change and also the elements of engineering and law. This provided a very rounded qualification and area of study. Equally, the people themselves seemed passionate about the subject that they taught and when I started the course, I wasn't disappointed in the

> range of study available by itself and the range of study beyond the actual core lectures themselves.'

As a footnote to this planner's educational experiences, the academic institution of Newcastle University, School of Architecture, Planning and Landscape promotes itself as one of the top educational institutes to focus on the built environment. The ethos of the school is imbued with a philosophy of sustainability, which emphasizes the linkages between education, practice and policy making, with an emphasis on making and sustaining places, which enhances life for present and future generations (www.ncl.ac.uk). It is also worth noting that Newcastle University opened a Sustainability Institute in 2008, which carries out innovative work on different aspects of sustainability (www.ncl.ac.uk/sustainability/about).

It is important also to highlight that the RTPI was formed in 1913 and its main aim and responsibility is to maintain professional standards in planning, by overseeing the professional certification of planners. This is done by promoting, developing and shaping policy and setting standards for education and professional development. In terms of defining planning, the RTPI (2014) identifies that planning has two main aspects to it as it involves:

- the management of competing uses of space; and
- the making of places that are valued and have identity.

The RTPI (2014) identifies that the broad range of activities of being a planner might involve:

- helping to design towns and cities, to include attractive buildings and public spaces, and promote healthy lifestyles;
- helping to regenerate socially deprived areas and creating new jobs;
- managing development and helping to create affordable housing;
- creating policies for managing traffic and providing sustainable solutions to our transport needs;
- improving energy efficiency and cutting carbon emissions in our homes, factories and businesses; and
- engaging communities to have a say in how their living space is developed and protected to improve their quality of life.

The RTPI not only works closely with universities, to coordinate linkages between theory and practice and the setting of professional standards but they also engage with wider pedagogy of planning and education, in their work with schools with the age group of 11–18-year-old students. This Future Planners Initiative informs curricula with issues related to concepts of planning, underpinned by principles of sustainability. The production of the film *Together, How Do We Plan Our World?* embodies the wider themes of planning and sustainability which underpin this book.

Q. How did your degree course prepare you for a career in planning?

A. 'I think the degree that I did, and degree courses given at that time, were very good at giving a broad overview of the history of planning, and the various strands that you need to pull together to design communities, and pull together new developments – even to the point of understanding some of the engineering principles of construction and very detailed design concepts of architecture. It was also good at flushing out the main political and social movements that have shaped the communities that we have today.

I think that one of the weaknesses, though, is that the theoretical coverage that we were given was very thorough but the reality of how decisions are taken and the political framework within which they are taken is very difficult to teach or to explain. It is only when you are actually in the workplace that you have to understand that more fully, and that is regardless of whether you are working in the private sector, working for a client or trying to get a permission through the system, or whether you are a planner working through a local authority assessing proposals and taking it through the political process. I think that is one of the difficulties that I faced once I had left college – to have to learn all of that machinery as it were.'

In terms of the evolution of the linkages between theory, practice and policy making, the current promotion of the planning course emphasizes the synergy between these different elements of pedagogy. Likewise the observations in Chapter 3, this volume, and particularly the London Women and Planning Forum, also emphasize the closer linkages between theory and practice in the evolution of the pedagogy of planning. Furthermore, the RTPI demonstrated above that they have a close linkage with working on the pedagogy of planning and universities' ongoing planning courses and there are closer synergies between theory and practice. This was confirmed in the answer to the next question.

Q. How have planning courses evolved?

A. 'What I do pick up is that planning courses have evolved and become more sophisticated to explain and deal more fully with the real world as it were: how decisions are taken and the consequences of decisions, how planners fit within both local authority contexts and also in the commercial world as well. Certainly, when I speak to graduates they seem to be much more sophisticated in their understanding of all that than certainly I was and my colleagues probably were. Even though we did a year out as part of our course, which was intended to prepare you for your final career, because of the wide range of planning environments, it depended very much on where you did your year out as to whether that was really going to prepare you for a career.

Now, there seems to be a stronger emphasis about preparing you for work, rather than just giving you a good education – giving you a degree which allows you to understand how we have got to where we are now in terms of the planning world that we see around us. But I suppose it is probably the same for most other qualifications as well, e.g. law or medicine or whatever. I imagine over the same period they have equally had the same emphasis on preparing students for the world of work.'

This is consistent with the discussion, in Chapter 3, this volume, on the notion of the 'real world' and the balance between theory and practice. The next question reveals the importance of a longer historical perspective on the inception and evolution of the development of the system of planning within a particular nation state, with key historical events influencing the style of planning. The RTPI suggested that the term 'town planning' was first used in 1906, in the UK, and that after the passing of the Housing and Town Planning Act of 1909, local authorities for the first time were able to plan how land would be developed in their local authority area (RTPI, 2014). This was an important point in the development of the social history of planning and development in the UK.

Q. As part of your studies who were the 'big names' in the history of planning who inspired you and why (and have continued to be an inspiration)?

A. 'Probably the obvious ones, and I would not say that they inspired me but I recognized the contribution to the world that we see around us today. What struck me were the developments that had happened in public health in the Victorian period (in the UK) and the names that they spawned later on, particularly Ebineezer Howard [see Osborn, 1946] and the Garden City Movement. But in many ways some of the more mundane changes in social structures and the engineering period certainly laid the foundations for the way in which our towns operate today. Even some of the by-laws that were passed which addressed some of the appalling housing conditions that people found themselves in during the Industrial Revolution certainly made a big impression on me. Then as you understood how things had changed over the 19th century, you could then understand why a lot of our cities look the way they do today, for very basic reasons which were principles of disposing of sewage or ensuring that buildings had enough space for air and light between them. And as the century changed into the 20th century [there were] more enlightened approaches to building [with] new housing layouts which then led to the large surge of house building before the First World War and between the wars themselves. So you can then appreciate that up and down the country very similar layouts and solutions to providing housing were brought about because of the common legislation and thinking that had influenced them in the first place.'

In terms of relating these insights to other cultural contexts, consider how other key names in the history of planning have significantly influenced the pattern of development and the living environments of communities.

Q. Please could you provide a quick overview of who these people were and what their contribution was to how planners think and practice?

A. 'But of course the names and the individuals that shape planning and shape the way things change did not finish in the 1960s – they carried on and individuals were involved in planning of the New Towns in the 1960s and in the 1970s. Although they were not particularly big names that people were aware of, they were individuals and groups of people who worked hard for aspirations for new ways of living. The people who were involved in Peterborough, Milton Keynes and Cumbernauld, places like that, were working at the edge of what people knew in those days about how to bring together communities and it was very much unchartered territory. Some of it didn't work, some of it showed even with the best plans or intentions it isn't possible to create communities from scratch, whereas others have been more successful and have provided opportunities for people to live quality lives in terms of their accommodation and their access to facilities and to employment.

I think Milton Keynes is a good example of that and the plans that they worked on with that city, which has now become one of the unsung heroes of planning. Certainly that project was used as a model throughout the world as to how you could create a quality space if you were very careful about how you segregated traffic and people and made sure that you provided facilities at the hearts of communities and also made sure that you created transport links as well.

Coming nearer to the present, I think one of the very influential politicians we have had is Michael Heseltine and his very influential role in bringing investment back to the heart of cities. Because until then and the 1970s, the emphasis had been very much on the communities outside of cities, new towns expanding cities on their edges and the centres of cities had often been largely forgotten about. People saw the problems that had been created and the decay that we had in the heart in many of our cities, but the policies did not seem to be in place to do anything about them. And it is fair to say, regardless of his political persuasion, Heseltine made ground-breaking urban interventions, particularly in places like Liverpool, but other cities as well. And he recognized that the same sorts of investments and the sorts of zeal and drive which had created the new towns should be harnessed and we should start to do the same thing within cities. There have been a lot of controversies about some of the solutions that were put forward, particularly the simplified planning zones where planning controls were largely taken out of the hands of local authorities, and I think it is fair to say that the investment that was put into those areas and the very clear purpose and vision that people brought to those areas has now [been] seen to have borne fruit.

A good example of that is the area around Canary Wharf [Docklands in London – see Chapter 3, this volume] and again a debate around whether that helped the local community. But if anyone has compared that area to what it was like at the beginning of the 1980s [they] would recognize the employment that is now available and the change that has happened in terms of the environment has been delivered. And [it] looks as though it will sustain into the future and continue to generate prosperity in what was a completely decayed part of London, forgotten about and largely without purpose in terms of its original role as industry and port-related activity. All of that had gone.

Moving further on, John Prescott – people should recognize his contribution. Again not everyone has agreed with the way that he has approached things, but again he had a vision to provide more housing and to try and 'kick start' the economy and bring investment back into the centre of our great industrial cities, many of which had lost their way, particularly in the 1960s and 1970s. He produced a task-force again and managed to harness various government departments and budgets to allow local authorities and local communities to regenerate their own communities. They are the sort of people who were influential when I went through my studies and went through my career. Some [were] more controversial than others but all of them wanted to make a difference. And I think that is what we need to do and see that we can make a difference and people are clear about what they are trying to achieve and are able to coordinate resources and policies, then some very significant changes and improvements can be made.'

These responses demonstrate how national government, particularly within the political arena of urban policy, have an important role in setting the policy priorities, in which planning and planners have to respond. The period of time incorporates the political administrations of the Conservative government, which was in power from 1979 to 1997, in four successive political administrations.

Michael Hesletine was a Conservative cabinet minister who created urban regeneration policies, largely in response to widespread urban riots in 1981, not least in Liverpool but in other urban centres around the UK. John Prescott was Deputy Prime Minister in four successive New Labour administrations, from 1997 to 2010 and was responsible for empowering local government functions and the development of regional government. The chronology of the career of the interviewed planner straddled this period of time, a time of political restructuring of post-industrial cities and also working in local authority structures which were abolished as part of reforming local authority structures in 1996, by the then Conservative government.

Q. Please can you provide a chronology of your career as a planner?

A. 'In terms of the chronology of my career, I have worked in various local authorities, together with doing a postgraduate course at Wye College in Kent which is a branch of University (Imperial) College London, which was a very specific course on ecological principles, landscape design principles and management of environments. Jobs in local authorities included working on mineral planning, working on reclamation schemes of former mining [areas], working in a multidisciplinary team which included architects and engineers, as well as some ecological expertise and [I] brought my recent qualification to bear on that as well. I worked as a planner within that team, preparing management plans for country parks and other natural areas, and areas both within towns and cities and within the urban fringe as well, and [I] worked on access strategies as well bringing in cycle and pedestrian routes.

I then went to Canterbury City Council where I still am. I have had many different roles within the authority, originally within planning policy and working on the Local Plan and also working on physical regeneration projects, the establishment of the management of country parks, and giving advice on planning applications and the policy side of those. Progressively I have moved into more of a managerial role, firstly managing the Local Plans Team and then added to that were the economic development functions, so working with local businesses, a service which provided business support and did work to get European money for regeneration and business support for the team. Then the tourism function was added, so I managed all the visitor information centres and all the work on promotional material for the district and working alongside other tourism organizations. Then, following another restructure, [I] managed the detailed development management service, so that is all the work around processing and assessing planning applications and all the planning appeals and the enforcement of planning regulations. My division also then included all the engineers, all the coastal and river flooding issues and that section provides services to adjoining authorities in terms of flood defence assessment and design. Also added to the section is the building control function and that is the detailed consents around construction methodologies, so making sure all the foundation specifications are met and all the technical specifications for buildings are done properly. The section also includes the Housing Strategy section and they are the people who produce the affordable Housing Strategy and work with Housing Associations to provide affordable housing for local people. I think that the way in which my career has evolved shows the role planners often take [is to] manage a wide range of services, because the education and work that you do prepares you for working in lots of roles where you need a systematic and planned way of operating. And often planners are put in charge of those sorts of wider technical services, and rather than just the professional role we take on the physical development side of things. I think that it also shows a strong linkage between the different technical disciplines which make up the framework within which towns and cities are planned and managed. In conclusion therefore, I suppose my career has moved from very detailed operational roles when I first came out of university, working on implementation schemes and managing contracts, through working on policy, and then into a managerial role, where much of my time is spent on managing the day-to-day running of the service. [This is based] around managing people and budgets and also the relationship with politicians and local groups and interests, as much as the more technical or professional role of planners which you may see more in the private sector. So my role now is really more of a managerial one, setting the framework for the division to then operate and deliver its services and to make sure that all that is managed properly and I am held responsible for the performance of that division.'

The range of posts described above, and particularly the current role and responsibilities of this planner, demonstrates the complex multidisciplinary operational, managerial and strategic elements of their role and provides insights into the combination of daily tasks they are engaged in, which make tangible the broad statements of what is contained within 'planning processes'.

Q. How has planning/the role of the planner/your role as a planner changed over the last 30+ years in the UK?

A. 'I think in terms of how planning [has] changed, it has moved more towards the planning system enabling things to happen. A few years ago, planning was something to regulate and control and [now it is] very much seen [to be] more about managing and facilitating, particularly in terms of housing and employment. There has also been a huge amount of legislation over the years and also an understanding to assess impacts before they actually happen. I am not saying that that didn't happen before, but there is a more coordinated way of doing that now, particularly things like strategic environmental impact assessments [which] are designed to look at a series of impacts, to look at habitats, etc. And that has to be done in a certain methodology before developments can be assessed, and indeed there are a number of documents and assessments [that are required] before particularly major applications are even registered.

So I think that there is a much more precise framework around major proposals now than there used to be. I think that one of the criticisms is that [it] has led to delays for lay people to get involved in the process as a lot of those studies are incredibly technical. It has also meant that there is a growing industry evolving to supply that information and many professional firms now exist to do those impact [assessments] and many planning firms who exist [to take on this sort of work] on behalf of their clients. Interestingly, throughout the rest of the world this sort of framework is also evolving and therefore a lot of the firms who are established in the UK are trading abroad. The system that we have here is seen by many as a good system and therefore the competencies of planners trained and operating here is valued elsewhere.'

Q. Highlight what the main challenges are which face the planner in the UK today.

A. 'In some ways it is an extension of what I have just been talking about. It is trying to manage that interface between central government policy and local aspirations – so my role often is to try and explain to people what we can and can't control. In an effort to try and free up the housing market, for example, the government has relaxed some of the planning regulations that we have, particularly around permitted development that people can do. That has caused some problems because it is not always clear what people can and can't do and often it causes significant neighbour disputes. On a bigger scale it goes back to some of the issues around the true ability of local communities to determine development levels, or how they want to assess planning applications, whether they want to give it permission or not. Having said that, the ability to set the agenda still is with local communities. If they want to have a strategy which promotes growth and tries to produce additional jobs and houses, they can do that, provided they can make sure that all impacts are taken into account and things like the road network can cope or they put in place new transport links to deal with [any problems caused].

Over the last 30 years I think the role of the planner in local authorities has probably diminished: 30 years ago the Chief Planner or Head of Planning would have been quite an important role within the organization, but I think over the years that has changed and been absorbed into more multidisciplinary departments. Perhaps that status has changed but that is probably the same for a number of professions within local government: in the old days the Town Clerk, the Borough Engineer and Chief Environmental Officer were very important local roles [but these have] diminished and changed to a more managerial role. Therefore, probably the role and influence of local authorities more generally has diminished. I think the other change also over this time has been to diminish the role of planning authorities because of the large amount of central regulation; much of what we do or have done has been dictated by central government or by regional organizations, and that has led to a lot of frustration within local communities because they feel that they are not in control of their overall destiny. That carries on to the present day – there has been a lot of talk of localism and the ability of local communities to set the agenda in terms of development and that has been very much on the basis of making sure the starting point is to provide sufficient housing to meet the census projections. So we need to make sure that is the starting point, and [if] we are not able to meet that number of houses per year then we need to have very good reasons why. And that obviously gets challenged at a public enquiry and if a local authority can't show that it is doing that, then any plans that it is preparing are less likely to be approved.

That is at the more strategic plan-making arena but also on a plan-making application basis we also need to make sure we take decisions on planning applications which are absolutely in accordance with the National Planning Policy Framework. And again in some ways particularly local politicians feel that their hands are tied very often because the agenda has already been set. Therefore they need to make sure that they are demonstrating that they have gone along with that agenda and have taken decisions in accordance with it, because quite clearly if they have taken decisions at odds with national policy, and if they refuse a planning application, then it is quite a simple job for a developer or agent to take that application to appeal, and if they can demonstrate quite clearly the local authority has not followed government policies then they will be able to get that refusal overturned and get consent.'

This section has demonstrated how the macro-political and economic environment has, over the last 30 years, led to successive national government administrations dictating the evolving dynamic of local governance and has dictated that the role of the planner is increasingly to pursue economic growth policies and strategies and approval for private housing initiatives.

Q. How do the principles of sustainability inform the planning process locally?

A. 'One of the main challenges is to incorporate the principles of sustainability into planning policies. Strictly speaking, and according to the planning legislation, you have to consider the social impact, the environmental impact and also the economic impact – it's those three strands all the time. That's enshrined in legislation and that goes to the heart of sustainability. But locally there's a big community angle to this, and as a political organization we have to be very conscious of what local communities want. Sometimes it will be environmental things, sometimes it will be more social facilities, sometimes it will be the need for jobs. We have to balance all of these issues. But at the end of the day the Council is held to account particularly for planning decisions which will often be taken – especially if they go to appeal – not necessarily taking on board local views, but taking on board national policy, which at the moment is very much promoting economic activity, housing delivery. Some of these are concerns of local groups who may feel they're very important but if there are national policies which say something different, we will have to try and respond to that as well, because if we don't, eventually our decision might get quashed somewhere else. We do need to respond to local views and concerns, but if we know that they just won't get supported elsewhere, we have a difficult task of explaining that to local groups and at the same time try and get the best for them, but fit[ting] within national policy. We need to balance all that out when we come to our decisions.'

It is important to emphasize that from a wider UK context and experience of other planning services over an extensive period of time, planners view the planning role in local authorities as an important process for advancing the goals of sustainability. They argue that planning processes lend themselves to incorporating the social, economic and environmental dimensions of sustainability particularly through the community-based consultative and participative processes of the Local Plan.

Public participation in planning

Regarding the process of planning the interviewed planning officer highlighted that:

> 'Over the period that I have been working in planning there has been a much greater recognition that the public need to get involved and should be involved in taking decisions, and public participation is very much at the heart of the system now. There has always been consultation on proposals but it is much more systematic now and also [it is] even enshrined in some of the regulations that there should be consultation, particularly before applications arrive with the planning authority. We call that the pre-application phase and developers and agents who are putting forward developments need to show how they have engaged with people and,

importantly, that it demonstrates that it is meaningful engagement.

For people trying to therefore assess the planning application when it comes in, all this has made life more complicated and it has meant that the resources needed probably need to be more significant. Although ironically at the same time resources have had to be cut back in the public planning sector and that is probably one of the big challenges today and certainly has been evolving over the last few years. But overall the process of looking at major schemes and assessing them as planning applications has become a lot more sophisticated and complicated.'

Sustainability and the natural environment

Consistent with discourses of sustainability and the nature of the physical environment and the dilemma of how to view natural resource use and preservation, the interviewed planner highlighted that:

> 'I think beyond that, another big change over the time I have been a planner is to emphasize the role of ecology and the natural world, as it were, in our assessment of proposals. I think when I first started the battle over recognizing the importance of urban or rather built conservation had largely [been] won in the '70s and '80s and historic buildings and conservation areas were very much central to what was happening in important historic areas, but the natural world had not been recognized to the same extent. So over the period I have been working in planning, the need to do habitat assessments and all [the] habitat regulations that have come in and the very detailed work that we need to do to assess the potential impact on vulnerable species has been a major change. [This is] not always viewed positively by everybody and I think for some groups and individuals it has led to frustrations that development is held up by the presence of one or two key species on a site. But overall I think it is a good thing to recognize that the quality of anyone's environment is not just the built world, it is also the natural world and the way in which we interact with that natural world brings about the quality of life that we all enjoy ... I mean overall people recognize that the management of our towns and communities and natural areas is something which needs to be worked at, it isn't just something [that] just happens and the planner's role within that has waxed and waned over that time.'

Case Study: Planning Challenges for Tourism in Canterbury

The ancient city of Canterbury, in the county of Kent in the UK, brings together a number of key considerations which are at the heart of this book. It provides an example of a popular destination to balance the needs of tourists and pilgrimages but first and foremost to respond to the needs of local residents. In the response to the question about their duties and responsibilities, the planning officer responded that:

> 'My service covers all the planning decisions that are taken by the authority – that's all the processing of planning applications and also the preparation of the planning strategy. My role also covers economic development and tourism activity, the countryside and conservation function and building control. Also policy on transport.'

(Continued)

Case Study. Continued.

This demonstrated the complexity of the role of this planner given the broad range of functions which have both strategic and operational elements but also that tourism is one element of many different planning priorities.

Balancing the planning needs of residents and tourists

Similar to the historic city of York, transport is a key consideration in planning for residents and tourists in Canterbury, with the interviewed officer stating: 'There are a lot of pressures on the city. It's a historic city which was never really designed to take the pressure in terms of traffic that it does now.' This challenge included managing the motor car within the city and motor vehicles generally. In this respect the planning officer highlighted that:

> 'One of the biggest challenges for the city is transport in the city. Obviously Canterbury is a centre for activity for a much wider area, whether it's shopping or legal or financial services or all the tourism pressures that we get, so we have a lot of traffic trying to access the city. We have worked hard to find new ways of tackling this – worked closely with the bus operators to have an attractive alternative to using the private car – provided park and ride sites on the outskirts of the city and also worked to deliver cycle lanes and walking routes. But there is still a lot more to do.
>
> Because Canterbury has such an international reputation and because it's such a popular place to visit and to study, these transport pressures are more acute than they would be in another city of the same size. If you speak to people across the world, they've probably heard of Canterbury and yet, in actual fact, it's not a very big town, with a population of just less than 50,000, and yet it supports a shopping and tourism industry which you might find in a town four times the size, and because of the universities, again, the services that are delivered here are probably more akin to a town [that's] much bigger. So because it's such a popular destination or a popular place to be, it has to cope with all these pressures and I think that's probably at the heart [of the problem] we face. And you see some similar situations in other popular and vibrant places like Winchester or other cathedral cities, so it's not unique to us but because we are smaller than, say Winchester or Exeter or places like that, I suppose it's more acute and it's more pronounced here. Whereas if you look at other towns of a similar size, which don't have those attractors, they maybe manage more easily with those sorts of pressures.'

In articulating the key visitor issues in more detail this officer stated that:

> 'We get over 6 million day visits a year – but it's a good position to be in. I think if you talk to residents or visitors or people studying here, they also like the place as a place to live – it's got a very attractive environment and they wouldn't want to change this attractive environment, and for visitors it also makes it a popular destination. We obviously have to deal with the consequences of that pressure and I think these are some of the consequences that have arisen, certainly over the last 30 years or so. Also its position near France as well – because Canterbury is literally within a couple of hours of Continental Europe it's easy to do as a day trip, and it's on the route through to London as well, so we get a lot of visits here which are part of a

(*Continued*)

Case Study. Continued.

> package of visits where people will come from Europe through Canterbury and then on to London or other destinations.
>
> So it is a very popular place, but you know, equally with that comes pressure; trying to work out where you park all the coaches, where do people park if they're arriving by car. We would like to convert those day visits to people staying overnight and supporting the night-time economy – shopping and restaurants, etc. But I think one of the issues we do have is that it is in many ways seen as a day trip or a transitory trip through to somewhere else. So that's something else we have been working on to try and convert that expenditure and make it more of a benefit to the local economy.
>
> In terms of tourism trips and trying to get people to stay, we've tried to do something as obvious as trying to get more hotels built. We've tried to work with hotel developers to get them to build hotels in the city. You'd think that would be a really straightforward thing to do in a place like Canterbury but the investment model for hotels is that you need to be able to fill those bedroom spaces during the week as well as at the weekend. During the week often it's to do with 'business tourists', in other words business people coming here to carry out various transactions and meetings, whereas at the weekend it's more people coming as tourists to visit. So that's been quite tricky.'

In terms of different partners who might bring about these changes in the development of the accommodation sector the officer stated that various transport organizations might be able to bring about changes, stating that:

> 'Well the bus operator, Stagecoach is one of them, Network Rail and Southeastern in terms of rail transport. Kent County Council is the Highway Authority so they're responsible for the investment in the roads. Some of the other things are more directly in our gift; the Planning Strategy is down to us – the Local Plan and the policies on things like trying to encourage hotel operators. But at the end of the day it requires someone else, i.e. a private investor or a developer or someone with land, to make that happen. We can only take it so far and have policies and encouragement in place, but it requires someone else's business decision to actually do it. Part of the issue with the planning process in this country is that it's typical of a mixed economy: it's neither completely private sector laissez-faire, nor is it completely a state-directed, centralist model, it's a little bit of both and at times it falls down because you can't make people do things and invest.'

It is clear that one of the key planning challenges is to plan effectively for tourist and pilgrim visitation to Canterbury's World Heritage Site.

World Heritage Site

In a contemporary sense, Canterbury is still one of the prominent locations for pilgrimage with the cathedral designated as a World Heritage Site. The United Nations Educational, Scientific and Cultural Organization (UNESCO) recognizes nearly a 1000 World Heritage Sites which are special sites of global significance that require protection for the benefit of current and future generations. 'UNESCO designated the Canterbury World Heritage Site on 9 December 1988 ...

(Continued)

Case Study. Continued.

and the site includes Canterbury Christchurch Cathedral and Precinct, St Augustine's Abbey and St Martin's Church' (Canterbury City Council, 2014b). The inclusion was partly based on the significance of the origins of Christianity in the UK and because of the status of Saxon and Roman architecture. UNESCO has issued a Statement of Outstanding Universal Value (SOUV) to reiterate the importance and criteria of designating the site and inviting public comment from residents of Canterbury about the site. Canterbury City Council has also made provision with policies within the Local Plan (supporting the World Heritage Site Management Plan) which protect the site through a development 'buffer zone' to preserve the integrity of the site and also to ensure that there is appropriate management, promotion and interpretation of the site (Canterbury City Council, 2014b).

CONCLUSIONS

Planning as an institution in the UK has a long heritage and has had to respond to different challenges and this chapter has provided a timely reminder of the vital role of planning and planners in society, offering the reader the contemporary voice of the planner and highlighting a complex series of considerations that the planner has to address in 21st century planning systems. This was highlighted by the interviewed planner who stated that:

> 'It's a very mixed bag of different people getting involved and we sometimes sit at the heart of a lot of different conversations. So we're having to talk with voluntary groups about very local, specific issues and then try and talk with the County or Government about much more strategic things. That's sometimes quite a tricky one to square up as well. In some ways we have to act as a translator between those different groups and try and act on behalf of the community from Government and also try and act on behalf of the wider community when we have local interest groups saying they want certain things to happen in certain areas; we sometimes have to try and take a wider view than that as well. It's quite challenging.'

This challenge is further made complex where the planner works:

> 'in a world in which the boundaries have become a bit blurred between physical regeneration and trying to promote new employment opportunities and also provide new skills for people working/living in areas of economic decline, so planners have actually moved into different areas other than just physical regeneration and development, and often you will find planners working in economic development or in tourism or in the management of natural areas. So I think that broad range of roles has evolved over those 30 years as well.'

QUESTIONS

1. Within other cultural contexts who were the most influential architects and planners in urban design within a named nation state?
2. How does the role of the planner help to negotiate the different interests that emerge within local governance?
3. Identify in a named destination how the role of the planner and the planning system can help to plan tourism in line with the needs of local residents.

BIBLIOGRAPHY

Boddy, M. and Parkinson, M. (eds) (2004) *City Matters: Competitiveness, Cohesion and Urban Governance*. The University of Chicago Press, Chicago, Illinois.

Brindley, T., Rydin, Y. and Stoker, G. (1996) *Remaking Planning: The Politics of Urban Change*. Routledge, London.

Campbell, S. and Fainstein, S. (eds) (1996) *Readings in Planning Theory*. Blackwell, Cambridge, Massachusetts.

Canterbury City Council (2014a) Introduction to Planning Policy. Available at: http://canterbury.gov.uk/planning-policy-introduction (accessed 25 July 2014).

Canterbury City Council (2014b) The Canterbury World Heritage Sites. Available at: http://canterbury.gov.uk/planning/conservationworldheritagesites (accessed 25 July 2014).

McCabe, A. (2010) Below the radar in a Big Society? Reflections on community engagement, empowerment and social action in a changing policy context. *Third Sector Research Centre Working Paper* 51. Consulting InPlace, Birmingham, UK.

Osborn, F.J. (1946) *Garden Cities of To-morrow*. Faber and Faber, London.

Royal Town Planning Institute (RTPI) (2014) Available at: www.rtpi.org.uk (accessed 6 July 2014).

Rydin, Y. (1993) *The British Planning System: An Introduction*. Macmillan Press, Basingstoke, UK.

chapter 5

Planning for Tourists and Residents in an Historic Destination – York UK

Ian Tempest*

City of York Council, York, UK

LEARNING OBJECTIVES

- To understand the development and governance of tourism and comprehend that town planning is as much about influencing behavioural change as it is about technical solutions.
- To understand that the fundamental problems of planning for tourism (in a congested historic environment) requires a variety of solutions from different sectors.
- To recognize how public sector investment in traffic management and public realm improvements can generate a direct benefit in private sector investment.

INTRODUCTION

This chapter describes how town planning and tourism initiatives in the historic city of York have developed over the past 50 years, to make the city one of the UK's leading destinations for some 7 million domestic and overseas visitors. It shows that tourism has evolved from the 1960s, as a modest contributor to the

*ian.tempest1@btinternet.com

York economy, to where it is today as one of the most important industries. These changes were facilitated through land-use planning interventions by local (municipal) authorities, and innovations in the organization and promotion of tourism and destination management by the public and private sectors. The different challenges that have emerged today will be outlined, not least that its very popularity as a visitor destination brings its own pressures in terms of the balance of interests between residents and visitors, continuing traffic and public-realm management issues and tourism's relative and absolute growth within the York economy.

A case study highlights the most complex destination management issue in an historic city, that of transport planning in an environment where traffic and pedestrian flows are constrained within an enduring medieval street pattern. In 2015, traffic management remains one of York's most difficult and controversial challenges.

HISTORIC BACKGROUND

York is a near 2000-year-old city, at the heart of Yorkshire, 200 miles north of London, at the confluence of the rivers Ouse and Foss. It has a significant archaeological and architectural heritage, and ecclesiastical, museum and archive collections showcasing many of the key periods in English and European history. It is in the top ten of the most popular and famous cities in the UK, for British and overseas staying visitors. The city centre follows a substantially medieval layout and the city wall is nearly 5 km in circumference. With a population close to 200,000, York consistently attracts some 7 million visitors into a compact 85 ha city centre with heritage a principal visitor driver.

Known as Eboracum, York was founded in ad 71 by the Romans and today is evidence of all of the key periods of European and English history, including Saxon, the Vikings and Normans through to the social and cultural influence of the Georgian period in the 18th century. By the 19th century and for the next two centuries, chocolate making and the railways were major employers. Today, there are thriving universities, financial and public services, including health services and education, a growing base of science-based companies, construction firms and substantial employment in retail and tourism. It has been less affected by large-scale redevelopment than most English cities with a rich heritage and extensive city walls, containing 1200 listed buildings of architectural and historical interest in three conservation areas and an Area of Archaeological Significance with 2000 years of history (Addyman, 1973).

HISTORY AND DEVELOPMENT OF TOURISM IN YORK

This historic environment has long been a fascination to visitors travelling from a distance, including pilgrims (with its ecclesiastical pre-eminence), authors, painters and others, with mass tourism a relatively new phenomenon. There is certainly an argument that leisure tourism (people coming for a visit rather than to work or worship) dates at least from the 18th century. The Theatre Royal, racing and Georgian buildings were used as venues for amusement and use of the Assembly Rooms for dancing started to emerge at this time. Stagecoach services from York to London were in operation in 1706, a 4-day journey that with the investment in the railways, in 1840, was reduced to 11 h 15 min, and direct services ran between the cities in 1850.

By the mid-20th century, although York Minster, the Castle Museum (founded in 1937), long-established annual race meetings and the medieval-origin York Mystery Plays (revived in 1951) were important attractions to visitors, tourist numbers only gradually increased to around a million a year by the end of the 1960s but with limited hotel accommodation, few visitor attractions and no annual calendar of events or festivals. Heritage as a tourism asset, with the narrow streets and ancient buildings, was viewed more as an encumbrance to the free flow of traffic.

THE 1960S – PLANNING CONSIDERATIONS

What was the context of tourism in the 1960s, and how was this economic sector affected by planning considerations at the time? The 1947 Town and Country Planning Act required local planning authorities (LPAs) to carry out a survey of their area together with a plan 'indicating the manner in which they propose that land in the area should be used'. York Corporation (the local authority's name at the time), the LPA for York, published its York Development Plan in 1951, based on assumptions of population increase, and identified land for residential and industrial land uses (Lichfield and Proudlove, 1976).

It acknowledged the need for 'the preservation of buildings of special architectural and historic interest', through transport plans focused on diverting through and cross-town traffic from the Central Area, but LPAs had very limited statutory requirements, certainly in the area of heritage and conservation before the 1960s (Willmot *et al.*, 1959; Lichfield and Proudlove, 1976). York had long had specialist, philanthropic bodies interested in conservation and heritage, including the Yorkshire Philosophical Society (founded in 1822 and responsible for establishing the Yorkshire Museum and Gardens) and others through to the York Civic Trust (founded in 1946) (York Civic Trust, 2014). York Corporation had been suspicious of some of this interest, fearing that conservation would inhibit growth and prosperity. In 1966, Richard Crossman, the then Minister of Housing and Local Government, announced the idea of pilot projects of four historic towns to consider how to reconcile preservation and progress. This was based on his concern at rampant commercial development, which failed to reflect the design, materials and scale of surroundings and was destroying the beauty and character of many historic towns. The York Civic Trust welcomed this initiative and invited Crossman to choose York for one of these projects. In May 1966, it was announced that Lord Esher, the former Lionel Brett, who had been an architect and town planner, would undertake a 'Study in Conservation' of York. The York Civic Trust offered to share its proportion of the fee, a gesture that resulted in an initially reluctant Corporation to support the project. Esher's report highlighted York's built environment as 'a great cultural asset and, with the growth of tourism, it is increasingly an economic asset as well' and its principal objectives were that:

- The commercial heart of York should remain alive and able to compete on level terms with its neighbour cities – old or new.
- The environment should be so improved by the elimination of decay, congestion and noise that the centre will become highly attractive as a place to live in for families, for students and single persons, and for the retired.
- Land uses which conflict with these purposes should be progressively moved from the walled city.
- The historic character should be so enhanced and the best of its buildings of all ages so secured that they become economically self-conserving.
- Within the walled city the erection of new buildings of anything but the highest architectural standard should cease.

Published in 1968, it created a vision which influenced many future policies which were adopted and included a vision for conservation and the relocation of industrial land outside the city centre, replaced by largely residential development. Not all of Esher's proposals were accepted – he recommended huge multi-storey car parks, which did not happen – but he advocated traffic restrictions within the city centre (Lord Esher, 1968). Detail of traffic management will be covered in the case study section later in this chapter.

In 1966, the Corporation produced a Guide for Developers and inaugurated a Town Scheme, for making grants to listed buildings and in 1968 a Conservation Area, aimed 'to protect the central historic core of the city', along with 34 other smaller conservation areas with special architectural or historical interest protected by extra controls (Alan Baxter and Associates, 2011; City of York Council, 2011b).

TOURISM IN YORK BY THE END OF THE 1960S

By the end of the 1960s, accommodation stock was predominantly in bed-and-breakfast and guest-house properties (with 540 hotel bedrooms; now there are 7606 bedrooms and over 17,000 bedspaces). Employment in York hotels was limited and in terms of retail, many shops were closed on Mondays and on Wednesday afternoons – a complaint of visitor surveys, along with traffic, litter and lack of signposting (both entering the city and to specific attractions in the city) and restaurant and tourist information facilities. Over half the visitors were from Yorkshire, 85% came in the 6 months of the summer and only 2.5% came from abroad (the percentage now is 19%). The city was dominated by school visits – trains and coachloads of school children – and York is still popular for school visits today (Lord Esher, 1968; Visit York, 2012). York's main attractions in the 1960s were York Minster (880,000 visitors in 1967, though facing substantial renovation by the end of the decade) and the Castle Museum (657,000 visitors in 1966). Of other attractions, the Yorkshire Museum had suffered under-investment for years (York City Council, 1971). By 1970 it was estimated that York was attracting just over a million visitors a year with an income to the city (excluding accommodation spend) of £2.85 million (Feinstein, 1981; York City Council, 1989).

PROGRESS IN THE 1970S AND 1980S

Along with linkages made between environmental improvement and tourist visitation, Esher recommended the appointment of a

Director of Tourism, encouraging and providing for overnight stays and he saw the potential for conference tourism, floodlighting, river spectacles, late opening of shops in summer and better tourist information. Developments started to happen. York Corporation set up the Department of Tourism in 1969 and established a Council Tourism Committee, by which time York Corporation had become York City Council operating a Tourist Information Office and the publication of annual tourism statistics (York City Council, 1989). By 1980, visitor numbers had improved to 2.8 million people and £10 million spend (York City Council, 1989), due to a combination of improved marketing opportunity through York's first dedicated Council Tourism Department and investment in new attractions and hotels – the National Railway Museum (1975), Viking Hotel (1969) and the Holiday Inn (1971) were the most significant, but the completion in the early 1970s of long-running works inside York Minster, and the creation of an archaeological dig at Coppergate (enjoyed by over half a million people from 1976 onwards) were also important (Addyman, 1973; Addyman and Gaynor, 1984).

By the 1980s, responsibility for tourism promotion and the Tourist Information Centres lay with the Marketing and Communications part of the Council, but the industry was not seen as one of the city's economic drivers. A new Economic Development Unit (EDU) was established in 1984 and joint research with the University of York concluded that some 4000 full-time equivalent jobs were associated with the tourism and hospitality industries (Barnett, 1984). Its key brief was to look at other aspects of the city's economy, in particular how to diversify from the traditional railway and chocolate industries through a combination of inward investment, small business grants and services, and encouraging developments such as, by the 1990s, a science park at the university.

THE CREATION OF YORK VISITOR AND CONFERENCE BUREAU

The Council's enthusiasm for promoting tourism was limited: while running two Tourist Information Centres it contrasted with active tourism businesses in York, who had a long-established membership organization, the York and District Hotels and Restaurants Association. An example of typical key proactive personalities, who make things work, was Donald Shepherd, Chairman of Portakabin, who in 1985 became Chairman of the York Chamber of Trade. He created a steering group of like-minded business leaders to focus on tourism and, by 1987, formed the York Visitor and Conference Bureau, 'with the objective of marketing the city as a tourist destination to provide direct benefits for the York business community'. Without City Council financial support, funding came from membership fees, accommodation, conference and group commission, merchandise sales and other sponsorship-raising activities. Business membership grew from 200 in 1990 to 700 in recent years (York Visitor and Conference Bureau, 1990). By the end of the 1980s the growing value of tourism was estimated as being worth £55 million to the York economy (York City Council, 1989). This was spurred by the creation of the Jorvik Viking Centre, attracting nearly a million in its first year of operation (1984–1985).

THE DEVELOPMENT OF YORK'S TOURISM OFFER – ACCOMMODATION, ATTRACTIONS AND EVENTS

Through private-sector investment of £200 million in hotels over the past 40 years, hotel accommodation bedspaces rose to nearly 6000 by 1988, and by the 1990s new budget hotels opened on the outskirts, and half a dozen within the city walls from 2000 onwards, with the first five-star hotel (the Cedar Court Grand) opening in 2010. Accommodation stock by the end of 2013 was running at over 17,000 bedspaces (Visit York, 2013b).

However, the biggest progress was with attractions and with events, with the creation of the National Railway Museum on a huge new site in 1975, and the Jorvik Viking Centre on the site of the Coppergate Dig attracting over 900,000 visitors in its opening year in 1984, and creating a paradigm shift in how York marketed itself. Other attractions began to open, such as York Dungeon (1986), and other historic properties and further facilities and attraction investment in recent years has culminated in the opening of the York Chocolate Story (2013). In addition, there have been multimillion-pound improvements to York racecourse, worth £58 million to the York economy in 2011 through its racing, conference and banqueting services (Centre for Economic and Social Research, 2011).

In the 1960s, the main annual events were horse racing and some minor traditional civic ceremonies and every few years the York Mystery Plays took place as a distinctive retelling of Bible stories. It has only been more recently that a calendar of events has started to take shape, with developments in the public realm, crucial for creating traffic-free spaces for events during the heart of the day. The economic impact of events and festivals is substantial: over the 9 years of Illuminating York (a lighting festival every October half-term) the amount generated has been estimated to be over £19 million, while the annual contribution of St Nicholas Fayre and pre-Christmas events and promotions is in the tens of millions (Visit York, 2012). In bringing together, as one audience, the residents and visitors, this is one of the best ways of ensuring community value from tourism.

THE ORGANIZATION OF TOURISM IN YORK – CHANGE IN THE 1990S

While the York Visitor and Conference Bureau was successfully promoting York tourism, the Council was still responsible for the Tourist Information Centre through its Marketing and Communications department. The EDU started to develop closer links with the industry in 1990, with the establishment of a tourism 'Employer of Distinction Award' designed to promote good employment practice in the tourism industry. From the Council's point of view the award, promoting good rates of pay and working conditions in the tourism industry, was addressing concerns about the quality, image and reputation of the industry as an employer, while for the Bureau, it was a good way to link tourism interests with Council priorities (York City Council, 1990).

Of more long-term significance was a York Tourism Strategy report, commissioned by the Council in 1994, from Touche Ross, which sought to identify how to create a deeper and more wide-ranging working relationship between the Council and the tourist industry. It provided the most in-depth statistical research into real economic value of York tourism so far. This revealed job numbers (8397), the economic value (£205 million) and the total number of visitors (nearly 4 million). Visitor spend was broken down into individual components – accommodation, food and drink, shopping, visitor attractions, etc. – highlighting the value of staying-visitor spend per head. This evidence was timely for tourism, coming at a period when the railway and confectionery manufacturing industries were going through further retrenchment (terminal in the case of carriage building). The need for tourism to contribute as strongly as possible to the city economy was clear – but it was also important to the Council, given the city's income from car-parking and coach-parking charges and income to the city's own Castle Museum, City Art Gallery and (following local government reorganization in 1996) the Yorkshire Museum (Touche Ross, 1994).

This report led the Council and various bodies to form a partnership arrangement – First Stop York. It was designed to make the most of the strengths of the associated organizations – not just the Council's EDU and York Visitor and Conference Bureau (which was soon to become York Tourism Bureau), but also the regional (Yorkshire) tourist board, the Chamber of Commerce, smaller membership organizations and the government body the North Yorkshire Training and Enterprise Council (TEC).

The Council, the Yorkshire Tourist Board and the North Yorkshire TEC funded the partnership and it enabled funds to be accessed from other sources, such as £270,000 of funds through the European Commission for areas affected by defence cutbacks, the Learning and Skills Council (£50,000 towards a training brokerage project) and, later, £3.6 million from Yorkshire Forward (the regional development agency). In year one some £700,000 of activity was coordinated across a range of projects including marketing and promotion, research and the promotion of training. During the 13 years of the partnership (1995–2008) direct funds and those levered in were used to develop further marketing activity, including: (i) European and other international campaigns, conference marketing, rail-based promotions, shopping promotions and trade shows; (ii) York's first tourism website (launched in 1998); (iii) public relations activity for the first time to augment paid-for advertising; and (iv) initiating ongoing research including an annual visitor survey and economic impact model to enable regular monitoring of performance as well as an annual report and newsletter to the industry at large on behalf of the partners. Funds therefore went on specific actions and to cover some of the additional staff costs to implement them.

Funding was also available for tourism partners for the first time, to carry out some direct destination management activity (which had formerly been carried out by the Council and the private sector alone). This work included:

- support for events, including designing new events such as the Residents' Festival (1996), a full Christmas marketing campaign (1997 onwards), the York Festival of Food and Drink (1997) and others to form a genuine year-round calendar of events;
- physical signage (brown signs on trunk roads) and interpretive signage on the city walls;

- trails publications (e.g. themed walking routes around the city) and dedicated print for coach operators;
- training courses for front- and back-of-house staff (funded by the North Yorkshire TEC and, later, by the Learning and Skills Council);
- a major lighting project with funds from Yorkshire Forward's Tourism Investment Fund, including architectural lighting schemes and an annual lighting festival (Illuminating York); and
- one-off projects including support for the Royal Ascot in York (2005), promotion of York's award for European Cities Marketing (2007) and the relocation of the Visitor Information Centre (2010).

By 2007, pressure was growing from stakeholders for changes to management arrangements for tourism, from the more informal First Stop York partnership approach, to the creation of a company limited by guarantee which would incorporate the Bureau and seconded staff from the Council's tourism team in the EDU. The new company, Visit York, would continue to be membership led, would be answerable to a Board with a private sector Chair and three Councillor-Directors as well as others from the private sector. It was part of a 'hub-and-spoke' arrangement for the Yorkshire Region, at the centre of which would be a revamped Yorkshire Tourist Board (Welcome to Yorkshire – established in 2009) alongside six sub-regional bodies.

While other sub-regional partnerships and structures fell by the wayside, and the key supporter Yorkshire Forward was abolished along with England's other Regional Development Agencies, after the 2010 General Election, Visit York has continued, and Welcome to Yorkshire has evolved into a strong regional tourism organization, leading on high-profile Yorkshire-wide activity, including the Tour de France 'Grand Départ' in July 2014.

VISIT YORK'S ACHIEVEMENTS SINCE 2008

Much has been achieved by the establishment of Visit York, continuing the deepening relationships with the Council, the enduring strength of membership and the capacity to secure funds from other sources. Achievements include:

- Access to Regional Growth Fund moneys with VisitEngland has ramped up promotion activity, including 'York's Wrapped Up', an off-peak short-break promotion with East Coast (rail company).
- A transformed website that is accessible in more than 20 languages, with foreign language translations and a chance to create tailored itineraries. Now there are over 2 million visitors to visityork.org annually (up 30% a year).
- Development of new social media network – Facebook, Twitter, etc. – complementing print, online and direct marketing.
- Ninety press visits in 2012–2013, including international publications.
- City-centre retail – a dedicated market research and marketing campaign.
- Active contribution to Welcome to Yorkshire and regional marketing activity.
- The Visitor Information Centre has doubled its footfall (to 450,000–500,000 visitors a year) and doubled merchandise sales since relocating in 2010.

- There are now a record total of 700 business members and there has been a growth in membership and member services including a large increase in member events and record entries in York Tourism Awards.
- Support for and promotion of events which result in large visitor numbers and have a major economic impact.

The improved promotion of York has created the climate for some £100 million of investment since 2010, including a five-star standard hotel, investment in the Yorkshire Museum, Jorvik and York Minster, and new city-centre retail and (especially) restaurant facilities, etc. The ultimate result is a 37% increase in visitor spend between 2008 and 2012 and a growth in the proportion of overseas visitors in that time (from 15% to 19%) (Visit York, 2013a, b).

CURRENT POSITION – NEW TOURISM STRATEGY (2013 ONWARDS)

Since the 1980s there have been a number of tourism strategies designed by the Council alone, by outside consultants (such as Touche Ross) and by tourism partners. The latest strategy for York has been produced by the City of York Council and Visit York (2013), with aspirations to double the value of tourism in 10 years (to create a £1 bn tourism industry in York) and to create additional high quality jobs by addressing all aspects of tourism, including:

- marketing via all platforms;
- product development/investment of all kinds – existing and new;
- business engagement;
- public realm investment to improve the appearance of the city centre and the quality of signage;
- skills and training development;
- events and festivals development; and
- technological innovation.

The areas to be addressed to deliver the growth targets – with the fine detail still being worked on – will include:

- tackling seasonality – strengthening promotion of York, and the visitor offer, in the winter period;
- encouraging more business visitors for conferences and meetings;
- extending the length of stay for both business and leisure customers;
- increasing the spend of domestic day and staying visitors; and
- increasing the numbers of overseas leisure and business visitors.

Discussions are currently taking place between the Council and Visit York as to how to deliver and implement this strategy, and to strengthen links with arts and culture – now such a strong part of York's visitor offer, including:

- nearly 50 attractions – and some £200 million investment since the 1980s;
- an extensive calendar of events and festivals – York as City of Festivals – enjoyed by hundreds of thousands of residents and visitors each year;
- Reinvigorate York – a new strategic partnership between the business community and the Council that has made a £3.3 million investment in the public realm to reimagine the city's streets and open

spaces and enabling art and performance in public space;

- a unique heritage with a modern outlook (i.e. showcasing of innovation, events and festivals against the backdrop of historic buildings); and
- creative industries as the fastest growing sector of the economy – 3000 people working in over 250 creative media arts companies, with substantial investment by the universities (York is now a UNESCO City of Media Arts).

The new way of working is likely to extend the marketing of York by facilitating closer ties between businesses, city-centre management and cultural activities (City of York Council and Visit York, 2013).

NEW THINKING ON PUBLIC REALM AND STRATEGIC PLANNING IN YORK

Pedestrianization schemes in the 1980s were a successful transport measure, and incorporated other interventions in the public realm, for example new signage, bollards, tree planting and repaving former tarmac streets. Further, more piecemeal improvements to the city centre were implemented in the 1990s, including: (i) giving permission for pavement cafés (only possible for the most part because of pedestrianization, but very much a 'happy accident' rather than planned); (ii) installing dropped kerbs; and (iii) projects like architectural lighting schemes (via the Tourism Investment Fund) and the major paving scheme known as the Minster Piazza outside the Minster which had funding through the Heritage Lottery Fund and the Council.

Current ideas for a more systematic, deliberate approach to public realm improvements started with a report by Alan Simpson in 2010, *York – New City Beautiful: Towards an Economic Vision.* This report, as a recent Cabinet paper suggested:

> strongly advised that quality of place and a rich diversity of activity in reinvigorated streets and spaces do affect personal and business decisions. Improving the physical appearance of the city, improving retail and commercial opportunity and activity, ensuring better accessibility and improving image and perception are all important.
>
> (City of York Council, 2011d)

A non-statutory body advising the Council's Cabinet was a new strategic partnership – Reinvigorate York – which brought together representatives from the business community, including from retail and the visitor economy, with Council officers concerned with improvements to the public realm, redolent of pedestrianization schemes at the end of the 1980s. Its three main objectives included:

- to improve the circulation in the city centre (of pedestrians, cars, cyclists and public transport);
- to restore and enhance the city centre's public spaces and streets (street furniture, signage, paving, clutter, etc.); and
- to achieve other improvements that will help enhance the tourist, retail and cultural experience of visitors, residents and people working in the city centre (City of York Council, 2012).

Crucially, the economic and tourism case for investment in the city centre, improving movement and the quality of public spaces, is now fully recognized. The York Visitor Survey 2011–2012 found that the overwhelming top activity of York's 7 million visitors is 'to stroll around and enjoy the ambience of York'. The public realm is very much part of the visitor experience. It is also the canvas on which a lot of activity takes place, for example specialist markets (Parliament Street) and events and festivals. Issues include the navigability of the city centre (e.g. signage) and the mix between a number of smaller projects and the opportunity for major schemes, including the recent trial to close Lendal Bridge to private vehicles during 'pedestrian hours' and city-centre lighting improvements, which can have direct benefits on the evening economy.

These detailed considerations are features highlighted in the emerging Local Plan for York – a development plan for York covering the 15 years from 2015 to 2030. The Local Plan 'provides the spatial or planning expression of York's strategic plan' as highlighted in the Strategy for York vision, 'York: A City Making History' and includes the following aims:

- building confident, healthy and inclusive communities;
- being a leading environmentally friendly city;
- being at the forefront of innovation with a diverse and thriving economy;
- being a world-class centre for culture, education and learning for all; and
- celebrating our historic past and creating a successful and ambitious future.

(City of York Council, 2013b)

The Local Plan also needs to consider issues in the shorter term. The 'City Action Plan – The Strategy for Growth' for 2011–2015, particularly highlights two crucial issues relevant to tourism (further progress on the local plan will emerge in 2015):

- to tackle transport congestion and address travel issues in, around and across York; and
- to enhance York's heritage and creative and cultural sectors and use them as a driver for economic growth (City of York Council, 2011c, 2013b).

Strategic planning policies and practical planning activity are underpinned by the principles of sustainability.

PRINCIPLES OF SUSTAINABLE TOURISM IN YORK

The principles of a sustainable approach to tourism have been pursued by the Council and by Visit York and its predecessors for more than 25 years, the future needs of visitors and residents being taken into account in developing the needs of present tourism. York has focused on the following policies:

- Protection and enhancement of the historic environment – the very reason people come to York – through the creation of conservation areas and other conservation policies.
- Maintaining the built fabric of York (conservation issues – but also ensuring that empty buildings are put to new use through the granting of appropriate planning permissions). With regular pressures of cost on the

maintenance and upkeep of historic buildings and assets, tourism and the visitor economy has helped bring a number of empty buildings – disused offices, churches and other properties – back into use as assets for visitors, residents and the city at large. The Cedar Court Grand (built as railway offices – opened as a five-star hotel in 2010) is a good example of this.
- Embedding the principles of sustainability in the tourism strategy, by prioritizing actions to increase the number of higher-spend staying visitors, rather than low-spending day visitors.

The promotion of sustainable transport by Visit York includes:

- Marketing, including:
  - promotion of access to and within York by sustainable means (cycling, buses, rail, park and ride, etc.);
  - production of material in print and online (website is www.visityork.org); and
  - availability of an extensive selection of publications and other information relating to walking, cycling and public bus and train timetables in the Visitor Information Centre.
- Working with rail and coach operators to promote access to the city by non-car means – minimizing the impact of cars on the very environment that visitors enjoy.
- Promotion of events and activities by many organizations in the city that showcase alternative transport modes, including city-centre cycle races, York Cycle Rally, Way of the Roses Cycle route and the Festival of the Rivers.
- Promotion of iTravel York (the Council's transport microsite) on the tourism website.

In 2013–2014, 32% of visitors get to York by train, 12% by bus or coach and the remainder by car (Visit York, 2014). Less than 1% of visitors come to York by bicycle but more hire bikes when they get to the city, while the compact city centre coupled with the largest pedestrian scheme in the UK means that walking in the city centre is popular.

RESIDENTS' OPINIONS OF TOURISM

A crucial aspect of sustainable tourism is seeking to ensure that the interests of residents are fully taken into account, so that as far as possible they can benefit from tourism investment. There is little published research into the attitudes of residents of York to tourism before the 1980s. Apocryphal evidence suggested that many people wore T-shirts and badges in the 1970s with the legend 'I'm not a tourist I live here', but it is likely that tourism did not impinge much in the daily life and consideration of residents, just as it did not for the Council. In the 1960s and 1970s the great industries of York were still the confectionery and railway industries: the 1971 Census of Employment shows some 25,000 people worked in manufacturing, transport and communication – 9500 in cocoa and confectionery and 7500 in railway offices and the carriage works. In contrast 2300 people worked in hotels, inns, etc. (Feinstein, 1981).

Back in the 1980s the creation of foot streets and the implementation of park-and-ride schemes were specifically promoted by the Council at the time as a boon for residents. The potential tourism benefits were not highlighted in the coverage at the time. Work was carried out in the face of fierce opposition from city-centre traders, many of whom thought they would lose business. There has long been a perception of difficulties between residents and tourism, largely based on isolated comments in the local media rather than a more widespread problem.

In developing the new tourism strategy for York, in January 2013 more than 200 residents were interviewed or responded to questionnaire surveys on their attitudes to tourism, including both positive responses and criticisms. The summary of responses revealed that residents evidently have a very good understanding of tourism and the city's qualities in attracting visitors:

- Tourism ensures history/old buildings/small shops are preserved.
- Direct benefits to residents are understood – restaurants, pubs, theatres, attractions, cultural life, events, craft fairs, retailers, shops.
- Financial benefits include: jobs, economy, and money for locals.
- City is vibrant, exciting, buzzing, busy, better vibe, more fun, alive.
- A number especially said they like/welcome visitors, enjoy showing them round, love to see lots of people, it gives residents pride.
- A number said it provides their livelihood (i.e. they work in tourism or in businesses affected by tourism).

Residents' support is not unconditional as they have important concerns including:

- Crowds, overwhelming numbers (especially at weekends), traffic/parking, congestion affecting cycle access, race days.
- 'Hen and stag' parties – general drunkenness and antisocial behaviour.
- Some feel York is more expensive than other places – and there are too many tourist shops (so residents' choice is restricted).
- Poor information to residents (e.g. on events).
- Residents, like visitors, want to see public realm improvements, getting the evening economy going, etc. (City of York Council and Visit York, 2013).

The tourism partners have long seen the benefits in ensuring residents' support of tourism, hence one of the first projects the original tourism partnership, First Stop York, worked on was the Residents' Festival, which encouraged York's attractions to open their doors free of charge over the last weekend in January. Up to 30,000 visits are made each year to dozens of attractions which will be run for the twentieth time in 2015.

The tourism strategy, by highlighting the value of tourism rather than just aiming for more and more visitors, is a direct response to the 'too many crowds' issue, as is the aim to get more and longer stay visitors in the off-peak months (City of York Council and Visit York, 2013). Separately the Council is working with police and the licensing trade on such issues as tackling antisocial behaviour, and with arts and culture groups on stimulating the evening economy. The traffic implications of the growth of tourism remains a challenging issue for the city.

Case Study: Transport Issues in Planning for Tourism

Tourism requires, fundamentally, the movement of people to and within a destination. For York the issue is compounded by a largely extant medieval street pattern which, combined with the historic buildings and medieval city wall, means that transport and access issues are acute and intrude on people's living, working, shopping, leisure and visitor experience. The challenge of traffic management has been an issue for decades, with early work on a partially completed ring road even in the 1930s. From 1948 onwards the first concrete plans were drawn up and these were modified from 1967 to 1971. Proposals were put forward with the support of the Corporation at this time for creating an £8 million 3-mile dual carriageway inner ring road for York.

There was considerable controversy about this as the plans would have required the demolition of over 350 houses, churches and other buildings. Further, the devastating impact it would have on the city's archaeological remains and 18 archaeological sites, impacted by the proposals and concerns about other development plans, led directly to the creation of the York Archaeological Trust in 1972, which from its inception, began to organize a number of digs across the city, attracting crowds of interested enthusiasts.

At the end of 1972, there was a public inquiry and the Inspector came out in his report (in 1973) in support of the inner ring road scheme but the ultimate decision was down to the Secretary of State for the Environment (Anthony Crosland). His decision on the inquiry was not made until February 1975 (by which time an outer ring road to the south and east of the city was about to open by the end of 1975). In highlighting grave objections (the impact on houses, civic amenities and schools, hospitals and public spaces and the cost) he emphasized that there should be 'further measures of traffic management and restraint and greater use of public transport' (Lichfield and Proudlove, 1976).

It is notable in hindsight that even in the middle of this debate the Council had started to close city-centre streets to traffic during the heart of the day. However, little further action was taken to restrict city-centre traffic until the mid-1980s, when two principal measures were taken: (i) pedestrianization; and (ii) investment in park-and-ride facilities.

There was an extensive programme of pedestrianization, known locally as 'footstreets', from 1987, closing some 30 key streets to traffic during the heart of the day, and forming one of Europe's largest pedestrian schemes. This involved a combination of road closures to vehicles (including deliveries) from 11 a.m. to 4 p.m. initially, with different restrictions at weekends, coupled with some repaving and other improvements to public space (Fig. 5.1). As former City of York Council Cabinet Portfolio holder for Transport, Councillor Dave Merrett said in an interview: 'Despite strong opposition to pedestrianisation from traders worried about servicing problems and customers being unable to pick up heavier items' (Laycock, 2012) work started in 1987 to replace kerbs and pavements with

(Continued)

Case Study. Continued.

Fig. 5.1. City centre of York prior to pedestrianization (a) and post-pedestrianization (b).

(Continued)

Case Study. Continued.

cream-coloured concrete blocks, with the intention of creating 'a safe, pleasant and long-lasting walking environment'. Deangate (adjacent to the Minster) was closed to traffic in 1989, removing 10,000 vehicles a day. It succeeded in reinvigorating and improving the ambience of the city, including pavement cafés and especially in Parliament Street (fully repaved in the 1990s) created a significant event space to be used by events and festival organizers, such as St Nicholas Fayre, the York Festival of Food and Drink and the Residents' Festival.

Concomitant with the creation of 'footstreets' was a programme of park-and-ride investments on the outskirts of the city, to ensure people could get into the city while restricting vehicle access. This was complemented by matching changes to city-centre car-park designation and pricing and was a key element in restraining traffic growth, which has been adopted by other historic towns and cities experiencing similar types of problems.

While removing the commuter and long-stay visitor vehicles from the city centre, it is still attracting business and short-stay shoppers. The provision of park-and-ride sites on the periphery of York with high quality, high frequency services into town (*the carrot*) has been accompanied by a decrease in the number of Council-run long-stay parking spaces in the city centre, steep increases in the price of those that remain and the widespread introduction of on-street parking regulation (*the sticks*). Its success is based on providing the quality, competitively priced service, backed up by bus priorities to make it more attractive than using the car (Titley, 2001). Following the initial park-and-ride scheme at Askham Bar (1991), four others were built (Grimston Bar 1994, Rawcliffe Bar 1996, Designer Outlet 2001 and Monks Cross 2004) and further facilities were opened at Askham Bar and Poppleton in 2014.

This programme provided clear advantages for the visitor, for example safe, secure and free parking in locations close to the main trunk roads approaching York and easy and longer access to the city (City of York Council, 2011a). Above all, in removing traffic from the city centre it substantially enhanced the ambience and environment of the central area, complementing the well-established 'footstreets' initiative and making a superb environment for people to enjoy. Over 4 million passengers now use park and ride and York now has one of the largest pedestrian areas in Europe (City of York Council, 2013a).

Transport issues remain controversial with a recent attempt to restrict traffic over Lendal Bridge, on the west side of the city centre, with 50,000 penalties levied. The Leader of the Council, James Alexander, has admitted that while city-centre footfall seems to have held up: 'It is clear, however, that the trial has been polarising and we need greater consensus amongst residents and businesses over measures to tackle congestion'. The trial closure, which started in August 2013, ended in April 2014. An independently-chaired cross-party commission has been proposed to come up with more 'consensual suggestions' for tackling congestion (City of York Council, 2014).

CONCLUSIONS

The tourism landscape in York has been transformed over the past 50 years, with improvements to the city (footstreets, park and ride, pavement cafés) and transformations to the city's events programme and the Visitor Information Centre service. This progress has been linked to a much more professional and integrated approach to tourism marketing and promotion – York Tourism Bureau and Visit York have been at the heart of promoting and encouraging this progress working with partners such as the Council, Yorkshire Tourist Board (now Welcome to Yorkshire) and hundreds of businesses. As a result, there is a greater understanding by the Council and by the city at large of the importance of tourism to the city's economy.

Attitudes on land use planning have also moved on, from the York Development Plan in the 1950s with no reference to tourism to the emerging Local Plan for 2015–2030 which needs to incorporate ambitions to 'celebrate our historic past' and 'enhance York's heritage and cultural and creative sectors' (Willmot *et al.*, 1959; City of York Council, 2013b). These changes in approach have created an environment of investment in new and existing attractions, new hotels and accommodation, and retail and restaurant investment – both in the city centre and out of town. These have in turn produced more jobs in tourism (20,200 in the latest figures) and progress on skills and careers in tourism, increased visitor numbers and a changing visitor profile and greater but not unconditional residents' support for and appreciation of tourism.

Tourism has come a long way, but progress isn't finished yet by any means. Some of the themes in Esher's report (1968) – about developing the evening economy and improvements to street lighting – have still to be fully implemented, although plans are afoot in both cases. Traffic management remains a serious and contentious subject as in the recent debate about a trial closure of Lendal Bridge to private cars and other vehicles during the day.

Are there any lessons? One has to be careful to acknowledge the context of the past. To modern eyes, so used to footstreets and traffic restrictions, the idea that the city could contemplate a major inner ring road in the 1960s and 1970s seems unthinkable – but back in the day, the car was the symbol of personal freedom and road building was integral to the modernization of Britain. Many other places had ring roads – why not York? Furthermore, the planners at the time believed it would ease the congestion which they recognized as damaging York's interests. Similarly, the lack of attention at Corporation level to conservation issues in the 1960s was a matter of priorities at that time, as their focus was on getting York's economy to progress and grow.

On that note, the transformation in employment prospects is striking. From some 17,000 jobs in the confectionery and rail sectors in 1971 and 2200 jobs in hotels, inns, etc. there are now 6500 jobs in confectionery and rail, and 20,200 in tourism and associated industries (1:5 of the local labour force). More widely the trend in York away from a 'producing economy' (manufacturing and construction) in favour of services (private and public sector) seems to have been accelerating.

Perhaps from former dependence on the traditional economic sectors in the 1960s, York now needs to watch against too great a focus on tourism. Certainly the growth of creative and knowledge-based industries and financial services is being encouraged by the Council and the private sector.

One final thought – whose tourism is it? Heritage tourism – based on the built environment and conservation – began to grow in the 1960s and 1970s. Tourism grounded (literally!) in archaeology created a paradigm shift for York in the 1980s. Maybe a different angle could be taken in future, of tourism based around people and stories rather than buildings?

One angle in the evolving new tourism strategy is that of 'exploiting York's unique mix of modern day and historic entrepreneurship and radical reforming traditions (e.g. the York Guilds, the Quakers, the Rowntree legacy)' (City of York Council and Visit York, 2013). Organizations like the Rowntree Society (www.rowntreesociety.org.uk) and York's Alternative History – examining the history of radicalism in York (Furness, 2013) – present ideas that offer a real chance to address York's history in a very different way, rooted in the people of York.

QUESTIONS

- How closely should residents' perception of their own area match that portrayed in the planning and marketing of that place by tourism authorities and can these views be adopted in promoting the destination?
- How far can an historic visitor destination marry its heritage outlook with a desire to portray itself as innovative, creative, cutting-edge and forward-thinking?
- How much can an historic destination expect to change incrementally, and how much through a more dramatic paradigm shift? What are the pros and cons of different speeds of change? What role does planning play in managing these changes?

BIBLIOGRAPHY

Addyman, P. (1973) Rescue Archaeology in York. The Fourth Elgee Memorial Lecture, 1 February 1973. Paper published by Teeside Museums and Art Galleries Service, Middlesborough, UK.

Addyman, P. and Gaynor, A. (1984) The Jorvik Viking Centre. An experiment in Archaeological Site Interpretation. Available as a reprint from the *International Journal of Museum Management and Curatorship* 3(1), March, 7–18.

Alan Baxter and Associates (2011) *York Central Historic Core Conservation Area Appraisal.* City of York Council, York, UK.

Barnett, R.R. (1984) *Tourism's Contribution to Employment in York.* Department of Economics and Related Studies, University of York, York, UK.

Centre for Economic and Social Research (2011) *The Economic Impact of Horseracing in Yorkshire.* Centre for Economic and Social Research, Sheffield Hallam University, Sheffield, UK.

City of York Council (2011a) Local Transport Plan 2011–31 (LTP3) Executive Summary. City of York Council, York, UK.

City of York Council (2011b) York Central Historic Core Conservation Area Appraisal: Consultation Draft Planning Committee, Paper 15, June 2011. City of York Council, York, UK.

City of York Council (2011c) *York: The City Action Plan – The Strategy for Growth 2011–2015.* City of York Council, York, UK.

City of York Council (2011d) *York New City Beautiful: Towards an Economic Vision.* City of York Council, York, UK.

City of York Council (2012) Reinvigorate York – Investing £3,300,000 in the City Centre. Cabinet Paper 4 September 2012. City of York Council, York, UK.

City of York Council (2013a) Improving Movement and Public Realm in the City Centre. Cabinet Paper 7 May 2013. City of York Council, York, UK.

City of York Council (2013b) Local Plan Preferred Options. Main Documents. City of York Council, York, UK.

City of York Council (2014) Cabinet: Commission Set to Tackle Traffic Congestion in York. Press Release 25 April 2014. City of York Council, York, UK.

City of York Council and Visit York (2013) York's Tourism Strategy. Interim Document. City of York Council and Visit York, York, UK.

Feinstein, C.H. (ed.) (1981) *York 1831–1981 150 Years of Scientific Endeavour and Social Change*. William Sessions Ltd, York, UK.

Furness, P. (2013) Not in the Guide Books – Yet. *The Guardian*, 16 January 2013.

Laycock, M. (2012) York City Centre Celebrates 25 Years of Pedestrianisation. *The Press* [York newspaper], 4 June 2012. Available at: http://www.yorkpress.co.uk/news/9742653.York_city_centre_celebrates_25_years_of_pedestrianisation/ (accessed 14 November 2014).

Lichfield, N. and Proudlove, A. (1976) *Conservation and Traffic. A Case Study of York*. William Sessions Ltd, York, UK.

Lord Esher (1968) *York: a Study in Conservation*. Ministry of Housing and Local Government. HMSO, London.

Simpson, A. (2010) *York – New City Beautiful: Towards an Economic Vision*. Report by Urban Design Skills (UDS) with Professor Alan Simpson. UDS, London.

Titley, C. (2001) We're Jamming. *York Press* [York newspaper], 5 April 2001.

Touche Ross (1994) *York City Council. York Tourism Strategy*. York City Council, York, UK.

Visit York (2013a) *Annual Report 2012–13*. Visit York, York, UK.

Visit York (2013b) Record Tourism Results for York. Press Release. Visit York, York, UK.

Visit York (2014) *York – Key Tourism Trends*. Visit York, York, UK.

Willmot, G.F., Biggins, J.M. and Tillott, P.M. (1959) *York – A Survey in 1959*. By G.F. Willmot, J.M. Biggins and P.M. Tillott as an Editorial Sub-Committee of the Local Executive Committee of the British Association. British Association, York, UK.

York City Council (1971) *York Official Guide*. York City Council, York, UK.

York City Council (1989) Statistics: Tourism in York. Paper published by the Marketing and Communications Group. York City Council, York, UK.

York City Council (1990) Employment Conditions in the Tourism Industry. Paper presented to the Policy and Resources (Employment and Economic Development) Sub-Committee, 4 October. York City Council, York, UK.

York Civic Trust (2014) The York Knowledge: Planning and Conservation. Available at: www.yorkcivictrust.co.uk (accessed 1 May 2014).

York Visitor and Conference Bureau (1990) *Annual Report for the Year Ended 30th June*. York Visitor and Conference Bureau, York, UK.

chapter 6

Destination Planning for Disabled People

Craig Grimes*

Experience Community CIC, Huddersfield, UK

LEARNING OBJECTIVES

- To highlight that historically there has been a dearth of tourism literature related to disability and particularly applied to planning processes which fully recognize the needs of disabled people.
- To provide an understanding of the concepts of disability, universal design and accessible tourism as key emerging concepts and components in the analysis of planning in destinations.
- To highlight the innovations in planning for disabled people in the destination of Barcelona and emphasize how they benefit both residents and tourists.

INTRODUCTION

This chapter is written from a number of perspectives and, consistent with Part II of this book, it provides perspectives on planning and tourism from an 'insider's view' from people working in the industry. Specifically, this chapter provides a practitioner's view of planning for disabled people. I have worked as an entrepreneur for 12 years providing bespoke holidays for disabled people, particularly wheelchair users in a number of continental European cities, including Barcelona, which forms part of the analysis of this chapter. The case study is on the evolution of a destination that has been successful in designing an urban environment which has recognized the accessibility needs of disabled people.

*cgrimes@experiencecommunity.co.uk

The corollary of this, is that Barcelona is an accessible location, first and foremost, for resident communities and therefore for visiting tourists. As a wheelchair user, the work that I have done in Barcelona and in other destinations is designed to contribute towards positive changes for disabled people so that everyday experiences are not diminished and result in lack of access to spaces and places that are routinely used and enjoyed by other members of society.

I am also an academic and work with students in the pedagogy of disability and tourism, leisure and events. The concepts and ideas shared and discussed with students on definitions of disability and the nature of accessible tourism will preface case study insights on Barcelona. This chapter will also identify how my current wider practitioner work and accumulated experience in this field makes a significant contribution to the processes of planning for tourism for disabled people. The first section initially explores the historic lack of literature related to tourism and disabled people and positions this literature within the wider context of disability studies and considerations for planning processes.

It is important to note that the title of this chapter is carefully worded in using the terminology, 'destination planning for disabled people'. It highlights that planning considerations should, in the first instance, be about planning destinations for people. This will include different types of citizens, including individuals and groups who traditionally have been marginalized and whose needs have not been fully recognized and planned for. Therefore, well-planned destinations for local people provide well-planned civic spaces, which can also be used and enjoyed by a range of visitors from different cultural backgrounds. The second chapter in Part I of this volume used the phrase, moving 'from theory to practice' and from theory to praxis was used in Chapter 3. By the end of this chapter it is hoped that the discussion of theory will be related to practice. My response to the phrase 'theory to praxis' is that the work I do makes a positive contribution to changing views, behaviour and attitudes to disabled people in society. Theory to praxis identifies that positive change is possible and theory can inform that process of change. So praxis is about creating actions which can change how society 'works'. Crucially, how disabled people should have equality in access and opportunities to the everyday aspects of society, enjoyed by other people and which are an important part of realizing the process of social equality. This is in turn a crucial part of working towards sustainability which forms the key focus of this book and all members of society should have a say in it.

In making observations related to disabled people. I would like to emphasize that I am not a spokesperson for other individuals and groups in society which might be viewed as 'excluded others' (Prince, 2008, p. 1). However, the observations and insights in this chapter in informing policy making and planning processes for disabled people might also be germane to other marginalized groups in society. Particularly, if their needs are not fully met because the existing ways of doing things follows a status quo of how spaces and places have been designed traditionally both in urban and in natural environments.

'INVISIBLE' PEOPLE – IN THE LITERATURE

In academic libraries across the globe, which contain sections on tourism studies, historically you will find that there is a paucity of references related to disabled people and tourism. When disabled people appeared in the literature, for example on the coverage of tourism demand, they might have appeared in the sub-section on suppressed demand and described as people with mental and physical handicap. To be fair, the tradition of mainstream tourism books appear to have been written from a different perspective, and might not have felt it was relevant to recognize through a social approach, what Prince identifies as 'excluded others' (Prince, 2008, p. 1). In particular, disabled people, along with people from other sections of society, were also absent from academic coverage of planning and development and technical detail of tourist destination requirements. What has been missing is that the analysis of destinations did not cover an assessment of individual needs, let alone the collective needs of disabled people and by implication other members of society, which in their 'invisibility' fitted the label of the 'excluded others' (Prince, 2008, p. 1).

Additionally, books which were keen to provide knowledge on tourism management might be concerned with management perspectives on tourism, which do not incorporate wider more socially orientated impacts of tourism, let alone social science perspectives. Tourism as an industry might be viewed in more general economic and commercial perspectives. In contrast to this approach, Schwab and Dustin (2013) provide a different critique on the utility of leisure and tourism in society and consider their role in social and environmental justice. Their edited text includes perspectives on disability in a contribution from Devine and Piatt (2013) who provide a US context to explore rural environments and outdoor recreation. Using this context they explore the concept of disability, arguing that there is a moral obligation for people with disabilities to have access to these leisure experiences (normalized leisure) as an inalienable right, if people are not to be denied basic human rights. Furthermore, they argue that this basic human right is important in establishing 'social justice based on dignity and respect and difference', in short 'individuals with disabilities have the right to participate in the same activities guaranteed to others, without their disability being in the forefront' (2013, p. 17). They argue that the status quo in society is predominantly 'ableist' (with everyday activities and people doing leisure having to conform to an able-bodied society).

In line with observations within this section about the invisibility of disabled people, Devine and Piatt note that from a US perspective:

> Although social justice scholars have conducted inquiries and raised philosophical questions relative to the women's movement, inequality of access to resources (i.e. transportation, community events) by people from low socioeconomic groups, and the plight of people of color, little research has been conducted relative to individuals with disabilities.
>
> (Devine and Piatt, 2013, p. 22)

Consistent with these observations is the work of Prince, who argues that despite an emerging academic field of disability studies (albeit from a North American perspective) in the context of urban politics and the city, that disability is 'relatively overlooked and under-theorized as a form of economic and social organization. A site for policy making, and a fundamental spatial context of everyday human relations' (2008, p. 1).

As a tutor, practitioner, entrepreneur, academic and from personal insights into the experiences of disabled people in everyday life, I concur with Devine and Piatt's and Prince's work, that historically there has been a lack of academic material that I can use with students who are studying tourism to enrich pedagogy related to disability and tourism. Later in this chapter, I will provide an overview of the work of Buhalis (Buhalis *et al.*, 2012), Darcy and Dickson (2009), who appear to be part of a new generation of academics to have provided in-depth materials on the concepts of disability and also accessible tourism.

But with regard to the history of tourism and tourism planning, disabled people have largely remained invisible and emerge from time to time in piecemeal analysis that only adds to the sense of being part of Prince's 'excluded others' in society (2008, p. 1). Devine and Piatt (2013) identify that it is possible to take for granted what might be viewed as everyday activities but for disabled people may not be possible for a number of reasons. The central thrust of their work is that it is not enough for all individuals and groups in society to have inclusion in everyday aspects of society, not least leisure and tourism opportunities, but that it should involve people making these things happen to enable participation to happen. They apply a structuralist approach to a socially constructed analysis of society in which: 'not all members of society especially individuals with disabilities, are provided with equitable access to life's opportunities. Unequal power, oppression, unearned privilege, and imbalanced life experiences make social and environmental justice elusive' (Devine and Piatt, 2013, p. 19). Within a UK context in the 1980s and 1990s, Brindley *et al.* (1996) noted how a move towards an ethical and moral movement in planning was led by groups with religious convictions and the Prince of Wales. This was in part a response to disadvantaged groups of society that included the poor, unemployed people, the isolated and the elderly. According to Brindley *et al.* (1996) these groups are less likely to be the beneficiaries of market-led planning. They identify that it was during this period that New Right political ideologies heavily influenced approaches to market-led planning in countries such as the UK, the USA, Australia and New Zealand.

They characterize this approach as having a range of externalities stating that:

> the implications of this pattern of development are only becoming apparent. The social groups which the market does not cater for are not only socially excluded but also spatially excluded. The market-led approach does not provide space in the built environment for those who cannot afford to pay for it.
>
> (Brindley *et al.*, 1996, p. 179)

Their work was applied to all social groups, not disabled people specifically, but their work is consistent with Devine and Piatt who argue that there is a moral obligation to enhance people's life experiences regardless of race, gender or ability level (2013, p. 25).

The general observations on the literature highlighted within this section raises questions of terminology. Authors using the terms 'mental and physical handicap' highlight that while terminology matters, there might be cultural variation in wording used to describe the concept of disability. You might want to consider how this terminology refers to the cultural context of your global location and how terminology can be both inappropriately and inaccurately used. For example the term 'people with color' is a racialized description from a US cultural context which might not transfer to other cultural contexts. Therefore, the use of the term disability requires further scrutiny.

DEFINING DISABILITY

In emphasizing the issue of the social construction of definitions of disability, Devine and Piatt, suggest that the medical model highlights how society might wish to 'transform people with disabilities to more like people without disability' and they contrast this model of disability with a social model where 'disability is a social construction in which society should not try to cure and normalise abnormality' (Devine and Piatt, 2013, p. 19). Starkly, they emphasize that disability should not be a problem to be solved. But who exactly are these people with disabilities? What is a disability and who does it affect? The United Nations (UN) *Convention on the Rights of Persons with Disabilities and Optional Protocol* (United Nations, 2006a) gives the following definition of disability:

> Persons with disabilities include those who have long-term physical, mental, intellectual or sensory impairments which in interaction with various barriers may hinder their full and effective participation in society on an equal basis with others.
>
> (United Nations, 2006a, p. 4)

Therefore, the convention takes into account people with physical disabilities including those who need to use a wheelchair or walking aid (such as a walking stick, crutches or walking frame) – these people may be considered as ambulant physically disabled people. It also includes people with mental health needs (anxiety, depression, schizophrenia), intellectual or cognitive disabilities (Down's syndrome, Asperger's syndrome, learning disabilities) and sensory impairments (visual and hearing impairments). The convention also considers that the impairment should be long term as do some other national disability rights laws such as the UK's Equality Act 2010, which replaced the Disability Discrimination Act (DDA) of 1995. The Equality Act defines long term as 12 months or more, which means that if a person has a temporary disability such as a broken leg, they may not be covered by this particular legislation. In addition to this, the Act has 'special rules about recurring or fluctuating conditions, for example arthritis', which may lead to differences in interpretation of the law. The Convention and the Equality Act only cover what might be considered to be the traditional definition of disability, while people have many more health conditions and disabilities that may affect their ability to visit a particular destination.

Modelling disability: medical and social models of disability

However, the *World Report on Disability* (WHO, 2011b) bases its definition of disability on the International Classification of Functioning, Disability and Health (ICF), which was developed in 2001. It:

> defines disability as an umbrella term for impairments, activity limitations, and participation restrictions. Disability refers to the negative aspects of the interaction between individuals with a health condition (such as cerebral palsy, Down syndrome, depression) and personal and environmental

> factors (such as negative attitudes, inaccessible transportation and public buildings and limited social supports).
>
> (WHO, 2002)

The ICF classification makes the definition of disability much broader than that of the UN *Convention on the Rights of Persons with Disabilities and Optional Protocol* or the Equality Act. It refers to health conditions that cause limitations in activity and participation. This means that it also includes conditions such as food allergies, epilepsy and diabetes, which may limit activity and participation in certain situations. Furthermore, the definition of disability is strengthened through recognizing that personal and environmental factors play a large role in disabling people. If a restaurant does not provide allergy advice on its menu, this could stop a person with a food allergy from eating there. If the attitude of the restaurant owners changes and they provide the information the person with a food allergy requires, the establishment is no longer disabling.

The ICF recognized that there needed to be a change in the way that we look at disability and placed a larger emphasis on health and functioning. The classification works upon the basis that everyone has health problems during their lifetime and therefore experiences some kind of disability, even if just for a short period of time.

> Disability is part of the human condition – almost everyone will be temporarily or permanently impaired at some point in life, and those who survive to old age will experience increasing difficulties in functioning. Disability is complex, and the interventions to overcome the disadvantages associated with disability are multiple and systematic – varying with context.
>
> (WHO, 2011b, p. 7)

The fact that the ICF appreciates that disability covers a wide spectrum of health conditions and can be permanent or temporary creates the foundation for the content of this chapter. The traditional definition of disability has become a stigmatized label, yet if planners consider the needs of people under the ICF definition, practically every single person on the planet would benefit either directly or indirectly at some stage in their lives.

Disability population statistics

Due to the different methods of defining disability, population statistics can also be misleading, confusing and contradictory. Even within a town, region or country contrasting figures are available depending on who and how the statistics have been gathered. An accompanying fact sheet to the UN *Convention on the Rights of Persons with Disabilities and Optional Protocol* published in 2006 states that 'Around 10 per cent of the world's population, or 650 million people, live with a disability. They are the world's largest minority' (United Nations, 2006b, p. 1). The World Health Organization (WHO), using the same data, admitted that they had previously underestimated the number of people living with some form of disability as they based the figures on data gathered in the 1970s until the *World Report on Disability* (WHO, 2011a).

The WHO has readjusted the number of people who have some form of disability for their studies since the publication of the 2010 global population estimates. In current publications they state that more than a billion people, approximately 15% of the population, have some form of disability. However, there are other statistics cited in the report from the *World Health Survey* (WHO, 2004) and

The Global Burden of Disease (WHO, 2008) that state the figures of 15.6% and 19.4% of the population, respectively. These estimates also attempt to identify the amount of the population with very significant difficulties in functioning is 2.2% (WHO, 2004) and those with severe disability is 3.8% (WHO, 2008). These figures appear to be measured on the traditional view of disability and include conditions such as quadriplegia, severe depression or blindness according to the WHO report.

An ageing population

Many studies have shown that the world population is growing older and with this comes an increased incidence of disability as health deteriorates with ageing. 'The risks of disability ... increase dramatically with age. In the US, about 14 million adults aged 65 and older (40%) report a disability and about 3 million (roughly 9%) report more severe limitations in self-care activities' (Institute of Medicine, 2007, cited in Freedman, 2011, p. 1). Freedman also looks at data from the 2007 National Health Interview Survey (NHIS) in the USA where it was found that 64% of people aged 65 and over had difficulty with an upper or lower body movement (NHIS, 2007, cited in Freedman, 2011, p. 7).

Buhalis and Darcy (2011) conclude that the WHO are concerned about ageing due to the release of *Global Age-friendly Cities: A Guide* (WHO, 2007), which provides information about the ageing population, directions for urban planners and a check list so that progress towards age-friendly cities can be monitored. Indeed, the document cited by Buhalis and Darcy states that the number of people aged 60 and over will double from 11% in 2006 to 22% by 2050 (WHO, 2007). The guide also makes the link between older people and those with disabilities with regard to access to the physical environment 'an age-friendly city is not just "elderly-friendly". Barrier-free buildings and streets enhance the mobility and independence of people with disabilities, young as well as old' (WHO, 2007, p. 6).

However, the WHO's (2011b) *World Report on Disability* asks if our longer lives will be spent in good or poor health, which according to them is under considerable debate among researchers. On the one hand researchers argue that there will be a decrease in disability as life expectancy increases, termed 'compression of morbidity' and on the other an increase in disability with longer life expectancy – 'expansion of morbidity' (WHO, 2011b, p. 12). Another view is that:

> as advances in medicine slow the progression from chronic disease to disability, severe disability will lesson, but milder chronic diseases will increase ... In the United States, between 1982 and 2001 severe disability fell about 25 per cent among those aged 65 or older even as life expectancy increased ... Unfortunately, this trend may not continue in part because of rising obesity among those now entering older ages.
>
> (WHO, 2011b, p. 12)

One of the issues highlighted in research is that we do not know if the disability was acquired after individuals reached the age of 65 and over or if they were from earlier in life. There are also commentaries regarding the links between life-long health and disability in later life. The coincidence of childhood diseases, living standards, environmental conditions, personal choices and exposure to health risks may all affect health and the likelihood of becoming disabled later in life. 'Exposure to toxic substances at work or at home, arduous physical work, smoking, alcohol consumption, diet and physical activity may have long-term health implications' (WHO, 2011b, p. 11). Researchers do not yet agree on the complex correlation between disability and ageing and more research needs to be done in this area.

TOURISM DESTINATION PLANNING: ACCESS FOR PEOPLE WITH DISABILITIES

A useful point to begin the discussion of tourism destination planning for disabled people is to attempt to define exactly what accessible tourism is. Buhalis and Darcy attempt to move towards a definition stating that:

> Accessible tourism is a form of tourism that involved collaborative processes between stakeholders that enables people with access requirements, including mobility, vision, hearing and cognitive dimensions of access, to function independently and with equity and dignity through the delivery of universally designed tourism products, services and environments. This definition adopts a whole life approach where people through their lifespan benefit from accessible tourism provision. These include people with permanent and temporary disabilities, seniors, obese, families with young children and those working in a safer and more socially sustainably designed environments.
>
> (Buhalis and Darcy, 2011, p. 10–11)

Therefore, when discussing tourism destination planning, it is first imperative that the planners know exactly who they are planning for and what the visitor may both want and require during their stay. For example, planning a ski resort will require different infrastructure and services to a city break in Las Vegas. The ski resort will require infrastructure such as ski lifts, so that the visitor can reach the slopes to undertake their chosen activity and services such as ski hire, which would obviously not be appropriate in Las Vegas.

The same is the case for other market segments within tourism where the destination may cater for many different types of visitor. Take into consideration a destination such as Barcelona: it has a wide demographic of visitors including shoestring backpackers, business tourists, city-break tourists, high-end tourists and day visitors who arrive in the morning on cruise ships and leave in the late afternoon. Each of these market segments requires different infrastructure and services. The shoestring backpacker will want cheap, hostel-type accommodation and good public transport infrastructure, while the business traveller may call for mid-range hotels and a decent taxi service and the high-end tourist might demand private airport transfers or a limousine, a five-star superior hotel and pre-booked tickets to a performance at the Liceu. The issues of infrastructure, services and different market segments also exist within the community of people with disabilities, yet few planners appear to fully understand their needs, wants and desires. This leads planners and architects to sometimes overlook this important market, looking only to local and national building regulations that enforce the rights of people with disabilities to have access to the built environment.

When an individual is planning a holiday, they will generally first decide upon the potential destination(s) and then begin searching for suitable accommodation and the method of getting there. There are then factors that will play a role in the ultimate decision-making process. These may include the cost and availability of the accommodation and the method of travelling to the destination. For example, if a person decides to fly to a particular destination but they cannot find a price within their budget at the right time of year, they may decide to change either the destination, their chosen travel dates or both. The same is also true of disabled people and access and facilities at accommodation play a vital role in the decision-making process. In fact, some would argue that the availability of suitable accommodation at the right cost is more important than the destination itself, which is highlighted in the following comments:

> for people with disabilities a foundation of any tourism experience is having accessible destinations (Israeli, 2002) and locating appropriate accommodation from which to base oneself while travelling (Darcy, 2002). Quite simply, to stay a night away from their normal residence requires appropriate accommodation that allows access to a bedroom and bathroom as a base for their stay.
>
> (Darcy, 2009, p. 816)

Many Western nations now have disability discrimination legislation making it illegal to treat a person differently due to their disability. Many of these laws (such as the Americans with Disabilities Act in the USA and the DDA in the UK) also have strict building regulations to ensure that disabled people can access and use them (Darcy, 2009). The building regulations generally give specific measurements regarding the physical environment such as the maximum slope for a ramp, door and corridor widths and the amount of space required for a turning circle for a wheelchair user. There will also be regulations for people with sensory impairments and other disabilities such as contrast markings on doors and stairs for people with visual impairments and flashing lights on emergency alarms for people with hearing impairments.

For accommodation providers, this also means that there are implications when designing bathrooms and this is also the most important factor for a person with a disability when making their decision on where to stay. Unfortunately, legislation, the architect's design and the needs of people with disabilities frequently do not overlap sufficiently. In short, the architect's implementation of the legislation renders the bathroom difficult to use due to their lack of understanding of the needs of people with disabilities.

Within an adapted bathroom at a hotel it will usually have features such as rails on the walls known as grab bars to help people with mobility problems stabilize themselves or transfer to and from their wheelchair if they use one. Building regulations will typically dictate where they should be placed, at what height and how far away from the WC, sink or bathing facilities with some room for tolerance (e.g. grabs bars should be installed at a height between X and Y). Some regulations may indicate how bathroom installations should correspond to each other; however, much of the actual layout, fixtures and fittings are left to the architect.

Although more research is required in this area, anecdotal evidence suggests that most people with mobility problems need or prefer to use a wet-floor shower room or wheel-in shower. This type of shower is level to the floor with no lips in the floor to prevent access with a wheelchair or cause tripping for an ambulant person with a mobility disability.

There is usually a fold-down seat attached to the wall with grab bars for transfers and stability. However, in some cases the regulations do not state that it is mandatory to install a seat in the shower nor its dimensions. If the seat is not installed, or if the dimensions are too small, it makes the shower unusable for many people with disabilities. In the USA the building regulations state the number of rooms that should have a level-access shower dependent on the total number of rooms at the accommodation. In many countries this is not the case. For example, in the UK an adapted room will fully comply with the legislation even if it has a bath rather than a level-access shower, which would be unsuitable and unusable for many people with mobility disabilities.

Further to these issues, a country may have different legislations in place at national, regional and local levels. A good example of this is Barcelona where an accommodation provider must comply with national Spanish legislation, regional Catalonian legislation and local legislation from the city's town council. Regrettably, to add even more confusion for accommodation providers and architects the different legislations contradict each other. Due to there being three legislations to comply with, the accommodation must undergo three different inspections. An accommodation may pass one, but then have to undertake alterations at a cost to comply with the second or third inspection. After completing the alterations, the accommodation would no longer comply with the legislation they had initially passed.

Disability legislation regarding building regulations can be complicated and open to a certain amount of interpretation, which is not only confusing for architects and accommodation providers but also means that often disabled people do not know what adaptations to expect when staying at a hotel they have not previously visited.

While this section focuses on access within accommodation, suitable facilities need to be available throughout the flow of the tourist experience. Although it is beyond the scope of this chapter it should include other aspects of the entire trip such as: (i) pre-booking information; (ii) access to the mode of transport to get to the destination; (iii) transportation at the destination itself; and (iv) access to tourist visitor attractions and other points of interest.

UNIVERSAL DESIGN AND DESTINATION PLANNING

In the 1990s Ron Mace, a design pioneer at North Carolina State University 'coined the term "universal design" to describe the concept of designing all products and the built environment to be aesthetic and useable to the greatest extent possible by everyone, regardless of their age, ability, or status in life' (Center for Universal Design, 2008a).

> The intent of universal design is to simplify life for everyone by making products, communications, and the built environment more useable by as many people as possible at little or no extra cost. Universal design benefits people of all ages and abilities.
>
> (Center for Universal Design, 2008b)

Unlike the building regulations that are attached to disability legislation, universal design does not have predetermined measurements, it is a more abstract concept comprising seven principles. It requires designers, planners and architects to consider the principles and implement them as far as possible into the design of products, spaces and buildings.

THE SEVEN PRINCIPLES OF UNIVERSAL DESIGN

The seven principles of universal design are as follows (Center for Universal Design, 1997).

Principle 1: Equitable Use
The design is useful and marketable to people with diverse abilities.

Guidelines:

- 1a. Provide the same means of use for all users: identical whenever possible; equivalent when not.
- 1b. Avoid segregating or stigmatizing any users.
- 1c. Provisions for privacy, security, and safety should be equally available to all users.
- 1d. Make the design appealing to all users.

Principle 2: Flexibility in Use
The design accommodates a wide range of individual preferences and abilities.

Guidelines:

- 2a. Provide choice in methods of use.
- 2b. Accommodate right- or left-handed access and use.
- 2c. Facilitate the user's accuracy and precision.
- 2d. Provide adaptability to the user's pace.

Principle 3: Simple and Intuitive Use
Use of the design is easy to understand, regardless of the user's experience, knowledge, language skills, or current concentration level.

Guidelines:

- 3a. Eliminate unnecessary complexity.
- 3b. Be consistent with user expectations and intuition.
- 3c. Accommodate a wide range of literacy and language skills.
- 3d. Arrange information consistent with its importance.
- 3e. Provide effective prompting and feedback during and after task completion.

Principle 4: Perceptible Information
The design communicates necessary information effectively to the user, regardless of ambient conditions or the user's sensory abilities.

Guidelines:

- 4a. Use different modes (pictorial, verbal, tactile) for redundant presentation of essential information.
- 4b. Provide adequate contrast between essential information and its surroundings.
- 4c. Maximize "legibility" of essential information.
- 4d. Differentiate elements in ways that can be described (i.e., make it easy to give instructions or directions).
- 4e. Provide compatibility with a variety of techniques or devices used by people with sensory limitations.

Principle 5: Tolerance for Error
The design minimizes hazards and the adverse consequences of accidental or unintended actions.

Guidelines:

- 5a. Arrange elements to minimize hazards and errors: most used elements, most accessible; hazardous elements eliminated, isolated, or shielded.
- 5b. Provide warnings of hazards and errors.
- 5c. Provide fail safe features.
- 5d. Discourage unconscious action in tasks that require vigilance.

Principle 6: Low Physical Effort
The design can be used efficiently and comfortably and with a minimum of fatigue.

Guidelines:

- 6a. Allow user to maintain a neutral body position.
- 6b. Use reasonable operating forces.
- 6c. Minimize repetitive actions.
- 6d. Minimize sustained physical effort.

Principle 7: Size and Space for Approach and Use

Appropriate size and space is provided for approach, reach, manipulation, and use regardless of user's body size, posture, or mobility.

Guidelines:

- 7a. Provide a clear line of sight to important elements for any seated or standing user.
- 7b. Make reach to all components comfortable for any seated or standing user.
- 7c. Accommodate variations in hand and grip size.
- 7d. Provide adequate space for the use of assistive devices or personal assistance.

(Center for Universal Design, 1997)

'The implication of universal design is that access would become central to design, rather than an add-on' to comply with disability legislation (Buhalis and Darcy, 2011, p. 9). Darcy and Dickson (2009) examined statistics from the Australian Bureau of Statistics (ABS) to see what percentage of the population would benefit from accessible tourism. They found that 20% of people had a disability and that 17% of people were aged 60 or over, and of these 51% reported having a disability (ABS, 2004, cited in Darcy and Dickson, 2009) – Darcy and Dickson assumed that there was an overlap with the disability statistics and thus reduced the percentage from 17% to 8% for seniors. As of June 2008, 1.4 million, or 6.4%, of the Australian population were aged 0–4 years (ABS, 2008, cited in Darcy and Dickson, 2009). They assumed that at least half of this age group used prams, which gave them 3%. Together the three groups they had identified as having access needs gave a total of 31% of the population that would benefit from universal design. Finally, they identified that, although they did not have any population estimates, people with temporary disabilities and workers who would be left with a safer working environment would also benefit from universal design (Darcy and Dickson, 2009). These groups are illustrated in Fig. 6.1.

Darcy and Dickson have identified specific groups of people with access requirements and the percentage of the population this affects. However, the concept of universal design, as mentioned earlier in this section, 'is to simplify life for everyone by making products, communications, and the built environment more useable by as many people as possible ... Universal design benefits people of all ages and abilities' (Center for Universal Design, 2008b).

Although the seven principles of universal design were published in 1997, it has existed for much longer; there just wasn't a name for it. A good example of universal design in practice is at supermarkets, which arrived in and around towns and cities long before 1997. Upon arrival at the car park there are generally spaces allocated and simply signed for people with disabilities and families with young children providing a system that is simple and intuitive to use, while the other spaces are for the rest of the clientele. Most modern supermarkets now have automatic doors that are of equitable use for everyone and simple and intuitive to use. The automatic doors have not been installed solely for the benefit of people with disabilities; they are there because many

Fig. 6.1. Universal design and accessible tourism beneficiaries. (From Darcy and Dickson, 2009).

people will be using shopping trolleys and carrying heavy bags. Both outside and inside the doors there is size and space for approach and use, the approach will also generally be flat, reducing the physical effort required to get to them either walking while pushing a shopping trolley or using mobility aids. The automatic doors also have tolerance for error built into them. If they begin to close while a person is passing through them, they will open again thus avoiding injury. As illustrated in this everyday example, six of the seven principles of universal design have been incorporated into the supermarket before you even start your shopping.

Darcy and Dickson's (2009) figure (Fig. 6.1) shows that 'it is not only those with access needs who benefit [from universal design] but all users as a universally designed environment considers occupational and safety issues...' (Buhalis and Darcy, 2011, p. 9) However, the figure may also mislead some into thinking that only the five groups mentioned benefit from universal design in accessible tourism and that only 31% of the population have access requirements. When considering the supermarket example, everybody would struggle through the entrance doors with a trolley if they were not automatic. Perhaps another method of representing universal design would be as in Fig. 6.2.

These same designs are also being implemented at airports and some hotels to make designs easier to use when carrying heavy luggage. For example, automatic doors and elevators are not just useful for disabled people but also for people with luggage and pushchairs, as they do not have to struggle with opening them or climbing flights of stairs.

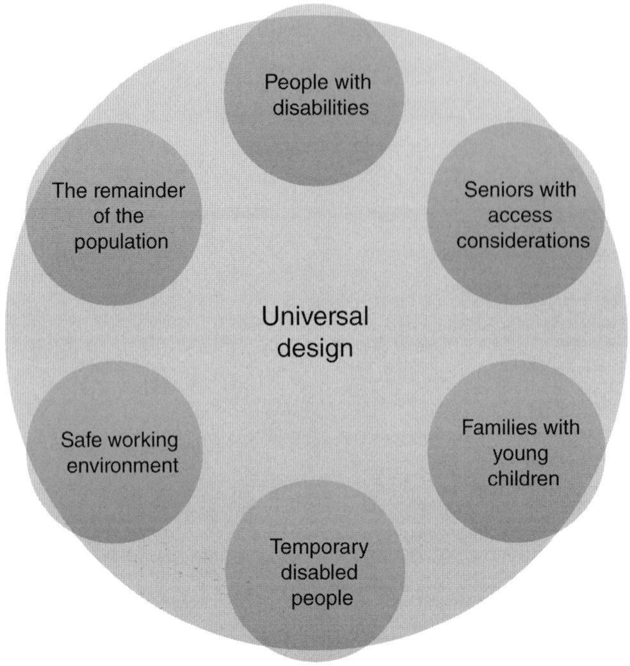

Fig. 6.2. Universal design and its beneficiaries.

PROVISION OF INFORMATION

Just as important as the infrastructure itself is the information supplied by tourism providers, tourist information boards, guidebooks and other stakeholders. In the past much of the information given to identify disabled access simply consisted of the international wheelchair symbol. Over time this began to change and varying symbols indicating various types of facilities for different kinds of disability came into play. In fact, so many different symbols were being used that it quickly became very confusing. In addition to this, tourism providers would use different symbols to represent the same facility thus creating a complex system that disabled people have to understand. While the use of symbols to represent specific facilities can be very useful they can also be supported by supplementary information.

Contextual information is where text is used to describe the access and facilities that are available for disabled people. For example, rather than just using symbols, an accommodation provider may expand on this by providing descriptive writing to explain different parts of the building such as the entrance, foyer, elevators, guest rooms and their bathrooms and other communal areas including restaurants and swimming pools.

Some tourism boards, including VisitEngland, have gone one step further than this, through making it compulsory for tourism providers to give an Access Statement if they wish to be listed with the organization on their website or in other materials on their accommodation and visitor attraction quality scheme. VisitEngland explain:

> An Access Statement is a description of a business' facilities and services to inform people with access needs. Access Statements allow for a written, descriptive approach to providing a wide range of information on accessibility. All areas of a business are described from car parking and arrival to toilets.
>
> (VisitEngland, no date)

They also offer an online Access Statement Tool that allows the user to select the type of tourism provider they are, which then gives them a suitable template to complete about their facilities. This information can then be typed into an online version that allows users to put their information on their own website and creates a PDF version for printing and Internet downloads for potential customers.

As well as giving written descriptions, Access Statements incorporate some basic measurements such as door and corridor widths and heights of steps and beds. However, much more statistical data can be specified to give even more detailed information so that an individual can decide if a tourism provider's facilities are suitable for their needs. Nevertheless, people using this information still have to interpret and understand it through building a mental picture of the establishment. One way of assisting this process is to make photographs and diagrams available to accompany the written text and measurements. Photographs can be annotated with useful information and diagrams, such as bathroom plans, and can also have measurements incorporated into them.

Besides photographs and diagrams, short videos can be of great benefit to help building a clearer picture of a tourism provider's access and facilities. Of course, some larger tourism providers may choose to pay professional film companies to make films for them, but this option does not have to be an expensive one. Most mobile phones and tablet PCs now have a built-in video camera, which can be used to create a Video Access Statement, recording the key features of a building or attraction. These can then be edited with free software on a computer (or indeed on a mobile phone or tablet App) and uploaded to video-hosting websites such as YouTube or Vimeo. Once uploaded many video-hosting websites allow the owner to add subtitles and embed them on to their own website using HTML codes or simple hyperlinks.

The obvious benefit to disabled people is that they can see first hand what the provider has to offer and make very informed decisions about their access needs based on the video, especially if it is provided in conjunction with the other elements as discussed above. Furthermore, there are also other benefits to the tourism provider that might not be immediately apparent. First, through hosting videos on much larger websites they have a much better chance of being listed by search engines and if this is combined with a good choice of keywords for the name and description of the film, it may help with search engine optimization. Similarly, if transcripts of the subtitles are available, this may help people to find it. To further increase the chance of the target market finding the video and other information about the tourism provider's access and facilities they can upload it to social media such as Facebook and Twitter. Finally, with many video-hosting providers the subtitles can automatically be changed into a variety of different languages and although the translation may not be perfect, it opens up the information to a much wider audience.

Case Study: Barcelona and Beyond

In 2010 the European Union launched its first Access City Award and although Barcelona didn't win, it reached the final four top cities out of the 66 that entered from 19 different member states (European Network for Accessible Tourism, 2013). However, Barcelona has not always been so inclusive for disabled people; it has been a long fought struggle for social change and equality that continues to this day. This section of the chapter looks at the processes and legislation that Barcelona has been through since the mid-1970s as well as the author's own experiences of living and working in the beautiful Catalan capital.

From 1974 social change was afoot in Barcelona, later known as the 'Rebellion of the lame' (Rebelión de los cojos) (Aragall i Clavé, 2003b) as disabled people, their families and friends began insisting on better rights. The protests reached their climax in June 1976 with a demonstration in Plaça Sant Jaume, the home of both the local and the regional governments, demanding employment, adapted transport, accessible streets and buildings, integrated education and a dignified pension (Aragall i Clavé, 2003a, b). Due to the continuous pressure and demonstrations from the protest groups a decree was finally passed in 1978 to remove architectural barriers from public thoroughfares. One of the breakthroughs of the decree was that from that point dropped kerbs were built on footpaths to allow better access for wheelchair users (Aragall i Clavé, 2003a, b). This was the start of Barcelona's long journey of strategic planning to become more inclusive and accessible for all.

In the same year, the Mayor of Barcelona met with representatives from disability groups to form the Municipal Institute of People With Disabilities in Barcelona (IMBD: Institut Municipal de Persones amb Disminució de Barcelona). The IMBD undertook many worthwhile projects including the integration of disabled people both socially and into employment. However, as Aragall i Clavé states the IMBD recognized by the late 1980s that 'it was necessary to bring about the integration of this group in society, in other words, it made it very clear the necessity to create accessible physical spaces (streets and buildings)' (Aragall i Clavé, 2003b, p. 8). In the early 1990s the IMBD signed an agreement with the College of Architects (Collegi d'Arquitectes) with a view to creating a benchmark for accessibility in architectural projects. By 1993, several major projects had been completed including Rambla Catalunya, which incorporated 'Kerb 120', a wide dropped kerb for all pedestrians, not just wheelchair users (Aragall i Clavé, 2003a, b). At the same time, 'the first low floor bus was unveiled, and 20 more were later acquired for use in the Paralympics' (Aragall i Clavé, 2003b, p. 10).

The 1992 Olympics and Paralympics in Barcelona 'acted as a definitive impetus' (Aragall i Clavé, 2003b, p. 10) to making the city more accessible and inclusive for disabled people, as an urban infrastructure with the necessary capacity to cope with both disabled and non-disabled athletes and visitors was required (Aragall i Clavé, 2003a, b). Part of the development was to create pedestrian zones in the Gothic Quarter of the city and

(Continued)

Case Study. Continued.

make improvements to the existing and new underground (Metro) lines. At the same time there were also trials to make the city's beaches more accessible and a Working Group for the Improvement of Accessibility (Grup de Treball per a la Millora de l'Accessibilitat) was established (Aragall i Clavé, 2003a, b).

By 1994 the local government and the IMBD had set up a new initiative, part of its remit was to 'Introduce the concept of Design for All, in other words, when thinking of accessibility, one is considering all the population, and not solely people with disabilities' (Aragall i Clavé, 2003b, p. 12).

The Design for All concept was broadened to encompass the whole of Barcelona under the title of 'City for All' (Aragall i Clavé, 2003b, p. 13):

> In short, the design of the City for All should take into account the diversity of the population:
>
> - People with permanent congenital disability.
> - People with permanent disability as a result of accident or illness during their life.
> - Temporary interaction problems with the environment.
> - Changes during one's lifetime (from birth to old age).
> - Persons with alterations to physical, psychological and sensory capacities.
> - Immigrants and tourists (culture, language and customs).
>
> (Aragall i Clavé, 2003b, p. 14)

This was a crucial time for the development of access and inclusion in Barcelona, although under the guise of different names, Design for All and City for All, are essentially the same concept as universal design as previously discussed in this chapter. The first Barcelona Municipal Accessibility Plan (Pla Municipal d'Accessibilitat de Barcelona) was developed in 1995 with the final objective of the city being 'totally accessible by the year 2006' (Aragall i Clavé, 2003b, p. 17).

Seven years later in January 2002, I moved to Barcelona for what was initially to be for 1 month; however, that wasn't long enough and I stayed for over 7 years, giving me an insight into the city and its accessibility from the perspective of a wheelchair user. One of the first things that impacted on my ability to live independently in the city was the lack of accessible public transport. Although some buses had lowered floors for easy access, they were few and far between and it was easy to wait at a bus stop and watch two or three inaccessible buses come and go before an accessible one eventually arrived. It was, at times, very frustrating, especially when time was of the essence and I had to be somewhere for a meeting. Secondly, although some of the bus fleet was by this time accessible, it did not mean that the bus stops themselves were. An accessible bus stop is where the kerb is higher than normal so that it is level with the vehicle's floor to allow a wheelchair user to enter the bus. So although some of the buses were accessible, the street infrastructure had not at that point been changed to allow easy access, this was especially the case outside the main tourist centre of the city.

(Continued)

Case Study. Continued.

Similarly, although the Metro line 2 and new lines had facilities for disabled people, other lines did not. The adapted lines typically connected the main sporting arenas and tourist hotspots and were designed for visitors, not people who actually lived in the city. Again, this was frustrating as a resident of the city, as it was not possible to travel to the places that I wanted to get to on the Metro. Another major issue both for residents and for visitors alike was the distinct lack of information regarding access in the city. Very little information existed about facilities for disabled people at museums, galleries or attractions and information relating to public transport was confusing, especially with regard to the Metro system as some stations were accessible for one particular Metro line, but not for another. It is interesting to note that although the concept of Design for All and City for All had been adopted, the problem of providing information hadn't been identified as in Principle 4 of the principles of universal design, 'Perceptible Information' (Center for Universal Design, 2007).

For this reason, I decided to begin providing information through a website called AccessibleBarcelona (Grimes, no date), which aimed to be a basic access guide to the city. The guide gives details about public transport including the airport bus, local buses, the Metro and even the Fenicular, a train that climbs Montjuic to the summit of the mountain where the Olympic Stadium is located. The guide also covers some museums and attractions as well as bars, restaurants and nightclubs that have facilities, including toilets that are suitable for disabled people (Grimes, no date), which were, at the time, difficult to come by. AccessibleBarcelona proved to be popular, especially with visitors to the city and requests for more information about suitable accommodation, adapted taxis and private airport transfers began to arrive. In addition to these requests people also had more specific needs and wanted to know where they could hire mobility aids and equipment such as wheelchairs, electric scooters, hoists and chairs for using in the shower. It was clear that there was an opportunity, not only to provide useful information, but to also create bespoke packages for disabled people who wanted to visit the city. Surprisingly, especially around the Christmas period, there would also be requests from local people who were residents in Barcelona for services such as adapted taxis and wheelchairs for elderly relatives, but they could only find the information through AccessibleBarcelona.

Consequently, I began researching accommodation providers that claimed they provided services for disabled people including 'wheelchair-friendly' rooms and visiting them to establish what they had to offer. The standards and facilities at different establishments was astounding, wheelchair friendly could mean anything from the building had an elevator and that was it, through to completely adapted buildings including bathrooms with wheel-in showers and grab bars to help people undertake transfers. There had to be a way of distinguishing between the different levels of facilities and so I used descriptive text, photographs and some basic measurements to help people

(*Continued*)

Case Study. Continued.

understand what was available at a given hotel. Over time this information developed further to include many more measurements and plans of the bathrooms so that people had a clear picture of the accommodation and could make informed decisions as to if it was suitable for their needs. The information provided gave people the confidence to book their holiday with AccessibleBarcelona and it slowly turned into a small destination management company for disabled people, one of the first of its kind for city breaks. The website also appealed to organizations from other parts of Europe and groups in northern Europe started to want to visit, which meant more research had to be undertaken to find wheelchair-accessible coach companies for week-long itineraries.

In 2003 Francesc Aragall i Clavé published *Barcelona. From the Accessibility Plan to the City for All* (*Barcelona. Del Plan de Accesibilidad a la Cuidad Para Todas las Personas*), which has been referenced in this section of the chapter. In it he refers to the fact that by 2003 a lot had been accomplished, but there was still a lot of work to be done if the city was to be fully accessible by 2006. By this time, there is a new recognition that information is vital for the city to be truly accessible and that technology should be utilized to its best potential. His reference to this is with regard to the citizens (residents), but it is also the case for visitors:

> When designing citizen information services, we cannot avoid the infinite opportunities that new technologies offer us, without forgetting that these services should be accessible and comfortable to use. Therefore, it is necessary to continue the search for new tools and new means to facilitate the relationship between the City Council and the citizen (Internet, Mobile Telephone, GPS, etc.).
>
> (Aragall i Clavé, 2003b, p. 34)

When 2006 finally arrived and Barcelona was scheduled to be a completely accessible City for All, it was apparent that there were still many problems for disabled people. Although the bus fleet was almost 100% adapted by this point, the bus stops still weren't and the Metro was far from being completed. In the same year The Grand Hotel Central, Barcelona opened after major renovations of the 1920s' building, which meant that it had to comply with national, regional and local disability building regulations. Although the guest bedrooms for people with disabilities passed an inspection under the national Spanish legislation, it failed under the regional Catalonian legislation. This was due to height measurements for some fixtures being slightly different for the two distinct sets of legislation. The hotel owners were forced to make a choice between move the fixtures and redecorate, pay a hefty fine or appeal. This example highlights some of the complex issues that planners, architects and developers were faced with, as outlined earlier in the chapter.

At this time I also became involved with Zexe, a project led by Antoni Abad, a digital artist who worked with disabled people to map obstacles in the city through the use

(*Continued*)

Case Study. Continued.

of mobile-phone photography. The participants were given mobile phones and were asked to take photographs of obstacles that they came across in their everyday lives. The photographs were then sent via text message along with the address where they were taken and were automatically logged on to a map of the city. They also identified that a large part of the access issues in Barcelona were not only due to permanent physical barriers, but also due to the lack of education and consideration of other residents of the city, such as parking vehicles on footpaths so that wheelchair users cannot get past. However, the participants also wanted to include photographs of examples of best practice, showing where architects and planners had thought through inclusive access. In total, 40 people took part in the 2006 project creating a map of 'inaccessible Barcelona' (Abad, 2006).

In 2008 Barcelona signed up to the UN *Convention on the Rights of Persons with Disabilities* and in 2010 representatives from the city's council made a presentation in Brussels entitled 'Barrier-free City for All'. The interesting parts of the presentation with regard to visitors from the city are 'Physical accessibility' and 'Mobility'. The presentation claimed that 98% of Barcelona's thoroughfares (streets, parks, places) had become accessible since the 1996–2006 Municipal Accessibility Plan, all six city beaches were accessible and that a Parks and Gardens Accessibility Plan had been created. The presentation also stated that with regard to public transport buses were 100% accessible and 85% of Metro stations now had lifts (Borda and Trujillo, 2010). The reference to 98% of Barcelona's thoroughfares being accessible is interesting as it does not take into account many of the barriers in the city, just the infrastructure that is owned by the council. Although the streets may be more accessible with more suitable surfaces and drop kerbs, there are still barriers such as steps and thresholds into businesses throughout much of the city, even on the city's main shopping areas. And again, the presentation refers to the 100% accessibility of the bus fleet, but makes no reference to the percentage of bus stops that have been improved to make alighting the vehicle easier.

Barcelona has been on a long journey to get to where it is today: the council have made a significant effort to work with disabled people's organizations, planners and architects to make the city more inclusive for everyone. The impact of years of legislation and integrated planning has ensured that it is now one of the most accessible cities in Europe, but as Aragall i Clavé suggested in 2006 there has been 'a lot accomplished, but a lot more to accomplish' (Aragall i Clavé, 2003b, p. 34) and this is still the case for Barcelona and many other tourist destinations.

CONCLUSION

Disabilities include a wide range of conditions and health problems, not just wheelchair users, and these individuals have varying abilities and needs. Their requirements are complex, which cannot all be addressed by the stroke of a large 'disability brush'. Yet, if society, governments and planners collaborate in an attempt to reach a final goal of creating universally designed destinations they would be useable by a much larger percentage of the population, both disabled and non-disabled alike.

As has been shown in this chapter through examining the case study of Barcelona, the driving force of change was initiated by disgruntled disabled residents, their families, friends and the wider community but the 'definitive impetus' (Aragall i Clavé, 2003b, p. 10) came from the planning of a large international sporting event, in this case the 1992 Olympic and Paralympic Games. This event placed international scrutiny on the city, bringing disabled and non-disabled athletes and visitors to the destination and the infrastructure developments that were implemented by the government have had a long-lasting legacy for both residents and visitors.

Through working with disabled people, including them in the planning process and implementing the principles of universal design, Barcelona has become a much more accessible city for all. However, accessibility is not merely concerned with altering and adapting environments nor creating new universally designed ones, without information the access and facilities on offer become irrelevant if people do not know about them. A major key to providing accessible destinations is the information itself and it is vital that it is provided in a variety of formats so that it is accessible and understandable to as many different people as possible. Therefore, if governments and planners truly want to create *destinations for all*, they must consult with local disabled people and their representative organizations and continuously update information to ensure that it is detailed and reliable to give everyone the confidence to visit.

QUESTIONS

1. What are the different definitions of disability?
2. How does the concept of accessible tourism provide opportunities for disabled people to engage in tourism?
3. Consider on your visit to a tourist attraction/destination, how far the principles of universal design have been applied.

BIBLIOGRAPHY

Abad, A. (2006) Barcelona*accessible. Available at: http://zexe.net/barcelona/message/index (accessed 27 August 2014).

Aragall i Clavé, F. (2003a) *Barcelona. Del Plan de Accesibilidad a la Cuidad Para Todas las Personas.* Ajuntament de Barcelona, Barcelona, Spain.

Aragall i Clavé, F. (2003b) *Barcelona. From the Accessibility Plan to the City for All.* Ajuntament de Barcelona, Barcelona, Spain.

Borda, R. and Trujillo, L. (2010) Barcelona and the UN Convention on the Rights of Persons with Disabilities. Presented in November 2010 at Eurocities – Workgroup Barrier-free City for All, Brussels, Belgium.

Brindley, T., Rydin, Y. and Stoker, G. (1996) *Remaking Planning: The Politics of Urban Change*, 2nd edn. Routledge, London.

Buhalis, D. and Darcy, S. (2011) *Accessible Tourism. Concepts and Issues*. Channel View Publications, Bristol, UK.

Buhalis, D., Darcy, S. and Ambrose, I. (2012) *Best Practice in Accessible Tourism. Inclusion, Disability, Ageing Population and Tourism*. Channel View Publications, Bristol, UK.

Center for Universal Design (1997) *The Principles of Universal Design*. North Carolina State University, Raleigh, North Carolina. Available at: http://www.ncsu.edu/ncsu/design/cud/about_ud/udprinciplestext.htm (accessed 29 April 2014).

Center for Universal Design (2007) *7 Principles of Universal Design*. North Carolina State University, Raleigh, North Carolina. Available at: http://www.ncsu.edu/ncsu/design/cud/about_ud/docs/use_guidelines.pdf (accessed 24 April 2014).

Center for Universal Design (2008a) *About the Center: Ronald L. Mace*. North Carolina State University, Raleigh, North Carolina. Available at: www.ncsu.edu/ncsu/design/cud/about_us/usronmace.htm (accessed 29 April 2014).

Center for Universal Design (2008b) *About UD*. North Carolina State University, Raleigh, North Carolina. Available at: www.ncsu.edu/ncsu/design/cud/about_ud/about_ud.htm (accessed 29 April 2014).

Darcy, S. (2002) Marginalised participation: physical disability, high support needs and tourism. *Journal of Hospitality and Tourism Management* 9(1), 61–72.

Darcy, S. (2009) Inherent complexity: disability, accessible tourism and accommodation information and preferences. *Tourism Management* 31, 816–826.

Darcy, S. and Dickson, T. (2009) A whole-of-life approach to tourism: the case for accessible tourism experiences. *Journal of Hospitality and Tourism Management* 16(1), 32–44.

Devine, M.A. and Piatt, J. (2013) Beyond the right to inclusion: the intersection of social and environmental justice for inclusion of individuals with disabilities in leisure. In: Schwab, K. and Dustin, D. (eds) *Just Leisure: Things That We Believe In*. Sagamore Publishing, Urbana, Illinois, pp. 17–26.

European Network for Accessible Tourism (2013) Accessible Cities. Access City Awards – Past Winners. Available at: http://www.accessibletourism.org/?i=enat.en.accessible-cities (accessed 23 August 2014).

Freedman, V.A. (2011) Disability, functioning, and aging. In: *Handbook of Aging and the Social Sciences*. Elsevier Science & Technology, Oxford, UK.

Grimes, C. (no date) AccessibleBarcelona. Available at: http://www.accessiblebarcelona.com (accessed 26 August 2014).

Institute of Medicine (2007) *The Future of Disability in America*. National Academy Press, Washington, DC.

Israeli, A. (2002) A preliminary investigation of the importance of site accessibility factor for disabled tourists. *Journal of Travel Research* 41(1), 101–104.

Prince, M.J. (2008) Inclusive city life: persons with disabilities and the politics of difference. *Disability Studies Quarterly* 28(1), 1–13. Available at: http://dsq-sds.org/article/view/65/65 (accessed 9 June 2014).

Schwab, K. and Dustin, D. (eds) (2013) *Just Leisure: Things That We Believe In*. Sagamore Publishing, Urbana, Illinois.

United Nations (2006a) *Convention on the Rights of Persons with Disabilities and Optional Protocol*. United Nations, New York.

United Nations (2006b) *Some Facts About Persons with Disabilities*. United Nations, New York.

VisitEngland (no date) Available at: http://www.visitengland.org/busdev/bussupport/access/info/Statements.aspx (accessed 22 August 2014).

World Health Organization (WHO) (2002) *Towards a Common Language for Functioning, Disability and Health (ICF)*. WHO, Geneva, Switzerland.

World Health Organization (WHO) (2004) *World Health Survey, 2002–2004*. WHO, Geneva, Switzerland. Available at: http://www.who.int/health- info/survey/en/ (accessed 9 December 2009).

World Health Organization (WHO) (2007) *Global Age-friendly Cities: A Guide*. WHO, Geneva, Switzerland.

World Health Organization (WHO) (2008) *The Global Burden of Disease: 2004 Update*. WHO, Geneva, Switzerland.

World Health Organization (WHO) (2011a) *Global Health and Aging*. WHO, Geneva, Switzerland.

World Health Organization (WHO) (2011b) *World Report on Disability*. WHO, Geneva, Switzerland.

chapter 7

Spatial Planning – Managing Tourism and Recreation in a Lived-in 'Protected' Area

Kathryn Beardmore*

Yorkshire Dales National Park Authority, Bainbridge, UK

LEARNING OBJECTIVES

- To realize that UK National Parks are very different to those found elsewhere in the world, being 'lived-in' landscapes in which the special qualities of the areas have developed through the interaction of people and nature.
- To understand that, having no direct power in terms of promoting and managing tourism, and land being in private ownership, the UK National Park Authorities must work in partnership with others to plan and influence tourism and recreation in their area.
- To realize that through developing a management plan for the area, the National Park Authority brings an integrated approach to balance the potentially conflicting needs of conservation, recreation and local communities.

INTRODUCTION

This chapter explores the key challenges involved in planning for tourism and recreation

*kathryn.beardmore@yorkshiredales.org.uk

in a protected rural landscape, specifically the Yorkshire Dales National Park, UK, and how these are addressed through a spatial, landscape-based approach. It looks at the concept of National Parks in the UK and crucially their planning and management.

The Yorkshire Dales was designated a National Park in 1954, and is an internationally recognized lived-in 'protected landscape'– International Union for Conservation of Nature (IUCN) category V. It covers 1762 km^2, has a population of 19,000, and accommodates approximately 5 million visitor days each year.

The area is located in the Northern Pennines of England and is remote hill country that rises to over 700 m. It is characterized by wild gritstone moors and limestone pasture where rivers have cut deep valleys (dales) with traditional meadows, and open areas of rough grazing (moorland) above. The area has been appreciated by visitors since the 18th century, with the tourism economy well established in the area long before it became a National Park. The area is valued by visitors for its distinctive landscape of limestone scenery, heather moorland, hay meadows, barns and walls, the opportunities it provides for many forms of outdoor recreation as well as the picturesque villages and hospitable residents that provide a warm welcome. It is a diverse area with busy honeypot sites as well as opportunities to 'get away from it all'. Visitor surveys show that the area is particularly appreciated for its beautiful scenery and peace and tranquillity.

The area is deeply rural, with the population of the largest settlement in the National Park only 2000 inhabitants. However, it lies within easy reach of some large urban centres including the conurbations of West Yorkshire, East Lancashire and the Tees Valley. The main challenge for the National Park Authority, in managing the area in accordance with statute, is to ensure that the opportunities for recreation and visitors' enjoyment do not undermine the very special qualities they come to experience. At the same time, the National Park Authority seeks to maintain the area's socio-economic well-being and ensure a prosperous future for the National Park and its communities. Much of the economy of the area is dependent on revenue generated through tourism and farming, though other employment sectors such as quarrying and education are also important.

NATIONAL PARKS IN THE UK

IUCN categorization

The IUCN is a global environmental organization, founded in 1948. It works with governments, organizations, scientists and local communities to conserve biodiversity (including geodiversity and landforms) and help address issues such as climate change and sustainable development at a global, ecosystem and local level.

The IUCN protected area management categories classify protected areas according to their management objectives. Category V is a protected area where:

> the interaction of people and nature over time has produced an area of distinct character with significant ecological, biological, cultural and scenic value: and where safeguarding the integrity of this interaction is vital to protecting and sustaining the area and its associated nature conservation and other values.
>
> (IUCN, 2014)

Many of the special qualities of the Yorkshire Dales are valued by visitors, particularly the distinctive landscape: hay meadows, heather moorland and the built heritage of barns and drystone walls (QA Research, 2013). These have come about through a combination of the underlying geology and centuries of farming and land management practices.

As 'lived-in' landscapes, National Parks in the UK are different to many National Parks found elsewhere in the world, which do not have human interaction and are IUCN Category II. For these National Parks, human activity is minimized in order to allow for 'as natural a state as possible' (IUCN, 2014).

The National Park concept as applied in the UK

It is generally accepted that the concept of National Parks originated in the USA, with the establishment of Yellowstone National Park, in 1872 (Dower, 1945, para 3). However, it was not until 1949, with the passing of the National Parks and Access to the Countryside Act, that they became established in England and Wales. It took until 2000 with the National Parks (Scotland) Act, for National Parks to be established in Scotland.

The idea of National Parks had been discussed since the 1930s, but their introduction only gained momentum after the Second World War. In 1945, John Dower was asked, by government, to produce a report on how National Parks could work in England and Wales.

In this report Dower said National Parks in application to England and Wales should be defined as:

> an extensive area of beautiful and relatively wild country in which, for the nation's benefit and by appropriate national decision and action,
>
> (a) the characteristic landscape beauty is strictly preserved,
>
> (b) access and facilities for public open-air enjoyment are amply provided,
>
> (c) wildlife and buildings and places of architectural and historic interest are suitably protected, while
>
> (d) established farming use is effectively maintained.
>
> (Dower, 1945, para 4)

Dower also recognized that:

> National Parks should be in a true and full sense *national*, if they are to be worthy of their name and purpose. This does not mean that *local* interests are to be disregarded. On the contrary, the well-being of those who live and work within them must always be a first consideration. But it does mean that their holiday and recreational use should be for people – and especially young people – of every class and kind and from every part of the country, indeed of the world. National Parks are not for any privileged or otherwise restricted section of the population, but for all who care to refresh their minds and spirits and to exercise their bodies in a peaceful setting of natural beauty.
>
> (Dower, 1945, para 13)

National Parks were seen as green lungs for the nation; important for open-air recreation and escape from urban life. John Dower's report was seen as a blue-print for National Parks in England and Wales and his vision is still relevant today.

The Environment Act and National Park Authorities

The statutory purposes and criteria for National Parks in England and Wales are set out in the National Parks and Access to the Countryside Act 1949, as amended by the

Environment Act 1995. The statutory purposes of National Parks are to:

1. Conserve and enhance the natural beauty, wildlife and cultural heritage of the area.
2. Promote opportunities for the understanding and enjoyment of the special qualities of those areas by the public.

A National Park is managed by a National Park Authority. The Environment Act 1995 makes it clear that a National Park Authority, in pursuing these purposes, shall 'seek to foster the economic and social well-being of local communities' within the National Park and shall 'co-operate with local authorities and public bodies whose functions include the promotion of economic or social development within the area of the Park' (UK Government, 1995).

National Park Authorities are similar to a local authority, with some of the same powers. However, they are not large and also, unlike a local authority, receive their funding from central government. To put this into context the government grant received by the Yorkshire Dales National Park Authority for 2014/15 was £4.2 million, and it had 97 full-time equivalent employees. A National Park Authority is governed by its members. The membership is appointed to represent local county councils, local district councils, local parishes and the national interest. Thus, on the Yorkshire Dales National Park Authority 16 members are 'local' representatives (i.e. six appointed by county councils, six by district councils and four by the parishes), and six are 'national' representatives, making 22 in total. The basic rules determining the size of these groups within an authority are that:

- Every district, county, or unitary authority with land in a National Park is entitled to appoint at least one member unless it chooses to opt out.
- The total number of local authority members must exceed the number of 'national' members (parish members are counted as national members because, although they are elected by parish councils, they are formally appointed by the Secretary of State).

It is the National Park Authority members that decide the overall aim and objectives and policy for the National Park to deliver the purposes of National Park designation outlined above.

Integrated management and the Sandford Principle

A National Park Authority brings integrated management to an area and, as part of this, balances the needs of conservation, recreation and the local community to achieve positive outcomes, in cooperation with those who own and manage the land. Managing a National Park can be challenging at times and often needs a proactive approach to prevent conflict arising. Conservation of the historic environment, geodiversity, wildlife and habitats and encouraging people to enjoy and learn from the countryside usually go hand in hand. However, if conflict does arise and cannot be managed through an integrated approach, the 'Sandford Principle' can be invoked. This is usually a last resort.

To help National Park Authorities make decisions between conservation and recreation, the National Parks Policy Review Committee made a recommendation:

'where irreconcilable conflicts exist between conservation and public enjoyment, then conservation interest should take priority' (Sandford, 1974). This is known as the 'Sandford Principle' named after Lord Sandford who chaired the Committee and is enshrined in statute (but now excludes the word irreconcilable). The Environment Act (1995) states:

> If it appears that there is a conflict between those purposes, [the National Park Authority] shall attach greater weight to the purpose of conserving and enhancing the natural beauty, wildlife and cultural heritage of the area comprised in the National Park.
>
> (UK Government, 1995)

Two well-known examples in the Yorkshire Dales National Park where this principle has been applied are:

- restricting climbing on particular crags, or parts of crags, away from the bird-breeding season (e.g. to allow the peregrine falcons to fledge at Malham Cove); and
- promoting alternative routes according to the time of year, on the Coast to Coast long-distance path, to prevent disturbance to breeding birds and footpath erosion.

This principle means that encouragement and promotion of the understanding and enjoyment of the special qualities must have regard to the first statutory purpose of conservation. While activities within National Parks are usually accommodated by cooperation, mutual understanding and good management, rather than by prohibition, if suitable forms of management are not possible or impractical, then the first objective of conserving the natural beauty takes priority.

National Park Management Plan

Section 66 of the Environment Act 1995 requires National Park Authorities to produce a National Park Management Plan and to review it at least every 5 years. The government's guidance on management of National Parks is explained in the UK Government Vision and Circular *English National Parks and the Boards* (Defra, 2010). In the circular it states:

> The Plan should provide statements of the relevant NPA's [National Park Authority's] policy for managing and carrying out its functions in relation to the Park and should reflect Park purposes, the duties of the Authorities and the need to engage other local authorities, statutory agencies and a wide range of partners, stakeholders and the community in the preparation and implementation of the Park Management Plan. They should set the context for the local development framework coverage of the Parks. It is essential that they are based on robust evidence and Government commends the practice of supporting the Park Management Plans with State of the Park reports and a range of more detailed documents dealing with specific topics, such as climate change, biodiversity, nature conservation, cultural heritage, forestry, dark night skies, recreation or management plans for particular areas.
>
> (Defra, 2010, para 135)

In essence, the National Park Management Plan shows how management of the area is joined up, in accordance with the purposes of the National Park, to achieve: (i) conservation; (ii) promoting understanding; (iii) access enhancement; and (iv) the socio-economic well-being of local communities. In addition, National Park Authorities are also the local Planning Authority, and produce a Local

Plan, in accordance with the National Planning Policy Framework and Planning Acts. The Local Plan sets out the detailed policies that are used to assess planning proposals for development and the use of land and buildings in the National Park, and sits alongside the National Park Management Plan.

POLICY AND STRATEGIES – A YORKSHIRE DALES NATIONAL PARK CONTEXT

Vision

Importantly, National Park Authorities have discretion as to how best to interpret their statutory purposes within their own local setting. The National Park Management Plan sets out ambitions for the National Park and how it will be looked after over the next 5–10 years. It provides the overall framework that guides the actions of all the bodies working in the National Park, including the Authority. To this end, the *Yorkshire Dales National Park Management Plan (2013–18)*'s vision and six ambitions are:

> Through their passion for this special place, local people and businesses will keep the Yorkshire Dales National Park a thriving area. Its unique cultural landscape will be treasured for its stunning scenery, exceptional heritage and wonderful wildlife, and every year millions of people will be inspired to visit or find out more.
>
> By 2040, it will be:
>
> - A distinctive, living, working, cultural landscape that tells the on-going story of generations of people interacting with their environment;
> - A friendly, open, and welcoming place with outstanding opportunities to enjoy its special qualities;
> - Home to the finest variety of wildlife in England;
> - Resilient and responsive to the impacts of climate change, storing more carbon each year than it produces;
> - Providing an outstanding range of benefits for the nation based on its natural resources, landscape and cultural heritage, and these underpin a flourishing local economy;
> - Home to strong, self-reliant and independent communities with good access to services.
>
> (YDNPA, 2013)

The six ambitions carry equal weight. For each one, the Management Plan then goes on to set specific objectives (53 in total) that the Yorkshire Dales National Park Plan partnership and many other organizations intend to achieve over the next 5–10 years.

Thematic strategies and plans

In the Yorkshire Dales National Park context the *Yorkshire Dales National Park Management Plan* is supported by a number of more detailed strategies relating to particular aspects of the management of the National Park, notably: (i) development; (ii) biodiversity; (iii) historic environment; (iv) woodland; (v) recreation and tourism; and (vi) interpretation. As with the Management Plan itself, a number of these strategies are strategies for the National Park, not just for the Yorkshire Dales National Park Authority (see Fig. 7.1). They are prepared by partnerships of local organizations, of which the Authority is just one member, through consultation with stakeholders.

As a 'lived-in' protected area, the National Park Management Plan and accompanying strategies recognize people are at the heart of

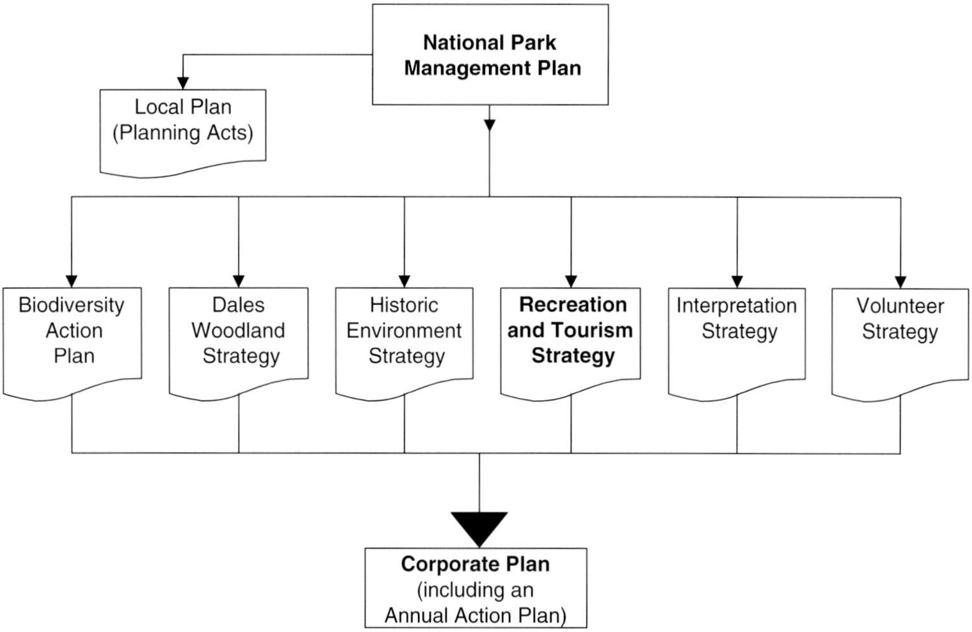

Fig. 7.1. The relationship between the Management Plan and other strategic documents. (Adapted from YDNPA, 2010.)

looking after the National Park: (i) the people who farm, manage and own the land; (ii) the people who live in and own the buildings; (iii) the people who own and work in the businesses; (iv) the people that give up their time to volunteer; and (v) the people who visit and love the Yorkshire Dales.

The detailed allocation each year of the Yorkshire Dales National Park Authority's own resources is set out in its published Corporate Plan. Among other things, this sets out:

- the Authority's priorities for using its resources;
- specific objectives that the Authority intends to achieve over the next 2–3 years – most taken directly from the National Park Management Plan; and
- actions that will be taken in the next 12 months to help achieve those objectives.

The Authority's workforce is structured, currently, into three Directorates. Supported by the Corporate Services Directorate (which includes Legal, Finance, IT, Websites, Communications, HR, Committee Services, Estates and Maintenance), the other two Directorates deliver the following main services. Conservation and Community Directorate:

- Land Management (farming, biodiversity, trees and woodlands);
- Sustainable Development (planning policy, cultural heritage, Sustainable Development Fund); and
- Development Management.

Park Services Directorate:

- Tourism (promotion, tourist facilities, National Park Centres, car parks, toilets);
- Access and Recreation Management (including the Ranger Service);

- Promoting Understanding (interpretation, Dales Countryside Museum, outreach and education); and
- Volunteering.

The Park Services Directorate is guided in its recreation and tourism work by the thematic strategy *Special Qualities, Special Experiences* (YDNPA, 2010).

Special Qualities, Special Experiences

The strategy *Special Qualities, Special Experiences* brings together the Yorkshire Dales National Park Authority's policies and principles for access, recreation, diversity, visitor management and sustainable tourism in the National Park. This strategy explains the interrelationships shown, diagrammatically, in Fig. 7.2 and outlines the National Park Authority's spatial approach to recreation and tourism management.

The strategy outlines:

- **The special environment and unique experiences.** The experiences the National Park has to offer through its special qualities and access and recreation opportunities, and the contribution it can make to other national concerns such as health and well-being, equality and diversity.
- **Access and accessibility.** The ease with which individuals can travel to and around the National Park to access the special qualities; encouraging visitors to think about their travel choices; information and signing to help visitors find their way around and gain enjoyment and understanding, in particular through interpretation using new technologies; and ways of increasing the diversity of the National Park's visitor profile.
- **Safeguarding the special qualities (including visitor management).** The spatial distribution of visitors and visitor facilities; managing people, including the numbers and timings of visitors to a site (where possible) to safeguard the environment; the quality of experience; and land management interests.
- **Sustainable tourism.** The economic and other benefits of people enjoying the National Park; ensuring that tourism is sustainable and provides economic opportunities without harming the special qualities; and is in accordance with the European Sustainable Tourism Charter principles (YDNPA, 2010, para 15).

Fig. 7.2. Sustainable tourism and the Yorkshire Dales National Park. (From YDNPA, 2010.)

CHALLENGES

Changes in visitor patterns over the years

Visitors were coming to the Yorkshire Dales long before National Park designation was confirmed. The area was made famous by writers and artists in the 18th century, during which time interest in recreational activities was growing. By 1781, there was already a tour guide to the caves in the Ingleborough and Settle area (YDNPA, 2001). As transport links improved, the numbers of visitors started to increase. Aysgarth became a staging post for coach travellers who would probably have wandered down to view the Falls while waiting for their horses to be changed. By the late 19th century, several railway lines had been built across the Yorkshire Dales, the most famous being the Settle–Carlisle line. Other railway lines were important for bringing visitors, including the Ilkley to Skipton line which called at Bolton Abbey, and the Wensleydale line between Northallerton and Garsdale (YDNPA, 2001).

The real boom in tourism took place in the second half of the 20th century with the rise in car ownership and increased leisure time. In recent times, a change in the way people take their holidays, with one main holiday plus a number of short breaks throughout the year has meant different visitor patterns for the Yorkshire Dales. However, it is still the opportunities for outdoor activities, alongside natural attractions such as Malham Cove and Aysgarth Falls and picturesque villages that draw people to the area (QA Research, 2013).

Despite these changes, the interdependence of public access, visitors, land management and economic survival of the local community were not really fully appreciated until the experience of foot-and-mouth disease in 2001 (Beardmore, 2013). The entire rights-of-way network was shut down to try and halt the spread of the disease, and the impact on the number of visitors in the area was immediate and devastating. For the first time it was demonstrated that the area's cultural offer, such as museums, galleries, local craft centres, visiting tea shops and driving around – in themselves major tourist activities – were not enough to sustain the local economy. Much of the tourism revenue turned out to be linked directly to recreational activities. People only visited the area if they could walk and ride in, and not just admire, the landscape.

The average length of stay for staying visitors, currently, is between 4 and 5 days. However, only 1 in 2 of the 5 million visitor days a year to the National Park is attributable to staying visits. Bounded by the Teesside conurbation to the north, York, Leeds/Bradford to the south, and Greater Manchester to the west, an estimated 9.5 million people live within an hour's travelling time of the National Park boundary (based on the 2011 census). Most people come for short breaks or day visits, mainly between Easter and October.

Monitoring of trends in visitor numbers and their economic impact on the area is carried out annually through the Scarborough Tourism Economic Activity Monitor (STEAM) model. Based on this model Yorkshire Dales National Park visitors contributed £200 million to the economy of the National Park and £500 million to the National Park's 'area of influence' in 2012 (Global Tourism Solutions, 2013).

The importance of stakeholders in managing recreation and tourism

One of the biggest challenges to taking a strategic approach to managing recreation in the National Park is the fact that 95% of the land is privately owned. The remaining 5% is owned by conservation bodies such as The National Trust, Natural England and a utility company (Yorkshire Water). The National Park Authority has a specific policy of not owning any land, and it also has no direct 'control' on visitor numbers. Its influence on the management of land and visitors can only be through partnership with others: organizations, local communities, farmers and land managers.

Similarly in terms of promoting and managing tourism, the National Park Authority has no direct powers or responsibility. It is not a Destination Management Organization (DMO) and does not have an economic development remit. This remit rests with the local authorities at a district and county level. Over the years there have been a number of partnerships established, with varying degrees of success, that have sought to bring together a range of tourism stakeholders around common objectives. Since the 1990s, the Yorkshire Dales National Park Authority has played an active part in these tourism partnerships for the area.

Visitor perceptions and characteristics

The National Park Authority regularly undertakes visitor perception surveys, (every 5 years a random, statistically valid, sample of >500 visitors), to understand the characteristics and views of visitors, and how these have changed over time. It helps to evaluate:

- why visitors come to the area (i.e. is it because of the area's special qualities, or for some other reason?);
- whether visitors are having a positive experience of the National Park and the National Park Authority's services; and
- the effectiveness of the National Park Authority's policies that underpin its work on sustainable tourism and visitor management.

As an IUCN category V protected area, how visitors experience the National Park and what they 'value', and how this in turn affects the resident population, and any improvements that can be made are important considerations.

Based on an analysis of visitor surveys, since 1994, there has been remarkable consistency in National Park visitor characteristics and perceptions (Beardmore, 2014). Most visitors are regular repeat visitors with only one in ten visitors interviewed being first-time visitors. The majority of visitors are couples, and do not have any children with them (>80%). There has also been consistency over the past 20 years in visitors' mode of travel, with 90% using a private car or van to get to, and travel within, the National Park on the day of interview. The use of public bus services among visitors during the summer has also remained consistent (3% of those interviewed), despite changes to public transport leisure services in recent years, while use of the train appears to have increased slightly over the last 10 years from 2% to 3% (QA Research, 2013). One noticeable change in 2013, compared with previous surveys, is that over one in ten visitors are now 'actively travelling' on foot or bike in addition to using their car to get to a location (QA Research, 2013). The survey did not ask why 'active travel' options were chosen by visitors but it could be a combination of: (i) it being a sustainable form of travel; (ii) it is healthy exercise; and/or (iii) the success of the National Park Authority's policy over the last 5 years or so, of producing and distributing information about active travel options and ensuring it is readily available through a variety of media, to help make choosing this option easier.

Over 80% of visitors go for a walk during their visit to the National Park, and satisfaction levels with the standard of rights of way remains very high. Since 2008, visitors have been asked (unprompted) what they see as the National Park's special qualities. The top four most frequent answers have been consistently: (i) natural beauty (78%); (ii) open space/remoteness (37%); (iii) peace and tranquillity (29%); and (iv) opportunities for walking (26%). Other answers specifically mentioned by more than one in ten respondents were hills/dales/valleys (17%), villages/traditional buildings (15%) and limestone scenery (13%).

In addition to the special qualities of the area, visitors were asked what they had actually enjoyed during their visit. The most frequent answers were similar: scenery/landscape; good walking; and peace and tranquillity. Very few visitors said anything had spoilt their visit. If they did mention anything it was usually the weather (9% of respondents).

Safeguarding the special qualities – why people visit

As the strategy *Special Qualities, Special Experiences* explains, people, places and the special qualities of the Yorkshire Dales are strongly connected and interdependent. The landscape, habitats, wildlife and archaeological features that combine to give the area its special character are actively shaped by the way the land is managed. In turn, the landscape, natural environment, cultural heritage and peace and tranquillity are *the* key attractions to visitors, who come to the Yorkshire Dales for recreation and to access this environment. These visitors form the basis for much of the economic well-being of the area.

People visit from many different places, with different interests and levels of knowledge about the area, and many undertake activities without obvious formal management or intervention by the National Park Authority. This is as it should be; understanding and enjoyment of the special qualities of the Yorkshire Dales National Park and the responsible use of the area is encouraged, with the aim of avoiding problems. It is important people do not undermine the very experience they come to enjoy, and equally, that the National Park Authority, through its visitor management and provision of visitor facilities, does not undermine those experiences for visitors.

Visitor management

As a 'lived-in' protected area the National Park Authority has little direct control on activities or visitor numbers within the National Park. However, it does manage visitors in the area indirectly through its policies on visitor management and having rangers 'on the ground'. The Ranger Service is very important in this context, providing a first point of contact for visitors and residents alike. It often means issues can be resolved quickly – as they arise – preventing disruption to the local community or damage to the area's special qualities.

Special Qualities, Special Experiences defines visitor management as 'the process of managing visitors so that they can enjoy the environment of the National Park whilst not harming the very qualities that make it special' (YDNPA, 2010, para 97). Intervention is only necessary where activities threaten the special qualities of the area or cause problems with other users or interests. Visitor management and the ways of influencing visitor behaviour and the choices visitors make can be through:

- promotion or non-promotion of areas or activities;
- targeting particular audiences, and/or type of promotion;
- raising awareness of users regarding the impact of visiting or activities;
- using a range of charges for facilities and activities offered;
- considering the type, number and locations of facilities provided and their opening times where relevant;
- Restrictions on what activities are allowed, where and when; and
- considering how accessible the environment is for different visitors (YDNPA, 2010).

Visitor management usually involves identifying the level of recreational use or development beyond which unacceptable consequences will occur, and managing this as far as possible by matching the provision of recreational opportunities and conservation of the environment with visitors' needs and aspirations. It is important to consider, on the basis of evidence, the 'impact' an activity is having, or is likely to have, on the environment and amenity of the area. Where this has, or is likely to have, an adverse effect on the special qualities of the National Park, then the National Park Authority considers the best management option available to achieve the outcome required.

Visitor management is relatively straightforward for new activities and/or new locations through either planning permission or influencing the activity before it becomes established. For example, following the introduction of the open access provisions of the Countryside and Rights of Way (CRoW) Act in 2005, the National Park Authority undertook positive visitor management, in Oxenber and Wharfe Woods for example. This is a Site of Special Scientific Interest, designated largely for its ground flora, and there was concern about possible degradation of what was previously a little-visited site. The National Park Authority developed three access points with interpretation to the site, and two well waymarked paths through the woodland. People counters were put in place and the site has been monitored. There are relatively high levels of use but the management methods have prevented any deterioration on the site while allowing people to enjoy it.

However, for well-established activities this can be more difficult. A notable example in the National Park requiring intervention in recent years has been the use of recreational motor vehicles on green lanes. This activity has been prevented through the use of traffic regulation orders (TROs) to conserve the special qualities of the area, particularly tranquillity and wildlife.

Case Study: A Spatial Approach to Planning and Management

In planning for visitor management in the area, the National Park Authority has taken a 'whole landscape' spatial approach – based on landscape units (Fig. 7.3). As explained in *Special Qualities, Special Experiences*:

> across the National Park there are quieter and busier areas, which for the most part reflect the relative remoteness and accessibility of different areas. Distinctions can be made between main and tributary dales, dale heads, uplands and watershed areas, and there is a general pattern that underpins the relative remoteness and wildness of individual areas.
>
> (YDNPA, 2010, para 106)

Special Qualities, Special Experiences sets out the relative sensitivities and opportunities to meet the needs of the access, recreation and tourism-related issues within these broad landscape units (YDNPA, 2010, Appendix A). By making the illustration used (see Fig. 7.3) generic, rather than specific, it can be applied to any dales and moor situation without difficulty. People experience landscapes holistically, and the illustration can be readily related to typical experiences and perceptions, so it is grounded in reality. It is used to identify opportunities and constraints for access, recreation and tourism throughout the whole landscape area underpinned by guiding principles and policies. The aim is to ensure

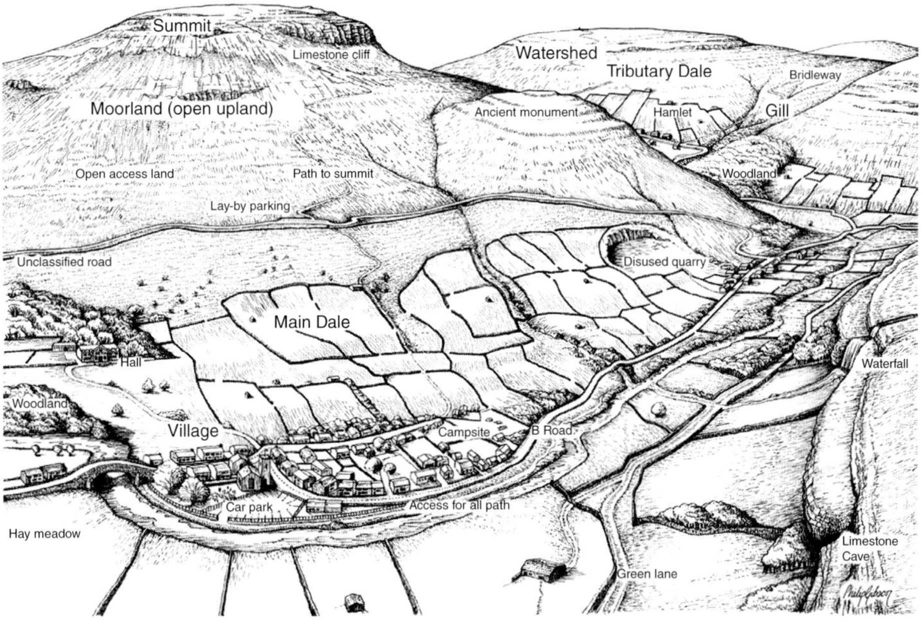

Fig. 7.3. Landscape units and features of a 'typical' dale. (From YDNPA, 2010.)

(Continued)

Case Study. Continued.

through careful visitor management that each area maintains its characteristics for which it is valued (i.e. tranquil areas remain tranquil, and busy areas are not stifled).

Relating this diagram back to the elements of Fig. 7.2, for example the experience of 'getting away from it all' will be most evident on the open upland. Here human influence will be minimized: there will be no formal car parking or waymarking of rights of way and signage will be minimal and only if necessary to resolve a specific problem. Access is likely to be on foot, cycle, horseback or by 'tramper' (for those with limited mobility). By contrast the main village in the dale-bottom will provide a very different experience with formal car parking, interpretation boards explaining features of interest, easy-to-follow waymarked routes and other visitor facilities.

In essence this reflects the way the different areas within the Yorkshire Dales have developed over time (YDNPA, 2010, para 111), and the strategy puts forward policies to retain this pattern and develop it further. Different visitor experiences are based very much on maintaining the characteristics that engender a strong 'sense of place' and feeling or association based on the busyness or quietness of an area and characteristics of the landscape units that make up that part of the Yorkshire Dales.

QUALITY OF EXPERIENCE

Through visitor surveys, the National Park Authority asks people to rate their overall satisfaction with their visit to the area. Table 7.1 shows satisfaction levels have increased significantly over the past 10 years and are very high.

Despite the challenges highlighted earlier, it is heartening to see that visitors have a very enjoyable and satisfying experience of the National Park. Many things contribute to this rating: (i) the high standards of tourism businesses and accommodation providers; (ii) the high standards of visitor infrastructure such as toilets, car parks, National Park Centres and a well-managed rights-of-way network maintained by the National Park Authority; and of course (iii) the amazing scenery and welcoming local communities.

WORKING TOGETHER

Dales Tourism Partnership and the EUROPARC Federation's Sustainable Tourism Charter

Since 2010, the National Park Authority has taken a lead role in tourism and established the Dales Tourism Partnership, and the Dales Tourism Business Network, when the Tourism Partnership for the Yorkshire Dales and Harrogate was dissolved. While the National Park Authority has no direct powers or role in managing tourism, only its statutory purposes of conservation and promoting understanding and enjoyment of the area's special qualities, cuts to funding of the district and county councils, and the demise of the Regional Development Agencies, meant that there were fewer resources available for others to undertake this work. The National

Table 7.1. Overall satisfaction rating with visit to the National Park (on a scale of 1, not at all satisfied, to 6, very satisfied).

	Percentage of respondents					
Year	**1 Not at all satisfied**	**2**	**3**	**4**	**5**	**6 Very satisfied**
2002	0	0	1	10	36	51
2008	0	0	0	5	21	73
2013	0	0	0	2	20	77

Park Authority recognized many organizations have an interest in tourism in the Yorkshire Dales, and a clear way forward for tourism was required to balance the economic benefits with the needs of the community and environment.

The Dales Tourism Partnership is a partnership between the Yorkshire Dales National Park Authority and Nidderdale Area of Outstanding Natural Beauty (AONB) (an AONB is a UK landscape designation recognizing landscape of a similar quality to a National Park). It manages tourism in the Yorkshire Dales geographical area with the help of a steering group of interested parties: local authorities, estates, conservation bodies and representatives of local communities and tourism businesses. The partnership is based on the EUROPARC Federation's Sustainable Tourism Charter model. The Sustainable Tourism Charter is designed to be a practical management tool which helps protected areas to continuously improve the sustainable development and the management of tourism, taking account of the needs of the environment, the local population and the local tourism businesses. The Charter's focus is on initiating and assisting a process of sustainable tourism planning through development of a strategy with the help of a forum.

The *European Charter for Sustainable Tourism* defines 'sustainable tourism' as:

> any form of development, management or tourist activity which ensures the long term protection and preservation of natural, cultural and social resources and contributes in a positive and equitable manner to the economic development and well-being of individuals living, working or staying in protected areas.
>
> (EUROPARC Federation, 2010)

In 2008, the Yorkshire Dales National Park Authority and Nidderdale AONB, two protected areas, applied jointly for the *European Charter for Sustainable Tourism*, and were successful in achieving the award. In 2013, re-validation for the Charter was confirmed.

The Dales Tourism Partnership recognizes that its role is to facilitate an environment that encourages and enables vibrant tourism businesses. Its vision can only be achieved through the tourism businesses in the area.

Working with businesses

The Yorkshire Dales National Park Authority, with Nidderdale AONB, has also set up the Dales Tourism Business Network to provide a way to communicate and work with the large number of small tourism businesses (less than four employees on average) in the area. Although there was some communication with tourism businesses prior to its establishment in 2010, it was on an occasional basis. The Network is run by one of the tourism businesses. It has grown rapidly and now has over 300 members. The aim of the Dales Tourism Business Network is to enable tourism businesses in the Yorkshire Dales to build on existing success, share good practice and improve performance where appropriate by capitalizing on the area's nationally designated environmental and recreational assets. It has provided an excellent forum for discussion.

A local distinctiveness project – Distinctly Dales – ran in 2011–2012 to actively engage tourism businesses through a series of events. There were 16 workshops held with over 220 local tourism businesses attending. The aim of these workshops was to help businesses use local distinctiveness and the area's special qualities as part of their marketing. To help, the National Park Authority has made available a range of resources including the Distinctly Dales Toolkit which pulled together the local legends, traditional food and special artefacts that make each of the dales unique.

The National Park Authority is keen that businesses and communities use the National Park logo to highlight their association with the National Park. To this end it has developed a series of adapted logos for use by businesses and individuals working within the area. It hopes that use of the logos will build on the strong, recognizable brand which celebrates the uniqueness of the National Park, as well as helping local businesses strengthen their own product.

CONCLUSIONS

Tourism is a vital component of the Yorkshire Dales economy, and the National Park Authority has a key role in providing facilities for the visitor – car parks, toilets, National Park Centres, and an easy-to-use rights-of-way network. Through visitor management it tries to ensure that the natural and cultural heritage of the National Park is protected. It also looks to support tourism businesses, and safeguard the interests of the local community.

Promoting sustainable tourism and working in partnership with Nidderdale AONB, other organizations and tourism businesses is central to management of the National Park. The strategic vision for this work is outlined in the National Park Management Plan, which includes agreed objectives. These are translated through the tourism and recreation strategies into coordinated actions across a number of organizations.

As *Special Qualities, Special Experiences* recognizes:

> the linkages, the mutual dependence and the opportunities that come from an integrated approach to the management of access, recreation, diversity, visitor management and sustainable tourism in the Yorkshire Dales. However, an 'integrated' approach will be successful only if it is applied consistently across the National Park, with consideration of the adjacent catchment areas. This is necessary to ensure that issues are not simply displaced and that a wider, more holistic understanding of the interactions and impacts of leisure and travel is managed and promoted.
>
> (YDNPA, 2010, para 169)

QUESTIONS

1. How should recreation and tourism development be planned and managed in the context of UK National Parks?

2. How can a spatial approach benefit visitor management in the Yorkshire Dales?

3. What role should the *European Charter for Sustainable Tourism* play in the development of sustainable tourism in the National Park?

BIBLIOGRAPHY

Beardmore, K. (2013) Living the dream? Active countryside tourism in the Yorkshire Dales National Park. In: *Conference Proceedings: Active Countryside Tourism*. Conference held at the International Centre for Research in Events, Tourism and Hospitality, Leeds Metropolitan University, Leeds, UK, 23–25 January. Available at: https://docs.google.com/file/d/0B6cNkxwXUXcPSGVRclY1emt-PQmM/edit?pli=1 (accessed 19 May 2014).

Beardmore, K. (2014) *Analysis of Yorkshire Dales National Park Visitor Surveys Report to Dales Tourism*. Yorkshire Dales National Park Authority, Bainbridge, UK.

Department for Environment, Food and Rural Affairs (Defra) (2010) *English National Parks and the Broads: UK Government Vision and Circular 2010*. Defra, London, UK. Available at: https://www.gov.uk/government/uploads/system/uploads/attachment_data/file/221086/pb13387-vision-circular2010.pdf (accessed 23 January 2014).

Dower, J. (1945) *National Parks in England and Wales Report*. HMSO, London.

EUROPARC Federation (2010) *European Charter for Sustainable Tourism in Protected Areas – The Charter*. EUROPARC Federation, Regensburg, Germany.

Global Tourism Solutions (2013) *STEAM data for the Yorkshire Dales 2012*. Report commissioned by the Yorkshire Dales National Park Authority, Bainbridge, UK.

International Union for Conservation of Nature (IUCN) (2014) Protected Areas Category V. Available at: http://www.iucn.org/about/work/programmes/gpap_home/gpap_quality/gpap_pacategories/ (accessed 23 January 2014).

QA Research (2013) *Customer Satisfaction Visitor Survey Research*. Report commissioned by the Yorkshire Dales National Park Authority, Bainbridge, UK.

Sandford, L. (1974) *National Park Policies Review Committee Report*. HMSO, London.

UK Government (1995) Environment Act 1995. HMSO, London. Available at: www.legislation.gov.uk (accessed 23 January 2014).

Yorkshire Dales National Park Authority (YDNPA) (2001) *Tourism – A Balancing Act*. Yorkshire Dales National Park Authority, Bainbridge, UK.

Yorkshire Dales National Park Authority (YDNPA) (2010) *Special Qualities, Special Experiences, an Integrated Recreation and Tourism Strategy – Policy and Principles*. Yorkshire Dales National Park Authority, Bainbridge, UK.

Yorkshire Dales National Park Authority (YDNPA) (2013) *Yorkshire Dales National Park Management Plan (2013–18)*. Yorkshire Dales National Park Authority, Bainbridge, UK.

PART III

Global Approaches to Planning and Tourism

This part of the book contains seven case study chapters, written by a range of academics from different global cultural contexts. These chapter essays explore unique insights on planning, tourism and sustainability from the destinations of Dubai, Portugal, the Lebanon, Hungary, Mauritius, Latvia and Peru. These chapters provide contemporary academic research insights into different processes and approaches to planning for tourism, and particularly evaluate the extent to which they are informed by the principles of sustainability.

- Chapter 8: *Planning and Developing 'Destination Dubai' in the Context of the United Arab Emirates (UAE)* – Filareti Kotsi and Ian Michael initially review the development of tourism in Dubai, which they characterize as a relatively young and modern tourism destination, within the wider context of the UAE. They consider the role sustainability plays in the development of a tourism destination and evaluate whether the staging of Expo 2020 can be used as an event to build a destination brand that exemplifies the principles of sustainability.
- Chapter 9: *Sustainable Tourism Development of Rural Destinations: the Schist Villages, Portugal* – Sofia Teixeira Eurico and Fernanda Maria Fernandes Oliveira highlight, initially, how the 'Schist Villages Programme', financed through European Union funding, has been developed since 2001 in a rural and socio-economic isolated area of Portugal. This chapter demonstrates how the creation of the 'Schist Villages' brand is based on a coordinated process of planning, involving different local, regional and national actors and their utilization of the principles of sustainability.
- Chapter 10: *Planning for Sustainable Tourism Development in a Context of Regional Instability: the Case of the Lebanon* – Nour Farra Haddad highlights how public and private sector stakeholders are struggling to develop strategic planning for sustainable

tourism in Lebanon. There is consideration of how domestic rural and nature tourism can be used as new niche tourism products, to help to develop sustainable tourism. The chapter also emphasizes that successive planning initiatives are hampered by the challenge of the insecurity in Lebanon, not least from the hosting of political refugees from neighbouring states.

- Chapter 11: *Understanding the Balaton: a Hungarian Case Study in Planning for Tourism* – Ágnes Raffay and Alan Clarke provide an account of the scale and complexities of governance around Lake Balaton, in Hungary. They critically assess the challenge of working towards sustainable development, within emerging governance structures, highlighting competition between different Destination Management Organizations (DMOs). In their assessment of this destination, they consider the challenges of planning multi-tourist activities in the Balaton area, given these governance complexities.
- Chapter 12: *National Approaches to Planning and Tourism in Latvia* – Dzintra Atstāja, Ivars Brīvers and Agita Līviņa provide an overview of the historical phases of the development of tourism in Latvia, evaluating how tourism governance arrangements in Latvia inform tourism policy, planning and management. They consider how tourism policy and planning, in Latvia, has incorporated the principles of sustainability and explore how niche forms of tourism, not least dark tourism, are part of the promotion 'Latvia. Best Enjoyed Slowly' which exemplifies the principles of sustainability.
- Chapter 13: *Beyond Enclave Tourism: the Road to Sustainability of Mauritius* – Girish Prayag profiles this small island developing state (SIDS) as an example of enclave tourism, which has limited economic links to the local community. Prayag considers the opportunities and constraints that enable sustainable tourism development to emerge. He particularly considers the consequences of non-participatory tourism planning and development processes on the residents' role in tourism.
- Chapter 14: *Cooperation, Local Communities and Planning of a Religious Heritage Site: Valle del Colca, Peru* – Dolors Vidal-Casellas and Neus Crous-Costa highlight the marginalization of rural areas in Peru, within the wider context of rapid economic growth in urban settings, with this economic growth stimulated partly by a growth in tourist visits surpassing 3 million in 2013. Specifically, they focus on the Colca Valley and describe the relevant stakeholders involved in the tourism industry from a social perspective. They evaluate the role of the local population in hospitality and tourism decision-making processes and related jobs, particularly in religious-based tourist sites.
- Chapter 15: *Conclusions: Diverse Paths to Sustainability* – Nigel D. Morpeth and Hongliang Yan offer a range of insights to emerge from the theoretical, practitioner and applied case study chapters from the range of academic authors. Morpeth and Yan finally provide ideas on future issues in planning, tourism and sustainability.

chapter 8

Planning and Developing 'Destination Dubai' in the Context of the United Arab Emirates (UAE)

Filareti Kotsi* and Ian Michael

Zayed University, Dubai, UAE

LEARNING OBJECTIVES

- To understand the development of a relatively young and modern tourism destination of Dubai within the wider context of the United Arab Emirates (UAE).
- To evaluate the role sustainability plays in the development of a tourism destination.
- To analyse the role of the Expo event in building a destination brand and exemplify the principles of sustainability.

INTRODUCTION

The intention of this chapter is to reflect upon the planning and development of tourism in the Emirate of Dubai, UAE using a variety of secondary data sources obtained from newspaper articles, government reports and international published research. The chapter begins by providing a brief overview of the human and physical geography of the UAE and Dubai, followed by the development

*filareti.kotsi@zu.ac.ae

and the role of tourism governance in Dubai, informed by the variety of secondary sources highlighted above, not least a variety of key academic studies on tourism. The principles of sustainability incorporated within the tourism governance are also highlighted in this section. More generally the analysis of sustainability and sustainable development within the UAE compares progress made within other parts of the UAE in contrast to Dubai. The rationale for choosing the case study of Dubai's Expo 2020 is that it demonstrates the capacity of this event to consolidate the branding of a destination but also the significance of this event to exemplify the principles of sustainability.

It is important to note that according to Balakrishnan (2013), research on the Middle East and North Africa (MENA) region accounts for less than 1% of all globally published research in business, management and accounting. For example, in the edited volume *Tourism in the Middle East: Continuity, Change and Transformation* (Daher, 2007) as well as in a more recent edited volume *Tourism in the Muslim World* (Scott and Jafar, 2010) there is no research related to Dubai or the UAE. However, this chapter uses additional literature to provide insights into the planning of a pre-eminent destination for tourism in the 21st century.

OVERVIEW OF THE HUMAN AND PHYSICAL GEOGRAPHY OF DUBAI

The UAE lies on the eastern coast of the Arabian Peninsula, bordered by Oman to the east and Saudi Arabia to the south and to the west by Qatar and Saudi Arabia. It stretches for more than 650 km, most is barren desert, and the total area approximates to 84,000 km^2. In the Musandam Peninsula, the creeks provide natural harbours, while the Hajar Mountains that rise to 2500 m extend through Oman to the east of the Arabian Peninsula. To the west, and within Abu Dhabi, the desert is interspersed with oases; to the south are high sand dunes stretching into the Empty Quarter. The climate is hot and dry with little rainfall, while the coastal plain experiences high humidity. Temperatures range between an average minimum in January and February of 10–14°C and an average maximum in July and August of 48–50°C (Dew and Shoult, 2000) (Fig. 8.1).

The UAE, formerly known as the Trucial States, is a young and dynamic country whose federal sovereign state was established in 1971. This 43-year-old country consists of seven emirates that were united by the late Sheikh Zayed bin Sultan Al Nahayan in a single state: Dubai, Abu Dhabi, Sharjah, Ajman, Umm al-Quwain, Fujairah and Ras al-Khaimah (the latter joined the federation in 1972). The UAE was formed from the group of tribally organized Arabian Peninsula sheikhdoms. In 1853, they signed a treaty with the UK, under which the sheiks (Trucial Sheikhdoms) agreed to a perpetual maritime truce where disputes among sheiks were referred to the British for settlement. The UK and the Trucial Sheikdoms established bonds in an 1892 treaty where the sheiks agreed not to dispose of any territory except to the UK and in return the British promised to protect the Trucial coast from aggression by the sea. In 1971, the UK announced its decision to end the treaty relationships with the seven Trucial Sheikhdoms that had been, together with Bahrain and Qatar, under British protection (Shoult, 2006). The withdrawal of the

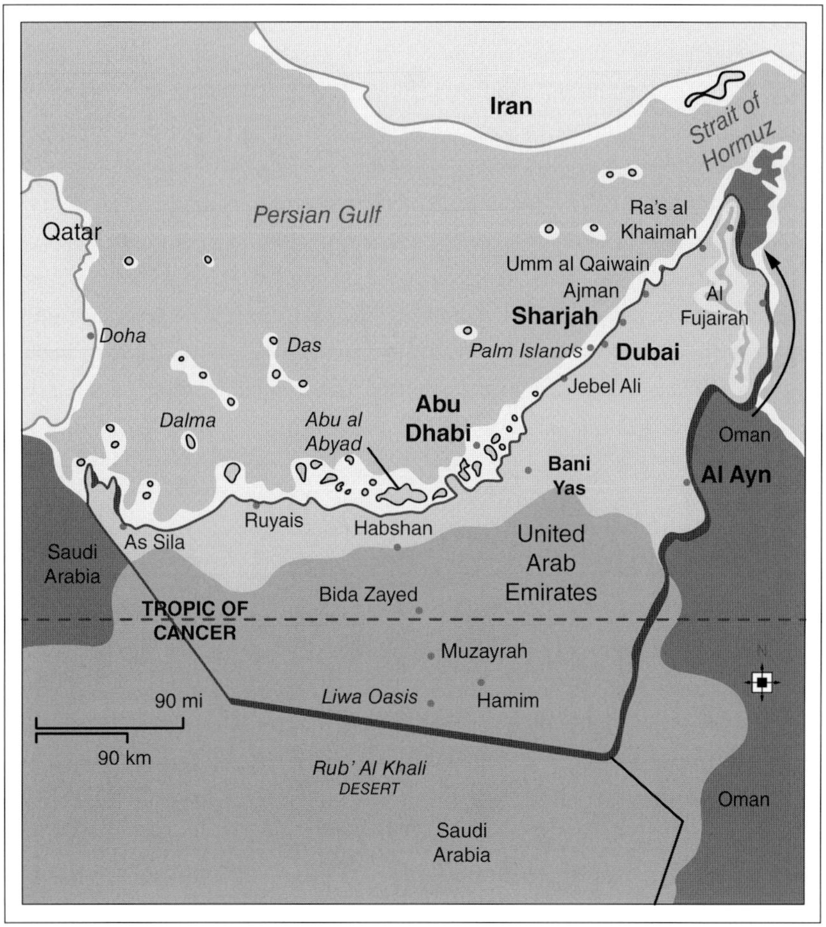

Fig. 8.1. Map of the United Arab Emirates (UAE). (From www.worldatlas.com)

British from the Persian Gulf led to the formation of the UAE.

With the collapse of the pearling industry in the 1930s, Dubai fell into a deep depression (Dew and Shoult, 2000); however, when in 1966 oil was discovered, the late ruler of Dubai, His Highness Sheikh Rashid bin Saeed Al Maktoum, utilized the oil revenues to spur infrastructure development in Dubai, such as schools, hospitals and roads as well as a modern telecommunications network (DTCM, 2014c). The UAE's primary natural resources are petroleum and gas; in fact the UAE is the world's eighth largest oil producer. More than 90% of these resources are found in the Emirate of Abu Dhabi. In addition to petroleum, the UAE extracts rock and sand, mostly for use in construction (Chronicles of the Middle-East and North Africa, 2014a).

Between 1968 and 1975 the city's population grew by over 300% (DTCM, 2014a) and in 2010 the UAE's population was estimated at 8,264,070 (United Arab Emirates National Bureau of Statistics, 2014). Dubai's population has crossed the 2 million mark and in 2013 touched 2,174,000 people (Nazzal,

2013). The emirate has not undergone a full census since 2005, but each year Dubai Statistics Centre releases population estimates. Native Emiratis are outnumbered in their own country with 15% made up of nationals, while the remaining 85% of the population is made up of expatriates. The nationals were originally Bedouins, principally desert-dwellers and sea-oriented people (Michael, 2013). Although no longer nomadic, much of the local population of the UAE is tribally affiliated. There are six major tribal groupings in the country – the most important is the Bani Yas and the two important ruling families come from branches of this tribe (Chronicles of the Middle-East and North Africa, 2014b). The country has one of the highest rates in the world with regard to the diversity of foreign nationalities and the diversity reflects a good investment environment. According to the Under-Secretary of the Ministry of Labour, a total of 202 different nationalities exist in the labour market (*Khaleej Times*, 2006). Over the years, not only have the numbers of nationalities increased rapidly, but others have appeared for the first time; there has also been a change in the nature of the jobs that expatriates perform (Hellyer, 2013). The population has a sharp imbalance in gender – men make up more than three-quarters of the population; the gender imbalance is generally attributed to the number of male expatriate workers who come to Dubai without their families. The three most populous neighbourhoods are industrial areas or areas providing accommodation for the labour force (Nereim, 2013).

Each of the seven emirates has its own ruler with the ruler of Abu Dhabi being president of the UAE, whereas the ruler of Dubai is vice president and prime minister. While the UAE has no political parties, the Supreme Council of the Federation is the highest political authority in the country and consists of all the rulers of the seven emirates. Each emirate retains substantial autonomy while a basic concept in the government's development as a federal system is that a significant percentage of each emirate's revenues should be devoted to the central budget (Shoult, 2006). Political order and continuity, alongside economic prosperity have been substantial factors contributing to the development of the country. It is a Muslim state with a degree of openness to outside influences (Henderson, 2006a).

UAE has witnessed rapid economic development and maximized the benefits to be derived from the country's substantial oil wealth. Dubai is distinguished as one of the trade centres in the Arabian Gulf region, highlighting a strategic competitive location in the middle of the UAE coast and linking maritime lines and transporting goods between East and West. In 1979, Jebel Ali Port, the biggest port in the Middle East opened and in 1985 Emirates airline launched. The same year Jebel Ali Free Zone was also inaugurated, attracting overseas investment and today Dubai has free zones throughout the city (Dew and Shoult, 2000). Dubai has also a Media City and Internet City consisting of IT (information technology) firms such as Microsoft, EMC corporations, IBM and media organizations such as CNN, MBC and Reuters (DTCM, 2014c). One of the defining characteristics of the country's economic development has been the government's far-sighted investment in high quality facilities, covering from port to airport developments, highway networks, and social and leisure amenities for the benefits of locals as well as expatriates (MacDonald, 2000a) (Fig. 8.2). For example,

Fig. 8.2. Dubai's skyline. (Photograph taken by Filareti Kotsi, 2013.)

Dubai Metro, the driverless metro rail network, declared by the Guinness World Records as the world's longest fully-automated metro network spanning 75 km, changed the way thousands of commuters travel within the city of Dubai (DTCM, 2014a).

The city is also home to two world iconic landmarks: (i) Burj Al Arab, the sail-shaped seven-star hotel built in 1999; and (ii) the world's tallest tower at 828 m (more than twice the height of the Empire State Building) – the Burj Khalifa, built in 2010 (Fig. 8.3). In 2006, the palm-shaped artificial archipelago of Palm Jumeirah – home to luxury beachside residences, apartment buildings and luxury resort hotels such as Atlantis – increased Dubai's shoreline by a total of 520 km. In 2008 Dubai International Airport expanded with the opening of Terminal 3, the world's single largest terminal building. Built for the exclusive use of Dubai's flagship carrier Emirates, Terminal 3 boosted the airport's capacity to 60 million passengers.

In 2013, Dubai's second airport at al Maktoum International Airport at World Trade Central opened and when all phases are complete, it will have capacity to handle 160 million passengers a year (DTCM, 2014a). In 2013, Dubai's airport surpassed Hong Kong's in traffic and became the world's third busiest after London and Paris (Cogswell, 2014). Passenger numbers at Dubai International have doubled from 24.8 million in 2005 to 47.2 million in 2010. The airport ranks fourth globally for international passenger and cargo traffic. Emirates' passenger numbers have increased sixfold over the course of a decade, making it the largest airline in the world in terms of international revenue passenger kilometres and flydubai is the world's fastest growing start-up low-cost carrier airline. By 2020 some 98.5 million passengers and

Fig. 8.3. Dubai's development seen from Burj Khalifa. (Photograph taken by Filareti Kotsi, 2013.)

4 million air freight will pass through Dubai airports. In fact, Dubai's airport has evolved from a small airstrip mainly serving as a re-fueling stop for a few airlines in 1960 to a global gateway for 150 airlines that is ranked among the world's top hubs (Dubai Airports Strategic Plan 2020, 2014).

The economy of Dubai has witnessed tremendous structural changes in the last three decades. Dubai's formula for development, such as visionary leadership, high quality infrastructure, an expatriate-friendly environment and zero tax on personal and corporate income, has turned Dubai into a business and tourism hub (DTCM, 2014c). Setting up the state-of-the-art infrastructure, as well as efficient financial markets and encouraging tax regimes, has shaped its good reputation in encouraging national and foreign capitals to enter successful investment ventures in different commercial, industrial and service fields. Dubai, aiming at the diversification of the sources of income, not only established industrial projects in the Jebel Ali Free Zone, but also boosted the private sector to set up manufacturing industries. It has also taken the lead among the other emirates in the volume of non-oil exports that constitute 78.4% of the country's total exports, as well as importing more than two-thirds of the UAE requirements of all kinds of consumable goods (Shoult, 2006).

In fact, no longer depending on oil, the internationalization of the economy through tourism has been a major element in the diversification process. Dubai's Strategic Plan 2015 leads the drive towards diversification; the strategy is to attract foreign capital into the industrial and other sectors including transport, petrochemicals, tourism, information technology, telecommunications, renewable energy, aviation and oil and gas services (UAE Interact, 2014). According to Hamad Buamin, Director General, Dubai Chamber of Commerce and Industry, the economic diversification helped UAE to survive the world economic crisis. During the crisis, finance was directly affected and construction and real estate sectors were impacted; despite the crisis, tourism remained strong throughout and the hospitality and tourism sectors performed well (Bitar, 2013).

HISTORY AND THE DEVELOPMENT OF TOURISM IN DUBAI

This section covers the history and development of tourism in Dubai from the birth of the UAE until 2014. A number of factors have made Dubai's transformation possible and in particular: (i) long exposure with foreign visitors through the city's trading heritage; and (ii) the vision of the ruling Al Maktoum family in recognizing the potential of the emirate's natural tourist assets, including strategic geographical location, winter sunshine, extensive beaches and desert scenery. Investment of both public and private sector, have also collaborated to ensure the creation of a first-class infrastructure, encompassing accommodation, aviation and recreation facilities. Tourism has therefore emerged as one of UAE's most dynamic growth sectors and it is evident in each of the seven emirates, but the industry in Dubai has become a major component in the city's economy (MacDonald, 2000b). Dubai has actually experienced a rapid tourism growth since the establishment of the nation. From having no hotels in the 1950s, by the late 1990s Dubai managed to enter the international high-end tourist market by investing in some of the world's most luxurious hotel projects. By 2000, 3.4 million tourists were coming to Dubai each year (Davidson, 2008). During the past 8 years, the increased number of visitors is strong evidence that Dubai has earned a reputation as a global tourism destination, since it doubled the number of tourists from 5 million yearly to 10 million tourists a year (*Khaleej Times*, 2013a).

Dubai was a fishing and pearl-diving village and through visionary leadership, it used its strategic location to grow into a trading centre, and has been transformed into an economic beacon for the Middle East (Balakrishnan, 2008). Balakrishnan has been a leading contributor in academic literature and scientific research within the MENA region and her work has also been specific to the Emirate of Dubai. Her paper 'Dubai – a star in the East' developed a destination branding framework and explains the importance of strategic vision and its competitive advantage. The chance discovery of oil in 1966 financed Sheikh Rashid Bin Saeed Al Maktoum's vision and gradually Dubai became the eighth most visited city in the world (Forbes Business, 2012).

Sharpley (2008) in his paper titled 'Planning for tourism: the case of Dubai', created a timeline from 1996 to 2010 and linked it to the growth of Dubai as a destination and policy creation during the 14-year period. This time period is broken down into two phases 1996–2000 and 2000–2010. The first phase saw the strategic creation of the Department of Tourism and Commerce Marketing (DTCM) established as the principal authority for the planning, supervision and development of tourism in the emirate. A number of organizations were active in promoting Dubai; however, it was not until the creation of DTCM that international marketing activities were coordinated through a single body (MacDonald, 2000b). Established in 1997, replacing the Dubai Commerce and Tourism Promotion Board that had been in operation since 1989, DTCM has played a pivotal role in Dubai's rise to prominence as one of the world's leading tourism destinations, welcoming more than 11 million visitors in 2013. DTCM markets and promotes the emirate's commerce sector and is responsible for the licensing and classification of all tourism services, including hotels, tour operators and travel agents. In addition to its headquarters in Dubai, DTCM operates 20 offices worldwide (DTCM, 2014b).

The first phase (1996–2000) saw the emergence of external consultants, pro-tourism policies, creation of facilities and attractions like the Dubai Shopping Festival. The second crucial phase between 2000 and 2010 experienced a sharpening and increase in focus on certain activities from Phase 1, like facilities and attractions, and enhanced activities and responsibilities for DTCM, devising more pro-tourism policies and introducing new activities and strategies such as creating structures and controls. Moreover, several incentives have been provided to draw people to the emirate such as the Dubai Shopping Festival, Dubai Summer Surprises, Dubai International Film Festival, Dubai International Jazz Festival, Arabian Travel Market, Dubai Air Show, Dubai World Cup (with a total prize fund of US$27.25 million; *Gulf News*, 2014), Dubai Desert Classic PGA Golf Tournament, Dubai Open ATP Tennis Tournament and many others.

A few researchers have focused on the social impacts of tourism development, for example Stephenson's and Ali-Knight's (2010) paper 'Dubai's tourism industry and its societal impact: social implications and sustainable challenges'. Ryan and Ninov (2011) examined the images held of Dubai by tourists in the area of Dubai Creek, while Ryan and Stewart (2009), in their paper entitled 'Eco-tourism and luxury – the Case of Al Maha, Dubai' examined the luxury site and its ecotourism status through its regeneration programme. Recently, Stephenson (2014) in his paper 'Tourism, development and "Destination Dubai": cultural dilemmas and future challenges' provided an assessment of Dubai's approach to tourism development, focusing again on its cultural implications. He emphasized the paucity of heritage resources within the tourism market as well as limited public knowledge of ethnic elements of the indigenous society.

However, a tourism body, The National Council of Tourism and Antiquities, was set up in 2009; its logo is 'seven emirates one destination'. The council is responsible for the creation of a federal level structure to regulate tourism departments in individual emirates and ensure uniformity and the successful implementation of a cohesive strategy to boost tourism throughout the UAE. This strategy is implemented by the consolidation of efforts between the tourism authorities in the UAE to identify the country as a leading tourism hub and coordinate the efforts to preserve and maintain the country's antiquities through unified external representation and co-operation with international organizations, governmental and non-governmental entities. In its endeavor to promote and showcase all seven emirates, it demonstrates its commitment in promoting the country's culture and heritage through the establishment of heritage villages and exhibitions in all tourist festivals in and outside the country, with a view to reviving the ancestors' heritage, promoting and improving it (UAE The National Council of Tourism and Antiquities, 2014).

Following Sharpley's (2008) timeline in relation to Dubai's growth as a tourism destination, a third phase ranging between 2010 and 2020 can be added. The emirate's development is planned following Dubai's 'Vision for Tourism'. Developed by DTCM under guidance from Sheikh Mohammad bin Rashid Al Maktoum, Vice President and Prime Minister of the UAE and Ruler of Dubai, Dubai's 'Vision for Tourism' aims to double its annual visitor numbers from 10 million in 2012 to 20 million in 2020 and the annual contribution made by tourism to the city's economy will triple. DTCM has set a number of objectives in order to achieve this target. First, Dubai wants to maintain the existing market share of the outbound tourism of all source markets. Second, it wants to increase awareness in a number of source markets such as Latin America, China and emerging economies of Africa. Third, DTCM wants to increase the number of repeat visits and length of stay, currently at 3.76 days (*Khaleej Times*, 2013a, b).

Sheikh Mohammed affirms that the UAE has succeeded in placing itself among popular tourist destinations on the global map of tourism, through launching ambitious developmental initiatives and high quality projects. He bears in mind that tourism is a broad service sector and that many government departments share this responsibility and join in developing it. The strategy behind the Dubai Tourism Vision for 2020 is based upon two central tenets: the first being to widen the range of tourism offerings across events, attractions, infrastructure services and packages and to enhance the overall visitor experience from arrival to departure. The second tenet is to adapt the marketing approach in order to showcase Dubai to a wider audience and grow the conversion rate of awareness to bookings (*Khaleej Times*, 2013a).

Moreover, DTCM in line with Sheikh Mohammed's statement that the UAE should be the world's leading family destination, will play a central role in coordinating the activities of industry and public sector partners to expand the existing attractions, events and experiences that cater to families. Dubai as an event destination is a second area of principal focus, with the aim of moving from a regional events hub to a global events and entertainment destination. Both the Dubai Events and Promotions Establishment – organizer of festivals such as Dubai Summer Surprises and Dubai Shopping Festival – and Dubai Calendar, were recently integrated into DTCM. The departments also worked closely with the Expo 2020 team to support and promote Dubai and win the bid to bring the World Expo to the MENASA (Middle East, North Africa and South Asia) region for the first time. The offer to business visitors is a third principal focus, with the strategy setting out plans for enhancing Dubai's status as a business destination. Already the Meetings, Incentives, Conferences and Exhibitions (MICE) capital of the region, DTCM will work with the industry in improving the business tourism offer to make sure that Dubai is the easiest and most effective place to do business and to adopt a more unified approach to business and leisure tourism (*Khaleej Times*, 2013a).

Dubai has a sense of unlimited possibility and a belief in the power to do anything imaginable. It now plans to build an enclosed, climate-controlled city featuring the world's largest mall, 'The Mall of the World' (Cogswell, 2014). It will take 10 years to complete and will occupy 4.46 million m^2 (48 million square feet), which will be covered by a glass dome. It will provide 100 hotels, it will house the largest indoor family theme park in the world, it will feature a 7 km promenade with retail shops, and a wellness and healthcare district (Maceda, 2014). The Mall of the World is the latest of the UAE's many drawing points to shift the world's centres of economics and culture to the east (Cogswell, 2014).

TOURISM GOVERNANCE IN THE UAE AND DUBAI

The concept of governance

The role of government in tourism, and the influence of state policy on tourism development have long been of interest to scholars (Jenkins, 1980; Jenkins and Henry, 1982; Hall, 1994; Hall and Jenkins, 1995; Bramwell and Lane, 2000; all cited in Hall, 2011). However, since the 1990s, there has been a gradual shift in approach in the tourism policy literature from the notion of government to that of governance (Greenwood, 1993; Hall, 1999; Yüksel *et al.*, 2005; Beaumont and Dredge, 2010; all cited in Hall, 2011).

According to Yee (2004), there is no single accepted definition of governance; the author provided a basic definition of governance as 'new governing activities that do not occur solely through governments'. Heritier (2002) on the other hand focuses on new modes of governance that include private actors in policy formulation and/or while being based on public actors are only marginally based on regulative powers or not based on regulation at all. The second broad meaning of governance is that it is used to denote a conceptual and theoretical representation of the role of the state in the coordination of socio-economic systems (Hall, 2011).

Treib *et al.* (2007) state that the development of an appropriate typology of governance is not based on a 'haphazard shopping list' that is compiled on all possible dimensions noted in the policy and planning literature that characterize decision-making processes, outputs and outcomes. Instead, the criterion for the frameworks of governance should be systematically based on defining what dimensions are included and which are left out. The most common focus of most contributions to the governance debate in public policy terms is 'the role of the state in society' (Pierre, 2000). Therefore, the core concept in governance in public policy terms is the relationship between state intervention (also referred to as public authority) and societal autonomy (self-regulation).

Governance and the UAE

The tourism board of Abu Dhabi, the Abu Dhabi Tourism & Culture Authority (TCA) is a very good example of private actors' involvement, within the tourism sector. TCA's first-generation stakeholder programme 'Partners in Progress' was introduced in 2009, during the authority's fifth anniversary year and marked the 'coming of age' of the industry as a whole. The overall programme represents the 'One Destination' approach the authority is adopting to build on the solid destination foundations already laid and to ensure the development of a sustainable tourism industry. TCA Abu Dhabi's Industry Development Committees (IDCs) are a key component of the authority's ever-expanding 'Partners in Progress' stakeholder programme. Six IDCs currently meet with TCA Abu Dhabi's Strategy team each month to identify sector challenges and opportunities, and to jointly forge solutions to stimulate progress and they include: (i) Hospitality; (ii) Aviation; (iii) Attractions; (iv) Tour Operators; (v) Business Tourism; and (vi) Retail (TCA, 2014a). Recently, TCA Abu Dhabi has organized a tourism education workshop under the title 'Identifying Community-based Intangible Cultural Heritage'. This workshop targets heritage enthusiasts, particularly teachers and researchers at the Ministry of Culture, Youth and Community Development, heritage departments, local institutions, individuals, non-governmental organizations (NGOs) and others who are actively involved in preserving intangible cultural heritage. The workshop is being supervised by Dr Nasser Ali Al Humairi, intangible heritage director and United Nations Educational, Scientific and Cultural Organization (UNESCO) international files coordinator at the TCA Abu Dhabi (*Khaleej Times*, 2014).

The Dubai Conventions and Events Bureau (DCEB), which is part of DTCM has introduced in 2011 the Tourism Ambassadorship Programme called 'Al Safeer'. This programme is a very good example of how DCEB has created and maintains a relationship between the various private actors (stakeholders) and the tourism sector with DCEB being the catalyst (DCEB, 2014). So herein are a few examples of how the emirates of Dubai and Abu Dhabi involve a range of private actors to create networks and relationships between various actors and stakeholders within the tourism and hospitality sector of the nation.

According to Sharpley (2008), the foundations of tourism in Dubai were laid out in the early 1980s when Sheikh Mohammed bin Rashid Al Maktoum, the former Crown Prince and now the present ruler of Dubai, set out a strategic vision for the emirate. Today, His Highness Sheikh Mohammed bin Rashid Al Maktoum, UAE Vice-President, and Prime Minister and Ruler of Dubai continues his role of providing strong leadership and a strategic vision for the city. Sheikh Mohammed plays a critical role in shaping the destiny of the tourism industry in Dubai.

The Emirate of Dubai released the highlights of the city's 'Strategic Plan (2015)', which is directed by the overall vision of Dubai's leadership, and has focused on the development of Dubai Strategic Plan along five sectors namely: (i) Economic Development; (ii) Social Development; (iii) Infrastructure, Land and Environment; (iv) Security, Justice and Safety; and (v) Government Excellence (Kotsi and Michael, 2014). According to the Dubai Strategic Plan 2015 (2014), the city's current GDP (gross domestic product) is very favourable since its strongest sectors by international standards happen to be highly conducive to future global growth. And these sectors include tourism, transportation, construction and financial services. Hence one can see a very close and strong link between the five sectors in the Dubai Strategic Plan 2015 and the tourism and transportation industry.

Kotsi and Michael's (2014) investigation further found that the government of Sharjah played a very proactive and positive role in strategy and policy formulation. In a recent press interview, Mr Mohammed Ali Al Noman, the Chairman of Sharjah Commerce and Tourism Development Authority stated that in accordance with the 'Vision' of the emirate's leadership, Sharjah's tourism policy has been consistently inspired and guided by the Emirati and Arab cultural heritage and legacy, and this was to remain the USP (unique selling proposition) and a distinguishing feature of the state.

Henderson (2006a) investigated tourism in Dubai and the barriers it had to overcome for the city to develop as a destination. Her focus for the investigation that formed her framework in analysing Dubai was on three variables, namely: (i) stability and government policy; (ii) accessibility and amenities; and (iii) attractions and promotion. Her research paper was mostly based on secondary data such as growth of the tourism industry, the various industrial sectors such as oil, trade, transportation, and logistics and how these contributed to growth and stability of Dubai and the UAE. Further, Dubai's tourism strategy and its promotion has changed since Henderson's 2006 paper was published, for example Henderson (2006b) states, 'much of Dubai's commercial interest has been in the luxury segment and the striking Burj Al Arab beachside hotel'. However, in 2009, Emirates Airlines and DTCM joined hands to create an advertising campaign called 'Fly Emirates, Meet Dubai' which showcased an expanded product 'Dubai' from not only being an exclusive destination for the 'rich and famous', but a destination that catered for a much broader market segment. Images of taking an inexpensive crossing of the Dubai Creek using an 'Abra' (boat), shopping at the gold and spice souks, and having a meal at a reasonably priced Pakistani restaurant (Ravi's) were all featured in this campaign. In 2010, the British Broadcasting Corporation (BBC) interviewed Michael, one of the authors of this chapter, about this campaign and he stated that a destination is much bigger and broader than a product or service, hence, there was a need to communicate Dubai's image to a much broader audience. In the interview he supported this new campaign, and the direction that the city was taking (BBC, 2010a). Since then, Dubai has seen a growth of three- and four-star hotels like Premier Inn, Holiday Inn Express, Citymax, Arabian Park and easyHotel.

The past few years witnessed a slow yet steady growth of research (especially secondary) on Dubai as a city destination and to a lesser extent, a mention at the most, of UAE as a nation. According to Nair (2014), Dubai has always been the 'go-to' destination for global brands looking to make a name in the Middle East marketplace. But, now, brands are seeing a Dubai exposure as not just confined to an entry point – in fact, the city is being raised to the status of a 'launch city', and on par with the status enjoyed historically by the likes of New York, London and Paris. As part of Dubai's Strategic Plan 2015, the government has highlighted Social Development and Security, Justice and Safety as two of the five sectors that have been identified by the Executive Council as part of the city's overall vision. This clearly proves Pierre (2000) and Treib's *et al.* (2007) suggestions, that governance should not be based on a 'haphazard shopping list' and should include a state society relationship.

Henderson (2006a) points out that stability in economic, social and political domains is essential for tourism as severe disturbance and volatility will deter many tourists, investors, and the industry of tour operators and travel agents in generating countries. The UAE and especially Dubai is seen as comparatively liberal and a cosmopolitan society, with little threat of civil unrest and low crime rates, with expatriates making up about 80% of the population. The creation of Free Trade Zones such as: Khalifa Industrial Zone (Abu Dhabi), Jebel Ali Free Zone – JAFZA (Dubai), Hamriyah Free Zone (Sharjah), Ras al-Khaimah Free Zone (Ras al-Khaimah), Ajman Free Zone (Ajman) and Fujairah Free Zone (Fujairah) are good examples of societal and commercial autonomy in the UAE.

The principles of sustainability and governance in Dubai and the UAE

The concept of sustainable tourism remains a fledgling industry in Dubai, and the Dubai government is beginning to take a closer look at its advantages. DTCM launched a 'Green Tourism Award' programme recently to encourage innovations in ecotourism (Government of Dubai, 2014). This initiative includes efforts to reduce the carbon footprint of the tourism sector's sprawling developments, and supporting activities that encourage a celebration of the UAE's natural assets. Some of the eco-friendly outdoor tourism activities and attractions that the various tourism bodies have created include hiking and camping in the Hajar Mountains, scuba and snorkelling outings on the UAE's eastern coast, and birdwatching adventures. With over 400 bird species in the UAE, there is scope for extensive birdwatching opportunities in places as diverse as Mushrif Park, Emirates Golf Club, Jebel Ali and Hatta, all located in the Emirate of Dubai (Frommers, 2011). The Emirate of Abu Dhabi government is also positioning the emirate as an innovator in renewable energy technology, launching the first ever World Future Energy Summit in 2009. This event is now being held every year during the month of January, drawing an international crowd and features exhibits on clean energy, and has witnessed a line-up of impressive speakers. In June 2009, Abu Dhabi was designated as the interim seat of the International Renewable Energy Agency (IRENA), and then at the first IRENA Assembly in April 2011, it was announced that Abu Dhabi would become the permanent seat of IRENA (International Renewable Energy Agency, 2014).

The National Council for Tourism and Antiquities, which was given the responsibility to actively represent the country at global tourism events, states on its website that the country's sustainable development strategy was initiated by HH Sheikh Mohammed Bin Rashid Al Maktoum, UAE Vice President, Prime Minster and Ruler of Dubai, and themed as 'Green Economy for Sustainable Development'. Through this initiative, the UAE aims to become one of the world leaders as well as a centre for the export and re-export of green products and technologies. The initiative consists of a number of fields. These include Green Energy, Investment Incentives, Green City, Climate Change Effects and Green Lifestyle, which form a vehicle that enables the country to realize its objective (UAE The National Council of Tourism and Antiquities, 2014).

In 2006, Masdar City was created as an emerging global hub for renewable energy clean technologies that positions companies in a place where businesses can thrive and innovation can flourish (Masdar City, 2011). The creation of Masdar is a very good example of how an environmentally sustainable initiative, in this case the creation of a unique city, is incorporated with the strategic vision of the Abu Dhabi government. A second case is now discussed, in this instance how the Abu Dhabi government through its tourism body got the city of Al Ain in the UNESCO World Heritage listing. The TCA's website clearly states that tourism is a vital component of economic diversification in the Abu Dhabi government's 2030 Plan. It further goes on to state that sustainable tourism development is integral to the emirate's commitment in building a sustainable future. Accordingly, TCA Abu Dhabi works towards the sustainable tourism development of the emirate; this has involved the government of Abu Dhabi and TCA being involved in the authority's annual sustainability report, creating and growing green tourism and hotel initiatives and getting involved in environment exhibitions and conferences (TCA, 2014b).

As part of its heritage and cultural tourism strategy, the government of Abu Dhabi submitted an application to UNESCO in 2009 to list the city of Al Ain as a UNESCO World Heritage Site (UNESCO, 2011). In 2011, UNESCO included the cultural sites of Al Ain namely: Al Hafit, Hili, Bidaa Bint Saud and Oases Areas as World Heritage Sites protected by the organization. According to Dr Sami El Masri, Deputy Director, Abu Dhabi Authority for Culture and Heritage, UNESCO's recognition of Al Ain was a testimony 'of an important culture and identity that has left its mark on the history of the world civilization and human contacts in the past and present day' (Ghazal, 2011). The BBC covered the story of Abu Dhabi's application in June 2010, where Michael, the co-author of this chapter, was interviewed (BBC, 2010b). During the interview, he mentioned that other places (destinations), in this instance, and for that matter, the entire UAE nation, could benefit from Al Ain's UNESCO listing. The examples of Masdar City and the UNESCO heritage classification of Al Ain are two good examples of how the government of Abu Dhabi has been proactive and been successful in creating a sustainable product, attraction and project for the UAE nation.

Based on this (the Abu Dhabi and Al Ain) experience, The National Council for Tourism and Antiquities's initiatives in recognizing HH Sheikh Mohammed's vision of a 'Green Economy for Sustainable Development' will spread to the other six emirates of the nation. For example, Dubai has the scope to enhance its extensive birdwatching opportunities in places as diverse as Mushrif Park, Emirates Golf Club, Jebel Ali and Hatta – these are all located in the emirate.

Since its inception, one of the strategies of the DTCM was to focus on the conservation of natural resources and sensible use of energy, water and gas, etc. as their excessive use causes degradation in the environment by disturbing the eco-balance in the atmosphere. DTCM began this in 2009 by officially launching the first-ever initiative to reduce carbon emissions in the hospitality industry by 20% by the year 2011, and further the Dubai Green Tourism Award was launched in line with the World Environment Day. These initiatives were in line with the vision and directives of His Highness Sheikh Mohammed bin Rashid Al Maktoum, to turn Dubai into a green and environmentally responsible community. As part of the Dubai Strategy Plan 2015, environment conservation and protection has been emphasized (Government of Dubai, 2013).

The Dubai Electricity and Water Authority (DEWA), formed in 1992, is an example of an organization's (community) involvement to help in environmental sustainability. DEWA has recently launched a campaign to reach out to commercial and industrial customers on the issue of power factor (PF) aimed at reducing the carbon footprint. DEWA comes up with interesting initiatives and information on its website such as: (i) Eco Appliances; (ii) Green Tips; and (iii) Peak Load Tips. The company also launched a series of specialized 'Your Decision' campaigns using efficient mass communication methods. This programme educated users to be careful when utilizing scarce energy resources, and it demonstrates DEWA's sustained efforts and its strong engagement in adopting creative ways to communicate conservation concerns with its varied stakeholders (Dubai Electricity and Water Authority, 2014).

Linking this to the case study of Expo 2020, it may be concluded that the city and the government of Dubai has precedence in terms of learning from the cases of Masdar and Al Ain, whereby the city and emirate can focus and achieve its objectives of being environmentally friendly and sustainable at Expo 2020. The emirate and the nation at large can also showcase the heritage and culture of its land and people during the event. Of course one has to be aware of maintaining a balance between a booming economy and environmental, socio-cultural and heritage sustainability.

Case Study: Dubai's Expo 2020

In recent years, sustainable tourism has become an increasing political leitmotiv for tourist destinations. Sustainable tourism principles are becoming progressively further integrated into strategic destination planning, since tourism activities have caused various environmental and socio-economic problems (Bramwell and Lane, 1993). The *National Geographic*'s Center for Sustainable Destination, states that sustainable tourism protects its product – the destination – it conserves resources and finally environmentally aware travellers patronize businesses that reduce pollution, waste, energy consumption, water usage, landscaping chemicals and excessive night-time lighting (*National Geographic*, 2013). The concept of sustainability is a key priority set by the Dubai government to achieve aggressive targets for its World Expo 2020. Some of the targets that are being focused on include generation of 50% of the energy on site from renewable sources in order to deliver a carbon-neutral Expo.

A world's fair, world exposition or universal exposition (Expo) is a large public exhibition and these exhibitions vary in character and are held in varying parts of the world, with Expo 2015 held in Milan, Italy. The crucial ballot to determine the winning bid to host Expo 2020 took place at the Organisation for Economic Co-operation and Development (OECD) centre after a series of 20-min presentations by the bidding candidate cities. The UAE's Minister of State and Managing Director of the Bid Committee, Reem Al Hashemi, highlighted the determination to deliver an inclusive Expo based on the principles of sustainability (Expo 2020 Dubai, United Arab Emirates, 2014a).

(Continued)

Case Study. Continued.

The theme of Dubai Expo 2020 'Connecting Minds, Creating the Future' was created to emphasize the importance of partnerships and innovation as the world's best hope to build a sustainable world today and in the future. Dubai's World Expo bid recognized the central importance of Mobility, Sustainability and Opportunity to the development of all countries and the bid committee included these three as sub-themes for the World Expo 2020. Accordingly, these three sub-themes would provide a unique platform for the global community to come together and explore creative and pioneering solutions to 'Mobility', 'Sustainability' and 'Opportunity' which were defined as lasting sources of energy and water, smart systems of logistics and transportation and new parts to economic development, all three identified as key drivers of global development (Expo 2020 Dubai, United Arab Emirates, 2014a).

An estimated total funding of €6.5 bn has been earmarked for this event with €5.2 bn anticipated as capital investment for the Expo's infrastructure environment. Dubai's economy is set to receive an added value boost of roughly €17.7 bn. More exciting, however, will be the innumerable opportunities that will be created boosting the growth of tourism and logistics in the UAE. An estimated 25 million visitors are set to arrive in the country, thus influencing tourism, trade, infrastructure and investment. Sultan Bin Saeed Al Mansouri, Minister of Economy, however, feels that this number is conservative and by the time people transfer the merits of Dubai by word of mouth the total could swell to 100 million. A total of 270,000 jobs will also be conceived to cater to an expanded economy (Chatterjee, 2013).

The timing of Dubai's Expo 2020 Theme Symposium was symbolic. It was during this week that HH Sheikh Mohammed bin Rashid Al Maktoum, Vice President and Prime Minister of UAE, and Ruler of Dubai, inaugurated a new solar power plant and which in turn coincided with the publication of the UAE's State of Energy Report. This State of Energy Report highlighted the nation's commitment to sustainability including a target to produce 50% of Expo energy requirements on site. The Dubai Supreme Council of Energy, in partnership with the United Nations Development Programme and the Dubai Carbon Centre, commissioned the State of Energy Report for Excellence (Dubai Carbon, 2014).

The sustainability aspect of the new booming Dubai is a cause for worry for various environmentalist groups. In a recent global analysis of ecological resource use, the UAE came top as the least environmentally friendly country owing to its heavy reliance on air-conditioning and desalinated water just to survive (in the summer it can be more than 40°C outside). According to the Royal Geographic Society, each person residing in Dubai uses twice the resources of the average Briton (Royal Geographic Society, 2014). In response to this, Mr Helal Al Marri, Director General of DTCM, CEO of Dubai World Trade Centre and a member of the Higher Committee for Dubai Expo 2020, stated that 'sustainability has informed every aspect of Dubai Expo 2020 from the overarching philosophy to the physical design of the site and infrastructure' (Expo 2020 Dubai, United Arab Emirates, 2014c).

(*Continued*)

Case Study. Continued.

In summary it is possible to speculate on the success of the brand image of the city of Dubai in securing the successful Expo 2020 bid and see *Ali's Film* that showed a young Emirati schoolboy Ali talking about Dubai's iconic landmarks as an important ingredient in sealing the bid for Dubai Expo 2020 (Expo 2020 Dubai, United Arab Emirates, 2014b). What needs to be determined in evaluating the success of Expo 2020 is how the city of Dubai can be sustainable in terms of its natural environment, heritage and culture, to enable it to be successful in future bids for other major events. This needs to be showcased to the world at Expo 2020.

CONCLUSION

The intention of this chapter was to highlight how a city like Dubai is creating 'waves' as a new and modern destination. Furthermore, aspects on governance and policy that can lead to a sustainable destination were discussed. Although the focus has largely been on the city (Emirate) of Dubai, at certain places Dubai's neighbouring emirates such as Abu Dhabi and Sharjah were mentioned alongside the city of Al Ain (that falls within the Emirate of Abu Dhabi) since these cities provide some interesting examples for the topic. For example, with Dubai being the selected city of Expo 2020, Abu Dhabi with its Masdar City and attractions in Al Ain now listed as UNESCO World Heritage Sites, these cities and the nation in general have become interesting cases of how these modern cities and a young nation can work hard and clever to create a sustainable and futuristic city and nation brand. The chapter also discussed the role of leadership and government policy in helping to develop a city and nation destination brand that can host major events of the world, sustain culture and the environment, and also become a modern city of the future. Many examples of government and the role of other actors such as industry partners have been cited to demonstrate how these relationships can foster a win-win situation for the tourism industry.

QUESTIONS

1. The last decade or so has seen the private sector in many cities and countries of the world playing a much more proactive role, be it in aviation and transportation, hospitality and tourism, roads and infrastructure, just to name a few. Has this changed and diminished the role of governments in governance and policy creation? Discuss this statement using the city of Dubai as an example.

2. What lessons can other (newer and upcoming) city destinations learn from DTCM in Dubai and the other tourism bodies in other emirates? Further, discuss the partnerships that government and the private sector could have to enhance a destination's brand image.

3. Do you agree with tourism researchers and writers when they say that government and industry bodies pay only 'lip service' to issues concerning the sustainability of tourism? Discuss using the UAE and the city of Dubai as background examples.

BIBLIOGRAPHY

Abu Dhabi Tourism & Culture Authority (TCA) (2014a) Stakeholder Relations. Available at: http://tcaabudhabi.ae/en/about/Pages/stake-holder-relations.aspx (accessed 28 June 2014).

Abu Dhabi Tourism & Culture Authority (TCA) (2014b) Sustainable Tourism Overview. Available at: http://tcaabudhabi.ae/en/tourism/sustainability/Pages/sustainable-tourismoverview.aspx (accessed 28 June 2014).

Balakrishnan, M. (2008) Dubai – a star in the east. A case study in strategic destination branding. *Journal of Place Management and Development* 1(1), 62–91.

Balakrishnan, M. (2013) Special issue on the MENA region. *Academy of International Business Insights* 13(2), 3.

Bitar, Z. (2013) Economic diversification helped UAE cope with global crisis. *Gulf News*, 9 September. Available at: http://gulfnews.com (accessed 15 June 2014).

Bramwell, B. and Lane, B. (1993) Sustainable tourism: an evolving global approach. *Journal of Sustainable Tourism* 1(1), 1–5.

Bramwell, B. and Lane, B. (eds) (2000) *Tourism Collaboration and Partnerships: Politics, Practice and Sustainability*. Channel View Publications, Clevedon, UK.

British Broadcasting Corporation (BBC) (2010a) Abu Dhabi Eyes World Heritage Boost. Available at: www.bbc.co.uk/news/10361685 (accessed 29 June 2014).

British Broadcasting Corporation (BBC) (2010b) Dubai Shows Different Face of Emirates in Advertising. Available at: http://news.bbc.co.uk/2/hi/business/8448768.stm (accessed 28 June 2014).

Chatterjee, R. (2013) Dubai to host World Expo 2020. *Gulf News*, 27 November. Available at: http://gulfnews.com (accessed 15 June 2014).

Chronicles of the Middle-East and North Africa (2014a) *United Arab Emirates*. Available at: http://fanack.com/en/countries/uae/basic-facts/geography-and-climate/natural-resources/ (accessed 30 June 2014).

Chronicles of the Middle-East and North Africa (2014b) *United Arab Emirates*. Available at: http://fanack.com/en/countries/uae/population/ (accessed 30 June 2014).

Cogswell, D. (2014) Can Dubai Move the Center of the Earth? Available at: http://www.travelpulse.com/news/destinations (accessed 15 June 2014).

Daher, R.F. (2007) *Tourism in the Middle East: Continuity, Change and Transformation*. Channel View Publications, Clevedon, UK.

Davidson, C. (2008) *Dubai. The Vulnerability of Success*. Hurst, London.

Department of Tourism and Commerce Marketing (DTCM) (2014a) *History: Tracing the Transformation from Fishing Village to a Global Metropolis*. DTCM, Dubai, UAE. Available at: www.dubaitourism.ae (accessed 28 June 2014).

Department of Tourism and Commerce Marketing (DTCM) (2014b) *Dubai's Department of Tourism and Commerce Marketing (DTCM)*. DTCM, Dubai, UAE. Available at: www.dubaitourism.ae (accessed 30 June 2014).

Department of Tourism and Commerce Marketing (DTCM) (2014c) *About Dubai*. DTCM, Dubai, UAE. Available at: http://business.definitelydubai.com/setting-up-a-business/about-dubai (accessed 28 June 2014).

Dew, P. and Shoult, A. (2000) *Doing Business with the United Arab Emirates*. Kogan Page, London.

Dubai Airports Strategic Plan 2020 (2014) *Connecting the World Today and Tomorrow*. Available at: www.dubaiairport.com (accessed 5 April 2014).

Dubai Carbon (2014) State of Energy Report. Available at: www.dcce.ae/energyreport (accessed 26 June 2014).

Dubai Conventions and Events Bureau (DCEB) (2014) *Al Safeer Program 2014*. Available at: www.dcb.ae/alsafeerpro.php?tab=2 (accessed 28 June 2014).

Dubai Electricity and Water Authority (2014) *Your Decision Campaign*. Available at: http://www.dewa.gov.ae/community/conservation/ydc/your_decision.aspx (accessed 28 June 2014).

Dubai Strategic Plan 2015 (2014) Available at: www.sclgme.org/Dubai%20Strategic%20Plan%20-%20 English.pdf (accessed 5 April 2014).

Expo 2020 Dubai, United Arab Emirates (2014a) Available at: http://expo2020dubai.ae/en/theme (accessed 26 July 2014).

Expo 2020 Dubai, United Arab Emirates (2014b) Available at: http://expo2020dubai.ae/en/multimedia/video/alis_film (accessed 26 July 2014).

Expo 2020 Dubai, United Arab Emirates (2014c) Available at: http://expo2020dubai.ae/en/hello_2020/article/heritage_meets_innovation_a_space_for_inspiration (accessed 26 July 2014).

Forbes Business (2012) Most Visited Cities in the World. Available at: www.forbes.com/2012 (accessed 28 June 2014).

Frommers (2011) Available at: www.frommers.com/destination/dubai (accessed 10 July 2014).

Ghazal, R. (2011) Al Ain wins World Heritage List honours. *The National*, 27 June. Available at: http://thenational.ae (accessed 28 June 2014).

Government of Dubai (2013) Available at: www.dubaitourism.ae/about/projects-initiatives (accessed 29 June 2014).

Government of Dubai (2014) Available at: www.green.dubaitourism.ae (accessed 28 June 2014).

Gulf News (2014) The winning horses at Dubai World Cup, 2014. *Gulf News*, 29 March. Available at: http://gulfnews.com (accessed 15 June 2014).

Hall, M. (2011) A typology of governance and its implications for tourism policy analysis. *Journal of Sustainable Tourism* 19(4–5), 437–457.

Hellyer, P. (2013) The UAE's expatriate mix reflects a much wider world. *The National*, 24 July. Available at: http://thenational.ae (accessed 15 June 2014).

Henderson, J.C. (2006a) Tourism in Dubai: overcoming barriers to destination development. *International Journal of Tourism Research* 8, 87–99.

Henderson, J.C. (2006b) Destination development: Singapore and Dubai compared. *Journal of Travel & Tourism Marketing* 20(3), 33–45.

Heritier, A. (2002) New modes of governance in Europe: policy-making without legislating? In: Heritier, A. (ed.) *Common Goods. Reinventing European and International Governance*. Rowman & Littlefield, Oxford, pp. 186–206.

International Renewable Energy Agency (2014) About IRENA: Future Headquarters of IRENA. Available at: www.irena.org/menu/index (accessed 28 June 2014).

Khaleej Times (2006) 202 Nationalities in labour market. *Khaleej Times*, 25 August. Available at: http://www.khaleejtimes.com (accessed 15 June 2014).

Khaleej Times (2013a) Shaikh Mohammed approves Dubai 2020 vision. *Khaleej Times*, 4 May. Available at: http://www.khaleejtimes.com (accessed 15 June 2014).

Khaleej Times (2013b) Mohammed approves new DM projects. *Khaleej Times*, 2 July. Available at: http://www.khaleejtimes.com (accessed 20 May 2014).

Khaleej Times (2014) Intangible heritage. TCA workshop from tomorrow. *Khaleej Times*, 23 August. Available at: http://www.khaleejtimes.com (accessed 20 May 2014).

Kotsi, F. and Michael, I. (2014) The role of 'vision' in creating city brands as tourist destinations. The case of Abu Dhabi, Dubai and Sharjah, U.A.E. Paper presented at the International Conference on Tourism Milestones – 'Preparing for Tomorrow', University of Sharjah, Sharjah, UAE, 31 March–2 April 2014.

MacDonald, P. (2000a) The commercial environment – internationalisation boosts non-oil sector diversification. In: Dew, P. and Shoult, A. (eds) *Doing Business with the U.A.E.* Kogan Page, London, pp. 26–29.

MacDonald, P. (2000b) Tourism development – the Dubai case study. In: Dew, P. and Shoult, A. (eds) *Doing Business with the U.A.E.* Kogan Page, London, pp. 30–35.

Maceda, C. (2014) Dubai's mall of the world to be built in 10 years. *Gulf News*, 9 July. Available at http://gulfnews.com (accessed 15 June 2014).

Masdar City (2011) *What is Masdar City?* Available at: http://www.masdarcity.ae/en/27/what-is-city-/ (accessed 28 June 2014).

Michael, N. (2013) Understanding tourist motivations. Emirati leisure travel to Australia. PhD thesis, La Trobe University, Melbourne, Australia.

Nair, M. (2014) Global brands preferring Dubai as a launch city. *Gulf News*, 1 April. Available at: http://gulfnews.com (accessed 15 June 2014).

National Geographic (2013) Available at: www.travelnationalgeographic.com/travel/sustainable/about_geotourism.html (accessed 30 June 2014).

Nazzal, N. (2013) Dubai population jumps 4.8 per cent to 2.17 m. *Gulf News*, 23 August. Available at: http://gulfnews.com (accessed 15 June 2014).

Nereim, V. (2013) Population of Dubai rises 5 per cent to 2.1 million. *The National*, 13 February. Available at: http://thenational.ae (accessed 3 July 2014).

Pierre, J. (ed.) (2000) *Debating Governance: Authenticity, Steering and Democracy*. Oxford University Press, Oxford.

Royal Geographic Society (2014) Available at: www.rgs.org (accessed 15 June 2014).

Ryan, C. and Ninov, I. (2011) Dimensions of destination images – the relationship between specific sites and overall perceptions of place: the example of Dubai Creek and Greater Dubai. *Journal of Tourism Marketing* 28, 751–764.

Ryan, C. and Stewart, M. (2009) Eco-tourism and luxury – the case of Al Maha, Dubai. *Journal of Sustainable Tourism* 17(3), 287–301.

Scott, N. and Jafar, J. (eds) (2010) *Tourism in the Muslim World*. Emerald Group Publishing, Bingley, UK.

Sharpley, R. (2008) Planning for tourism: the case of Dubai. *Journal of Tourism and Hospitality Planning and Development* 5(1), 13–30.

Shoult, A. (2006) *Doing Business with the United Arab Emirates*. GMB Publishing, London.

Stephenson, M. (2014) Tourism, development and 'Destination Dubai': cultural dilemmas and future challenges. *Current Issues in Tourism* 17(8), 723–738.

Stephenson, M. and Ali-Knight, J. (2010) Dubai's tourism industry and its societal impact: social implications and sustainable challenges. *Journal of Tourism and Cultural Change* 8(4), 278–292.

Treib, O., Bähr, H. and Falkner, G. (2007) Modes of governance: towards a conceptual clarification. *Journal of European Public Policy* 14(1), 1–20.

UAE Interact (2014) Available at: http://www.uaeinteract.com/business/economy.asp (accessed 15 July 2014).

UAE The National Council of Tourism and Antiquities (2014) Available at: www.uaetourism.ae (accessed 28 June 2014).

United Arab Emirates National Bureau of Statistics (2014) Available at: www.uaestatistics.gov.ae/ (accessed 29 June, 2014).

United Nations Educational, Scientific and Cultural Organization (UNESCO) (2011) Available at http://whc.unesco.org/en/list/1343 (accessed 28 June 2014).

Yee, A.S. (2004) Cross-national concepts in supranational governance: state–society relations and EU-policy making. *Governance – an International Journal of Policy and Administration* 17, 487–524.

chapter 9

Sustainable Tourism Development of Rural Destinations: The Schist Villages, Portugal

Sofia Teixeira Eurico* and Fernanda Maria Fernandes Oliveira

Escola Superior de Turismo e Tecnologia do Mar, Peniche, Portugal

LEARNING OBJECTIVES

- To read and understand a tourism development plan regarding sustainability principles for a specific territory.
- To see why governance processes are a crucial part of the sustainable development of a tourist destination.
- To understand how local communities can and should be involved in designing tourism development planning.

INTRODUCTION

The chapter is structured in five main parts. In the first part, a global characterization of the schist villages is provided. Readers are presented with an overview of the physical and human geography of the schist villages in the context of the region and more widely in Portugal. The second part focuses on the history and development of the 'Schist Villages' project, considering the framework in which

*sofia.eurico@ipleiria.pt

it was developed and its main guidelines towards a sustainable development. The third part highlights extensive details of the structure and processes of governance, including key actors involved in tourism planning, policy formulation and the implementation process. The fourth part is directly related to the principles of sustainability and strategies used in this project to promote the sustainable development of tourism in this territory. Finally, a case study describes the network of schist villages' shops and their importance for development of local initiatives and the economy, as well as for the promotion of local products in national and international markets.

GLOBAL CHARACTERIZATION OF THE SCHIST VILLAGES

With the oldest borders in Europe, Portugal is situated in the extreme south-west of Europe and it is bathed by the Atlantic Ocean and the Mediterranean Sea. With a total area of 92,072 km^2, it is the third largest exclusive economic zone of the European Union (EU), having special rights over the exploration and use of marine resources of 1,727,408 km^2 of sea zone (Datacentro, 2012).

The Portuguese continental territory is divided into five main administrative areas, corresponding to Nomenclature of Territorial Units for Statistics (NUTS) level II: (i) North; (ii) Lisbon and Tagus valley; (iii) Centre; (iv) Alentejo; and (v) Algarve. The schist villages are part of the Centre region, which represents 31.3% of the Portuguese mainland and 23.7% of its population. It is thus an extensive region in the national context, but with a low population density and much diversified territory as it extends from the Atlantic Ocean to the border of Spain, including urban and rural areas.

The estimated resident population of the Centre region is 2,298,938 (Datacentro, 2012) out of a total of 10,487,289 for the national territory, with a population density of 81.5/km^2 (113.7/km^2 for Portugal) and with an ageing index of 164.5, which is considerably higher when compared with the ageing index of the country (131.1) and is also aggravated by the fact that the proportion of young people is also lower than the national percentage.

Moving on to the schist villages (Fig. 9.1), they occupy an area that is mostly mountainous, forested (eucalyptus and pine), crossed by important rivers and provided with a set of hydroelectric dams, which gives it a very specific identity in terms of natural resources. The accessibility to the region is difficult due to its meandering territory, narrow roads and lack of public transport. However, efforts have been made in the last decade to improve the accessibility, namely with improvements in the roads' surface and by adding better road signs.

Another feature of this territory is the abundance of granite and schist, which have been used for different purposes by its inhabitants and is clearly part of the built cultural heritage. Moreover, landscape differences are noticed throughout the year as seasons influence the diversity of fauna and flora according to the time of the year. Boars, deer and chestnuts are some of the most typical examples of the species and fruits one may find here, as its specific climate and geographical features provide a rich environment for a valuable and high level of biodiversity.

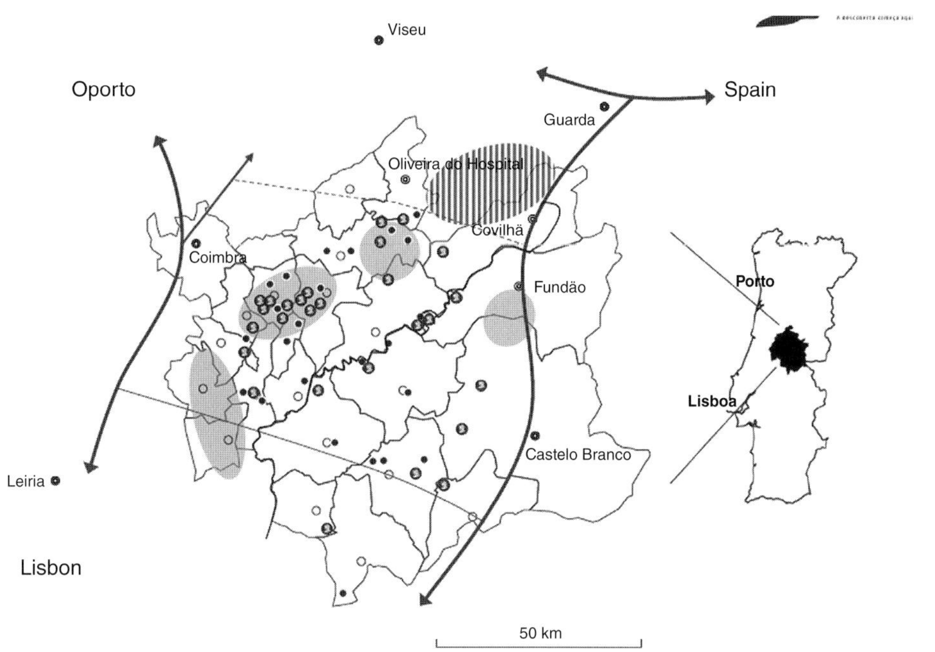

Fig. 9.1. Location of the schist villages. (From ADXTUR, 2009, p.12.)

Furthermore, it may be characterized by cold and rainy winters and hot and dry summers. These extremes in climate represent hard life conditions for inhabitants, demanding a serious and difficult adjustment to such extreme and opposite conditions. Difficulties in accessibility and the climate conditions are two of the main reasons that have meant that young people, in particular, have deserted the territory as they migrate to urban areas where there is a better chance of getting a job and increasing opportunities in different fields, namely education and culture and even health support.

THE 'SCHIST VILLAGES' PROJECT

The Commission for Coordination and Regional Development of the Centre Region (CCDRC) included in its strategic plan for the development of the territory the requalification of the schist villages, one of the most deserted and socio-economic vulnerable areas of Portugal. It applied for EU funds (through FEDER – the European Regional Development Fund) in order to improve conditions in 24 villages, distributed among 14 councils, making them more appealing in terms of housing, and also for the development of activities and business. By 2006, the requalification of several houses had been approved with financial support (Fig. 9.2).

After this first stage, a local development association – *Pinus Verde* – supported by the CCDRC, prepared a Global Plan for Sustainable Development of the Schist Villages, promoting actions such as meetings with communities in each village, the elaboration of a municipality plan for each village and

Fig. 9.2. Traditional houses in the schist villages.

research on technical architecture and traditional materials. The plan also considered the promotion and enhancement of the schist villages, considered essential for integrating these territories in the cultural agenda as tourist destinations.

The Global Plan for Sustainable Development of the Schist Villages assumed the following priority actions:

- to develop a brand (image) identifying and promoting the territory;
- to improve signing by placing information placards about each village and across the network of the 24 villages and also directional signage on roads;
- to refresh an existing magazine on the 'Schist Villages' but now aiming to encourage the public to visit and enjoy the villages as an emerging tourist destination;
- to establish a network of shops with the aim of commercializing the products produced by the inhabitants of the villages;
- to establish an agency for the tourism development of the schist villages, Agência para o Desenvolvimento Turístico das Aldeias do Xisto (ADXTUR), joining public and private entities that worked together to contribute to the development and promotion of the territory, heritage preservation and enhancement and improvement of living conditions for its inhabitants; and
- to construct a network of the schist villages.

The main strategic lines of the plan were therefore the sustainable development of a territory and the promotion of its endogenous values. It intended to: (i) fight the stigma of human desertion in the territory of the schist villages; (ii) reduce forest fires; (iii) answer and fight the private sector's inertia; and (iv) re-establish local identity, promoting its unique cultural and creative values, directly linked to local traditions. By promoting direct contact with the people and their culture, as well as advertising proper accommodation and services, the project's initiatives would transform this territory, enriched by the richness and diversity of its landscapes, into a distinctive tourist destination in Portugal. Moreover, the creation of the 'Schist Villages' brand, directly related to this specific territory and to its endogenous products, equipment and services, was crucial to consolidate the project nationally and to launch it in international markets.

THE PROCESSES OF GOVERNANCE IN THE TOURISM PLANNING AND MANAGEMENT OF THE SCHIST VILLAGES

This is a project of local and regional development guided by the principles of sustainable development of tourism. Within this framework, and taking into consideration the physical, social and economic characteristics of the territory, the 'Schist Villages' project was based on a properly phased and coordinated planning process in which all stakeholders were integrated: public and private (local, regional and national).

This section of the chapter aims to explain the whole planning and development process of the 'Schist Villages' project, highlighting the contributions of all stakeholders from the perspective of the concept and processes of governance. Indeed, the beginning of the network of the schist villages was based on a strict plan and a set of instruments that allowed a previous outline to define each stage of its development. Moreover, and in terms of governance, it was clear that one of the main areas of the network of villages was the social component which had to guarantee the involvement of residents in all future actions. They represented a key element in the strategic planning process and participated at an early stage in the preparation of the 'Schist Villages Programme' (PAX), when planners wanted to know their openness to the idea of having their villages renewed – from the point of view of architectural and landscape features. Their individual opinions and suggestions were considered as a priority in the PAX, and provided information on their acceptance of the project as well as their individual predisposition to invest in the upgrade of their houses. This procedure embodies the participatory nature that characterizes this sort of strategic planning (Veal, 1994; Hall, 2000), and it gives a voice to the residents in the planning process and allows the use of their input in the political decision-making process; furthermore, it reinforces the pillar of governance participation. Indeed, according to the International Fund for Agricultural Development (IFAD) the concept of governance:

> requires that civil society has the opportunity to participate during the formulation of development strategies and that directly affected communities and groups should be able to participate in the design and implementation of programmes and projects. Even where projects have a secondary impact on particular localities or population groups, there should be a consultation process that takes their views into account. This aspect of governance is an essential element in securing commitment and support for projects and enhancing the quality of their implementation.
>
> (IFAD, 1999, p. 3)

The first planning tool is the PAX, which is a document that identifies a set of actions aimed at the territorial development of the region. This instrument was issued by the CCDRC NUTs II and its implementation was supported by the European Regional Development Fund. In the PAX document, the villages considered for intervention and the objectives for each one of them were identified. The PAX implementation was phased in two distinct periods: first the rehabilitation of the villages, and second their promotion as a tourist destination. The rehabilitation of each village depended on the design of a Village Plan (PA), which underlined the major intervention works needed for the rehabilitation of private properties and basic infrastructures (public equipment and spaces). The PA was performed by the Local Technical Offices that were under the dependence of the municipalities. In methodological terms the village's plans included the opinions and priorities of reclassification obtained from the surveys conducted among residents, homeowners and local agents (if any) of each village. It also included a plan of action for the restoration of the houses' outside facades and roofs and for the improvement/implementation of basic infrastructures and rehabilitation of social spaces.

As far as determining strategies, measures and actions towards promotion, another planning tool was designed – the Global Plan for Sustainable Development of the Schist Villages. This instrument was prepared and sponsored by the local development association *Pinus Verde*, with the support of CCDRC, and in partnership with various municipalities, institutional and cultural entities and economic actors in the region (Carvalho, 2012).

The need to accomplish the objectives set by PAX and to financially guarantee their realization led to the development of a new planning tool – an Action Plan for 2009–2012 – that was presented by ADXTUR as an application of the Programme for the Economic Enhancement of Endogenous Resources (PROVERE). PROVERE supports actions based on institutional partnerships among private companies, but also among public administration entities such as city councils, business enterprises for local and regional development, and research and technology knowledge transfer units, among others. In terms of governance, the PROVERE Network of Schist Villages is a Strategy for Collective Efficiency (EEC PROVERE) which is promoted by a consortium of regionally or locally based institutions, including companies, business associations, municipalities, educational institutions, research and development institutions, regional and local development associations and other relevant institutions. It gathers a coherent and strategically justified number of initiatives, integrated into a programme of action, which looks for structural funds that cover the different projects transversally. The PROVERE Network of Schist Villages involved a partnership with more than 160 players in the area and received financial support under the National Strategic Reference Framework (NSRF).[1] The four main aspects of this PROVERE were (ADXTUR, 2009, p. 11):

- the growth and sustainability of the 'Schist Villages' brand and all the associated sub-brands (e.g. Schist Villages Shops, SVShops) in national and international markets;
- the use of brand awareness of endogenous resources to attract, guide and enhance new investments and investors;
- to strongly invest in innovation, human enhancement and cooperation as a way to create new services, new attractions, new investment opportunities and specific solutions for this region; and
- to strengthen and improve the system of governance of local partnership and deepen and enhance the technical cooperation model design for this effect.

As for the PROVERE's structure and methodologies that would monitor and evaluate the whole process, a model of governance was followed and it relied on the concept of 'democratic decision making'. It induced the participation and full involvement of all partners in the process of managing the partnership and the following actions took place:

- visits/meetings with potential institutional investors/partners;
- working sessions and strategic coordination between partners/sponsors;
- information sessions/public release; and
- thematic workshops (Restaurants, accommodation and local products; Tourism recreation; Hotels and huge investments; Institutional partners).

The implementation of the EEC PROVERE Network of Schist Villages has an evaluation/monitoring model which aims to:

- monitor indicators of regulation and strategic cooperation between the consortium and the realization of investment projects;
- periodically assess the progress made towards achieving the specific objectives of PROVERE; and
- review and audit entities managing the PROVERE implementation.

As an example, we present some results of one of the workshops that took place within the process of evaluation of the strategy and implementation of this EEC PROVERE that occurred in June 2013 in one of the villages. This evaluation process is being developed by the Portuguese Society for Innovation and this initiative involved 20 organizations that are part of the 'Schist Villages' network, including representatives of municipalities, tourist accommodation, restaurants, local associations, artisans and tourism recreation companies. The main purpose is to collect opinions of different stakeholders about the project, its progress and to gather contributions for a greater efficiency. From the dynamic participation some important information for the future of the project was collected.

For example, at the level of interactions between entities included in the consortium, some of the answers were (SPI, 2013, p. 4):

- 'There is a huge similarity between the institutions and there is a strong link with the territory.'
- 'Yes, we feel that we now have more value and ADXTUR team has favoured the connection and liaison necessary to the relationship between entities.'
- 'The strategy revolutionized the territory and ADXTUR reached private entities demonstrating an exemplary role.'

As for the appearance of new projects or activities, stakeholders also mentioned that with the EEC PROVERE Network of Schist Villages, there is now greater empowerment of agents to create and develop new projects that result from planning, that consider the territory features and some that were not considered and in the meantime are being taken care of. Moreover, promoters learned to network, discovered this learning channel and the results have been manifestly positive because they want to belong to the network, and do more through the method that is adopted by the network.

However, promoters also note that the projects and activities have not been developed according to the initial expectations. In fact, they were able to go further but this did not happen because there was a reduction in the rate of implementation of the EEC PROVERE and a decrease in the financial strength was felt. In this workshop, participants could also identify the main difficulties, weaknesses and problems within the EEC PROVERE (whether in politics, the operationalization of instruments, the governance, the commitment of the participants or any other field assumed as relevant). Beyond identifying problems, possible solutions to solve them have been proposed. For example (SPI, 2013):

- *problem identified*: increase communication between collective people (partners) and inhabitants; *solution presented*: to organize an open day with activities organized with the help of the local community of the village and promote and disseminate it to neighbouring populations;
- *problem identified*: lack of visibility and connectivity within the schist villages; *solution presented*: to implement signage recognized by the national authority responsible for the roads and highways of Portugal, the creation of a partnership in terms of telecommunications covering the network of schist villages and invest in media promoting it in the major nearby cities.

In terms of the level of satisfaction shown by the participants considering some of these items, it was found that 87.5% are satisfied or very satisfied with the role of the private sector and 70.1% are satisfied or very satisfied with the new relations of cooperation (created under the EEC PROVERE). As for the sustainability of implemented projects and of the established relationships, the degree of satisfaction is 70.1% and 45.8%, respectively (SPI, 2013).

PRINCIPLES OF SUSTAINABILITY WITHIN THE 'SCHIST VILLAGES' PROJECT

With the analysis of the entire process of planning and development of the schist villages, it is clear that the implementation of sustainability principles regarding the tourism sector was carefully followed. If we take a deep look into the leading organizations that act at the level of European policies for sustainability (European Commission) and specific guidelines for the sustainable development of international tourism (United Nations World Tourism Organization, UNWTO), one may clearly see that the 'Schist Villages' project has considered them. From the EU's perspective we will consider the nine principles for achieving competitive and sustainable tourism, presented by the Tourism Sustainability Group (European Commission, 2007) in the Report 'Action for More Sustainable European Tourism' and equally assumed by the 'Agenda for a Sustainable and Competitive European Tourism' (621 Final; COM, 2007). Because of its importance and wider perspective we highlight two of the set of 13 principles defended by UNWTO (UNEP/UNWTO, 2005) as far as sustainability and tourism are concerned.

Principles common to both approaches

Beginning with the principles that are common to both approaches and that should be respected by all actors involved, we have:

- The need to take a holistic and integrated approach in which tourism planning and development must be articulated with the whole range of public policies and private activities that affect the society and the environment.

Planning instruments described in the previous section of this chapter as well as the adopted methodologies reflect the integrative and participatory nature of the planning and development process of the schist villages. Moreover, regulations, policies and strategic plans set by the Portuguese government have determined that the project must fulfil strategic guidelines and mandatory rules and establish a close liaison with other areas of political action (mobility, urban planning, regional development, employment, environmental protection, tourism development, etc.).

- The importance of planning tourism in a long-term perspective ensures the sustainable development premise that cares for the 'needs of future generations as well as our own. Long term planning requires the ability to sustain actions over time' (European Commission, 2007, p. 5).

This project is a great example of phased and structured planning, contemplating a long-term implementation process. The vast structural weaknesses of the territory, together with the diversity and quantity of strategies and planned actions, imposed a long, slow timescale to ensure smooth integration and accomplishment of the project principles aimed mostly at safeguarding the region's social and environmental identity values.

- The need for gathering contributions that reflect all stakeholders' involvement in the process of policy decision making and on practical implementation. This principle of sustainability is related to the involvement and engagement of all stakeholders, whether in terms of local control or working together, and it may involve formal partnerships between public or private organizations or arrangements that stimulate common interests and promote the participation of all in the development and management of tourism projects.

This principle was largely implemented in the schist villages, as reflected in the scheme shown in Fig. 9.3 which shows the number and variety of agents that are direct partners of ADXTUR. Besides companies directly connected with tourism and public entities representing local government associations (sectorial and local development), some of the major institutions of higher education in the region were also part of this network.

- The importance of setting and respecting the limits determined by regulation.

Defining limits to development is a usual measure to guarantee a responsible and balanced use of resources (natural and cultural), equipment and infrastructures, to ensure high levels of tourists' satisfaction and respect residents' well-being. When considering private houses and public infrastructures' intervention (particularly in terms of roads, sanitation, water supply, natural gas, electricity and telecommunications) a set of regulations and restrictions determined by territory planning instruments had to be met. These instruments dictate rules that bind any type of intervention – public or private – and determine, for example: (i) the categories of land use; (ii) areas of protected cultural heritage and natural resources; (iii) types, areas and levels of construction; and (iv) the type of equipment and features created.

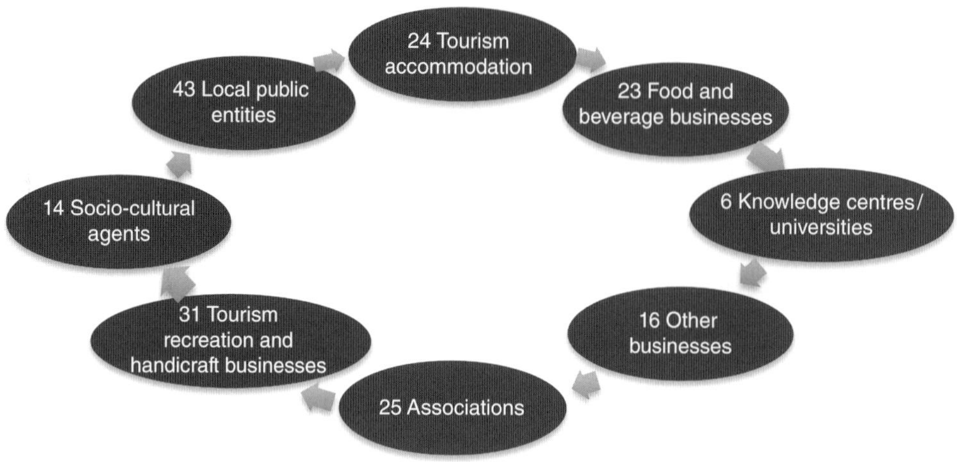

Fig. 9.3. Agência para o Desenvolvimento Turístico das Aldeias do Xisto (ADXTUR) partners.

- The 'precautionary' principle – which means 'putting in place measures to avoid damage before it occurs rather than trying to repair it afterwards' (UNEP/UNWTO, 2005, p. 17).

In addition to the general rules for planning, some of the villages have to comply with specific regulations if included in areas classified by Natura 2000 (11 villages), or if included in protected areas (one village) and in areas of reservoirs (two villages). The environmental fragility of these spaces, of their natural resources and habitats demand greater limits to growth, the processes require planning permission and more rigorous licensing and, in most cases, the preparation of an Environmental Impact Assessment, which materializes the precautionary principle. In fact, risk assessment is an important requirement of sustainable tourism development, so when facing situations where 'there is limited evidence about the possible impact of a development or action, a cautious approach should be adopted' (UNEP/UNWTO, 2005, p. 17).

- The 'continuous improvement (or monitoring)' principle.

Any planning process must consider the inclusion of a monitoring system that cares for the implementation of the various phases, objectives and planned actions. The whole process of planning and developing the schist villages has been founded on a set of indicators to identify and evaluate rates of completion of each project phase. As for the financial implementation of a great part of the 'Schist Villages' project, it was only possible by submitting it to a broad set of applications for national and community financial aid programmes that demand presenting diagnostic indicators for the *ex ante* situation and *ex post* evaluation.

- The 'user and polluter pays' principle.

This is the only principle for which, to date, there is no example of implementation. This principle argues that everyone (and every business) must be responsible for the environmental negative effects resulting from their actions, so the impacts of tourism should be imputed to the prices of tourism goods and services. This is the only way of having available funding to ensure the correction of the negative effects of the sector.

Principles only set out by the European Commission (2007)

There are still a few principles only set out by one of the approaches. In the case of the principles only sustained by the European Commission we have the following examples.

- The principle related to the achievement of an appropriate pace and rhythm of development, arguing that 'the level, pace and shape of development should reflect and respect the character, resources and needs of host communities and destinations' (European Commission, 2007, p. 4).

This was performed in the schist villages since the physical intervention ensured the preservation of traditional architectural design and created a contemporary image in the villages, both in the public equipment and spaces and in the private buildings. Comfort, quality and habitability of the property were secured using building techniques and materials and modern equipment which met the requirements and needs of tourists and residents.

- The principle which defends the use of the best available knowledge.

Policies and actions should be informed by the latest and best knowledge available. Information on tourism trends and impacts, and skills and experience, should be shared across Europe. One good example of its implementation on the schist villages is the ECO-ARQ_ project (Application of Eco-rehabilitation in the traditional architecture of the South-west Europe). The project is an initiative of transnational cooperation (supported under the SUDOE Programme – INTERREG IVB) of a partnership between seven local development agencies (one of them being ADXTUR) in five regions of Spain, France and Portugal, all of them already with extensive experience in projects whose theme is the preservation of heritage. The ECO-ARQ_ project: (i) brought together and capitalized on the know-how and best practices of eco-rehabilitation; (ii) designed a website (http://www.eco-arq.eu) to disseminate such information to the various actors involved; and (iii) encouraged their implementation as well as the development of a specific and viable tourism that enhances eco-rehabilitation.

Another example is the Heritage Schist, a project developed in the territory of the schist villages, which involves a partnership between the local development association *Pinus Verde* and the Museum of Røros (a Norwegian city listed as a World Heritage Site by UNESCO). It was based on the knowledge transfer among Norwegian and Portuguese craftsmen, through intervention in buildings in the schist villages in accordance with the philosophy of preservation (maintenance and repair) of assets deployed in Røros. The basic idea of this project was to provide an exchange of knowledge and best practices in maintenance and repair of cultural heritage in rural-setting operations, through a programme of workshops that consisted of activities for the conservation and repair of buildings of traditional construction which are built of local materials, the schist stone and wood. The knowledge and experiences developed are properly explained and documented in http://patrimoniodoxisto.blogspot.pt/ which is available to everyone.

Principles only sustained by UNEP/UNWTO (2005)

With regard to the principles referenced only by UNEP/UNWTO (2005) they are listed below and are complemented with examples of how they were achieved (or not) in the schist villages.

- Equating sustainability and quality – as defended by UNEP/UNWTO (2005, p. 16): 'quality tourism destination or product is one that addresses the full range of sustainability issues rather than simply concentrating on visitor satisfaction'.

Everyone should be involved in the development of tourism sustainability components including tourists, suppliers and the local community. Considering tourists, encouraging them to think and to value sustainability issues will contribute to the sustainability of tourism businesses but also to make them more responsible citizens. In terms of quality, the project has developed a set of initiatives designed to ensure the excellence in the equipment and infrastructures created in the marketed products and in the existing services. Some examples are discussed below.

The need to qualify residents for a better relationship with the outside (especially with visitors and tourists) led to the organization of various activities, many of which are framed in terms of training aimed at the population of the schist villages. This provided them with the necessary skills in areas such as tourist service, weaving, stonemasonry and gastronomy, thus enabling them to appropriately respond to higher levels of requirement and quality demanded by the establishment of the network of schist villages.

The development of the network of schist villages' shops, selling traditional products most of which are locally produced, was of extreme importance as it stimulated the traditional activities of the area and gave them a decisive role in the economic side of the project. Farmers and artisans got the opportunity to value their products and found themselves committed to ensuring the needs of the network of shops' stock too. Regarding the principle of quality, we also have to emphasize the fact that the products sold are genuine and selected for their quality. Moreover, the creation of the 'Schist Villages' brand aimed to identify the services, products and infrastructures that care for social and environmental values that identify this territory. Also in this context, we can mention the Charter of Principles, which underlines particular aspects related with economic transparency, commercial fairness and genuine quality.

From the point of view of equity, ADXTUR is the promoter of the project 'Equal Villages', approved by the Commission for Citizenship and Gender Equality within the POPH (Human Potential Operational Programme). The main objective is to make the enterprises of this territory more socially responsible. Thus, ADXTUR aims to contribute to the organizational change of its associates (involving all workers, men and women), helping these companies to integrate a culture of social responsibility in their management policies and practices. This achievement will contribute to the 'Schist Villages' network standing as a global project of sustainable tourism.

- Considering functional alternatives – This is a principle that defends the use of the most efficient development options, as referred to by UNEP/UNWTO (2005, p. 17): 'consideration should be given to whether the same function can be performed and the same result achieved by doing things in a way that has more positive and less negative impacts on resources'. In other words, 'preference should be given to those options that bring the least environmental and social impacts and the highest economic returns' (UNEP/UNWTO, 2005, p. 17). This principle is directly related with the one that highlights the importance of addressing global and local impacts (when making the decision on which development options to implement, considering those with a broader scope and that point out global impacts such as climate change). The principle also cares for the tourism development need of adapting to [external] changing conditions which can affect its performance and the level of its impact.

The schist villages respond to this principle by adopting strategies that value reuse, renovation, revitalization and redevelopment of an area – including resources, knowledge and heritage. The goal is to use what the territory offers, increasing local production and reducing the need to buy from other regions. With this approach, we are considering human resources, building materials and other raw materials, agricultural products, crafts and other traditional products, and the project succeeds in optimizing the territory resources, as well as reducing financial and environmental costs (by reducing the need for buying from other regions, impacts are minimized in terms of transport and the resulting greenhouse gas emissions, also the option to reuse old or traditional building materials is to minimize the production of waste).

By valuing the use of local products (whether in the schist shops, accommodation units or local grocery) the 'Schist Villages' network is promoting sustainable consumption, and putting forward production practices that minimize external environmental factors. However, in the field of sustainable consumption and considering that it does not only apply to individual consumption choices, namely visitors and tourists, but also to businesses and the implementation of green procurement in the public sector, one may say that a long path has still to be taken in order to fully accomplish this goal.

Finally, we can conclude by this analysis of sustainability principles within the schist villages, that there are many examples that illustrate best practices for sustainability of tourism in this territory. However, there is still work to be done, mostly if we consider the 'user and polluter pays' principle and the one regarding sustainable consumption. The difficulty of implementing these two principles is related, in the first case with cultural issues (payment for the use and/or for the protection of public goods is not a common practice in Portugal), and secondly with internal management sales' decisions of private enterprises and public entities. In this case, issues directly related to the non-existence of more environmental and socially responsible choices in the market and the costs that are implicit to the ones that are available, greatly influence managers' decisions.

Case Study: Villages' Shops

Gathered with the endogenous resources and products typical of the schist villages, the know-how of local craftsmen to work traditional materials with ancient techniques justified one more initiative within the project that established a network of shops that aimed to commercialize these products, which were therefore recognized with a seal that ensured quality. They were thought to affirm the 'Schist Villages' brand, through a commercial sub-brand, that of the Schist Villages Shops (SVShops).

The SVShops (Fig. 9.4) sell local products and provide goods and local services of selected quality (to the community and visitors). From traditional to modern, items are handmade by artisans and inspired by the people and culture of the villages. Many local products, such as cheese, wine, herbs, infusions, regional sweets, as well as the bread baked in the community oven, were exclusively used by the inhabitants of these 24 villages. Others are reinterpretations of traditional processes and materials but adorned with a contemporary design and inspiration, namely hats and scarves, suitcases, bags and wallets, rings and necklaces.

Moreover, the shops' image and interior decoration were created jointly by contemporary producers and old master craftsmen. The result was new designs for art pieces made with raw materials, including clay, basketry, wood and iron. These shops also assumed a character of the neighbourhood shop in the villages as sometimes, in smaller and more distant villages, they work as grocery shops too. So the shops have created a new dynamic in the life of the residents, who have an opportunity to sell their products and another gathering point to get together and somehow combat the loneliness and isolation that characterizes their daily routine and geography.

Fig. 9.4. Schist Village Shops' (SVShops) logo. (From authors)

The network of SVShops has clearly brought social and quality benefits with this strategy of marketing and promotion of local products and the villages themselves, as well as boosting the local economies. It has proved to be an effective means of affirmation of this territory par excellence and it has contributed to the certification of almost unknown products and producers. In fact, the qualification of all agents involved in this project, including traders, producers, sellers, merchandisers, promoters of shops and

(Continued)

Case Study. Continued.

tourist guides among others, as well as the qualification and certification of products has been a main goal of the shops' project. Furthermore, young designers/companies have been motivated to come to the villages, through means of financial support, to settle their business in the territory and an incubation system for these small businesses was created. With this system, it was possible to create lines of handmade products, developing new products that were shown and shared with the local artisans in the format of workshops and tutorials that allowed the locals to replicate them and work together with the young artists. By employing local people and encouraging the opening of small enterprises related to the schist shops, this project has attracted young entrepreneurs to the schist villages.

Central Purchasing (an online store) for the network of SVShops was created in order to act as a platform for the acquisition and concentration of products that are then distributed to the different shops. Central Purchasing is responsible for selecting the products through a process that cares for the accreditation of suppliers and guarantees enough supply to the different shops. Besides the shops that have opened in the schist villages, this network has been consolidated with the strategic opening of similar shops in important touristic cities of Portugal and in foreign cities such as Barcelona (Spain). Opening SVShops abroad is of utmost importance for brand internationalization. International shops will in fact be a permanent brand presence in foreign markets, providing an effective channel for promoting and facilitating access of the products and services offered by the network to international tourists. A mobile shop is part of the project and it enhances the marketing of local products nationwide and in international markets by its participation at trade fairs and events promoting tourism in Portugal.

A shopping guide to SVShops was also produced with information on products, artisans, community practices, a product portfolio, road maps, description of points of interest, entertainment and eating/accommodation information, among others. All communication supporting the network of SVShops, including the shopping guide, is multilingual and information is spread at foreign trade fairs and international shops.

Finally, we highlight the existence of the 'Schist Villages' guidelines which refer to the network of SVShops' responsibilities and works 'to protect fair trade as a way of contributing to the long-term development of this region of the country'. The guidelines are as follows:

- the development of local initiative and economy;
- a face-to-face approach based on trust;
- the establishment of a network of producers and craftsmen;
- transparency, dialogue and respect in commercial activities;
- a fair distribution of income between seller and producer;
- fair pricing of products; and
- differentiation through genuine quality.

CONCLUSIONS

This chapter focuses on a rural area in the centre of Portugal with high levels of human and economic desertion and a set of very attractive properties and environmental values in terms of tourism.

Since 2000 this territory has been submitted to a process of revitalization of a number of villages composed of schist houses and an engaging nature that is aesthetically very appealing to new tourism markets. This project is based on a working partnership involving all local and regional stakeholders, including local government, regional government bodies, community residents and local businesses. Furthermore, all stages of development were properly planned in a democratic way and articulated among the different sectors of political action. The aim is to create a dynamic region, where quality of life and well-being of residents is promoted and entrepreneurship is encouraged. All this is under the umbrella of sustainable tourism development.

This chapter identifies some of the good practices that make this project a case of success, illustrating how theoretical concepts and methodologies related to the concept of sustainable tourism can be well applied.

We leave you with some questions on this specific case study.

QUESTIONS

1. In the case of expansion of the network of Schist Villages Shops (SVShops) and a consequent increase in the demand for its products, will local producers be able to cope with a substantial rise in production?
2. What would the feedback from local stakeholders and community residents to a governance model such as the PROVERE be if the schist villages were a mass-tourism destination, with very strong lobbying interests?
3. Can we really consider this project economically sustainable? Or in order to be economically sustainable will it have to consider an increase in growth of companies and businesses that will somehow endanger the sustainability of the environmental values of the territory of the schist villages?

NOTE

[1] The NSRF is the framework for the implementation of EU policy on economic and social cohesion in Portugal 2007–2013.

BIBLIOGRAPHY

Agência para o Desenvolvimento Turístico das Aldeias do Xisto (ADXTUR) (2009) EEC Rede das Aldeias do Xisto – Candidatura PROVERE. Available at: http://maiscentro.qren.pt/private/admin/ficheiros/uploads/doc_estrategico_aldeias_xisto%20(anexo1_1_eec_rax_).pdf (accessed 15 May 2014).
Aldeias do Xisto (no date) Available at: www.aldeiasdexisto.pt (accessed 15 May 2014).
Carvalho, P. (2012) Património Cultural, Lazeres Turísticos e Territorialização das Políticas de Desenvolvimento: O Exemplo das Aldeias de Xisto. In: Carvalho, P. (coord.) *Património e Lazeres Turísticos. EUMED.* Universidade de Málaga, Málaga, Spain, pp. 25–36. Available at: http://www.eumed.net/libros-gratis/2012b/1225/index.htm (accessed 27 February 2014).

COM (2007) Communication from the Commission – Agenda for a Sustainable and Competitive European Tourism. 621 Final, Brussels, 19 October 2007. European Commission, Brussels.

Datacentro (2012) Available at: http://datacentro.ccdrc.pt/Viewer.aspx?D=DCAB89A60FFF316C1AD946AA2A836997AC65C6A87184BDC633146E4F7CA3B078 (accessed 15 May 2014).

European Commission (2007) Action for More Sustainable European Tourism. Report of the Tourism Sustainability Group, pp. 3–4. Available at: http://ec.europa.eu/enterprise/sectors/tourism/files/docs/tsg/tsg_final_report_en.pdf (accessed 10 May 2014).

Hall, C. (2000) *Tourism Planning: Policies, Processes and Relationships*. Prentice-Hall, Harlow, UK.

International Fund for Agricultural Development (IFAD) (1999) *Good Governance: An Overview*. Sixty-Seventh Session EB 99/67/INF.4, 8–9 September. IFAD Executive Board, Rome.

Sociedade Portuguesa de Inovação (SPI) (2013) Estudo de Avaliação da Estratégia e do Processo de Implementação das EEC-PROVERE – Relatório Workshop: Rede das Aldeias de Xisto. Available at http://web.spi.pt/provere/relatorios/WorkshopAldeiasdoXisto.pdf (accessed 30 May 2014).

United Nations Environment Programme/United Nations World Tourism Organization (UNEP/UNWTO) (2005) *Making Tourism More Sustainable: A Guide for Policy Makers*. UNEP Division of Technology, Industry and Economics (DTIE), Tourism Programme, Paris.

Veal, A.J. (1994) *Leisure Policy and Planning*. Leisure Management Series. Longman, London.

chapter 10

Planning for Sustainable Tourism Development in a Context of Regional Instability: The Case of the Lebanon

Nour Farra Haddad*

NEOS Tourism, Beirut, Lebanon

LEARNING OBJECTIVES

- To highlight a simple approach for comprehensive tourism planning in Lebanon and explore the ways in which tourism can be developed in a sustainable perspective.
- To explore and challenge the insecurity in Lebanon to develop sustainable tourism, taking into consideration the evolution of the tourism sector in Lebanon.
- To understand the existing planning strategies and initiatives related to the development of this sector.

INTRODUCTION

This chapter starts with a description of the physical and human geography of the Lebanon. Lebanon is a fascinating and confusing nation, steeped in history, a land of

*nour@neoslb.com

rich natural beauty, breathtaking scenery and cultural diversity. It is the only country in the Middle East where one can find a lifestyle, nightlife and entertainment facilities that is the same as the one encountered in Western countries and at the same time oriental traditions well rooted in the society. This section is followed by an overview of the development of the tourism sector in Lebanon through history to the strategies and policies that exist nowadays to support this sector that is suffering from the precarious political and security situation in the Middle East and especially in Lebanon. The promotion of domestic tourism is highlighted as a case study. Looking forward to re-establish Lebanon on the international touristic map when the situation will allow it, tourism governance is trying to develop and promote local tourism. A significant body of critical literature refers to domestic tourism as the sector of the tourism industry that can support and boost tourism in general, even in a context of insecurity and geopolitical instability. The conclusions drawn are presented as recommendations that respond to the challenges of national insecurity that have caused the tourism crisis, within the wider context of a preliminary analysis of the general contemporary tourism sector.

PHYSICAL AND HUMAN GEOGRAPHY OF THE LEBANON

Lebanon is located at the extreme east of the Mediterranean Sea and sandwiched between Syria and Israel/Palestine. The most astonishing geographical fact about Lebanon is its size: packed into 10,452 km^2, it extends from east to west averaging a distance of only 50 km and 225 km from north to south along the coast. Despite the fact that it is densely populated, it is ranked 125th worldwide for its population of 4 million inhabitants. Two mountain ranges run parallel to each other down the length of the country: Mount Lebanon and Anti-Lebanon. The Anti-Lebanon constitutes the natural border with Syria. Mount Lebanon is higher, runs close to the coast and is to host six ski stations allowing one of the clichés of the country to survive, demonstrating that you can ski in the morning and swim in the sea the afternoon.

It is its location which has always given it such key strategic importance through history as a part of the Fertile Crescent (an arc of land rich in water connecting Egypt and Iraq). All across the country are historical and archaeological sites, reflecting ancient and modern world history including Canaanite, Phoenician, Persian, Hellenic, Roman, Byzantine, Umayyad, Abbasid, Crusader, Mamluk, Ottoman and French Mandate sites. Lebanon offers a myriad of elements: from ancient ruins to well-preserved castles, caves and interesting karst formations, Christian and Muslim religious sites (Mosques, maqâms, churches, convents, chapels, etc.), beautiful beaches nestled along the Mediterranean Sea, exciting non-stop nightlife and discothèques, and mountainous ski resorts and mountain sports adventures.

The cosmopolitan flair of modern-day Beirut, the gastronomic capital renowned for the country's food, cuisine, Arak and wine, and an educated and outward-looking population complement a country that is both traditional and progressive in outlook. The tourism products which Lebanon offers are so varied, from coastal urban tourism to rural tourism, while the demand for niche products such as religious tourism and community and eco-based tourism is increasing.

Unfortunately, the bulk of the tourism benefits remain concentrated in the capital city, Beirut, and its suburbs, in addition to a very few highly renowned archaeological and religious sites, despite Lebanon's rural regions containing most of the prominent historical, archaeological and natural attractions. Tourism in Lebanon has traditionally been urban based (i.e. most tourists head for other urban areas and poles such as beach resorts, country clubs, casinos and restaurants). In recent years, however, tourism and recreation are increasingly practised in natural settings such as forests, protected areas, valleys and mountain areas. Reasons for this imbalance are many, the most crucial of which is the lack of reliable, quality hospitality services outside the urban cities, as there is a lack of support for the rural hospitality infrastructure (roads, signage, etc.). This is exacerbated by the absence of a national tourism strategy to support and develop the tourism sector in rural regions. Lebanon is very rich in resources that could be leveraged to extend economic development to all of its regions. Multiple points of interest can easily be reached from almost anywhere in the country, to spread the economic benefits of tourism to all rural areas.

HISTORY AND DEVELOPMENT OF TOURISM WITHIN THE LEBANON

Lebanon has a long-standing history of cultural tourism with interest in the oriental Levantine culture influenced by multiple references to Lebanon in the Bible, the writings of Homer, in the visits of many European Orientalists, scholars and poets, particularly Alphonse de Lamartine, Ernest Renan and Victor Guerin.

The Lebanon Tourism Service was originally created in the 1930s, within the Ministry of National Economy. In 1948, the 'Commissariat Général du Tourisme, de l'Estivage et de l'Hivernage' (CGT) was established and through the 1960s, CGT played a significant role in promoting Lebanese tourism. Major tourism projects were carried out, and Lebanese attractions were promoted worldwide. In 1959, CGT was annexed to the Ministry of Information, which became the Ministry of Information, Orientation, and Tourism (http://mot.gov.lb/Ministry/About). After independence in 1943, Lebanon enjoyed, a 'golden age', three decades of prosperity under a free market economy. Tourism, agriculture, education and democracy flourished, claiming for Lebanon the title 'Switzerland of the East', and for Beirut 'Paris of the Middle East'. However, the golden decades did not continue to thrive with the surrounding regional and international events, and crisis and discord in the Middle East. Since independence Lebanon has had a service and banking economy that relies significantly on tourism and it has long relied on tourism as an important source of national income, especially from residents of the Persian Gulf states and the immigrant community.

In 1962 the 'Conseil National du Tourisme' (CNT) was formed. It is a private association charged with conducting all tasks related to tourism promotion and investment in order to boost the tourism sector. The growth of tourism on a global scale led to the creation of the Lebanon Ministry of Tourism in 1966 (law no. 21/66, issued on 29 March 1966).[1] With the CNT the Ministry had the right to build-own-transfer (BOT) projects and public–private partnerships (PPPs). No other country in the world was taking these steps at that time. It was a 'golden age' for tourism, contributing to Lebanon's economy, accounting for almost 20% of national gross domestic product (GDP), in the two decades before 1975. Lebanon's service sector has traditionally been based upon tourism and banking, and the sector itself was the most important pillar of the economy until nowadays. Until the mid-1970s, Lebanon was a very popular destination for both travellers and businessmen.

The war in Lebanon (1975–1991) has not only resulted in hundreds of thousands of deaths, injuries and the displacement of locals in governorates across the nation, but it has also seriously damaged its economy and infrastructure and especially devastated the tourism industry, leaving guides, agencies, hotels, restaurants and shop owners without a source of steady income. Tourism plays a crucial role in the economy due to its strong linkages with other service sectors such as transportation, construction, real estate, banking and financial services and retail. These have all been badly affected by the war.

In the years since 1991, Lebanon has attempted to rebuild much of its physical and economic infrastructure but tourism has yet to return to its pre-war levels and the country re-emerge as a genuinely compelling and unique tourism destination, supported by dynamic tourism industry leaders from both the private and the public sectors. The tourism industry was slow to recover and continued to experience setbacks, as the country underwent profound political, economic and social transitions.

In 1992, a decree transferred the tasks of the CNT, including overseas promotion and execution of tourism projects, to the Ministry of Tourism. Significant private investments were made in the modernization and expansion of this sector and international hotel companies will return to Lebanon. Casino du Liban, which historically constituted a major tourist destination, reopened in 1996, claiming for Lebanon the title of 'Las Vegas of the Middle East'. Lebanon is the only country in the Arab world that has a casino, and it is also the only one that offers skiing and related winter sports activities. The government believed that, because of the return of peace and stability to the country and with the development of the necessary infrastructure, tourism will again contribute significantly to Lebanon's economy.

By 1998, tourism represented an important and growing economic sector, generating US$1300 million in net revenues, up from US$673 million in 1994. In 1995, services including tourism accounted for 16.4% of GDP (up from 14.3% in 1973). In 2001, the tourism sector was one of the fastest growing sectors of the economy, with the number of tourists visiting Lebanon between 1996 and 2000 growing at the rate of 14% annually. By 2003, the number of tourists travelling to Lebanon again topped the 1 million mark; however, most visitors were staying and spending money only in Beirut, while rural regions throughout the country were not reaping the gains of tourism's revival. In 2003 the tourism industry contributed US$6.374 bn to the Lebanese economy and in 2005 the receipts per capita reached US$1433.

The United Nations World Tourism Organization (UNWTO) held a regional workshop on tourism statistics in Beirut on 22–24 March 2004, in order to better coordinate the collection of reliable tourism statistics. For the past several years, there has been a steady 10% increase in the number of tourists.

In 2006 the Ministry of Tourism counted 373,037 admissions to the country's main ministry-run touristic sites. Tourism directly contributes between 10% and 12% of GDP (UNWTO, 2006), and employs around 300,000 people with one-third of these working in catering businesses. In 2006 the Ministry of Tourism predicted 1.6 million visitors that would generate some US$2 bn in revenue, and this reflected the major advance in economic and political stability achieved by the country before the outbreak of war in July 2006. This prediction was on track during the first half of the year, with record tourist numbers, registering a 49.3% increase over the 2005 figure. However, the war led to: (i) a dramatic decline in tourist numbers by 58.6% in the third quarter of the year; (ii) the cancellation of festivals and concerts; and (iii) the closing down of restaurants, hotels, nightclubs and beach resorts.

This war episode, known as the July war in Lebanon (Harb Tammouz) or in Israel by the 'Second Lebanon War', took place in Lebanon, northern Israel and the Golan heights. This military conflict started on the 12 July and ended in the 14 August by a ceasefire negotiated by the United Nations (UN). The conflict has killed about 1300 Lebanese people and 165 Israelis. It severely damaged the Lebanese infrastructure and displaced approximately 1 million Lebanese. Right after the official 'couvre feu' efforts focused on the regeneration of this sector; however, attracting tourists and investors proved to be very difficult. In 2009, Lebanon hosted about 2 million tourists, a record number, passing the previous 1974 record of 1.4 million tourists. The number of tourists grew by 39% over 2008, the largest increase in any country according the UNWTO. Lebanon was also featured by several international media outlets, including the *New York Times*, CNN and *Paris Match*, as a top tourist destination at the beginning of 2009. Lebanon's annual income from tourism reached $7 bn, about 20% of GDP, according to the Minister of Tourism. Lebanon experienced two unique years, in terms of number of arrivals in 2009 and 2010. It seemed the country was back to its heyday, with more than 2 million visitors in 2010 and global media praising the outstanding offerings of this tourism destination (Mitchell, 2009; Fabricant, 2010). Most of the increase is due to heightened political stability and security.

Then, Syria, Lebanon's neighbour, with very strong ties to its internal politics, fell into a civil war in 2011, which affected Lebanon's own security, and left tourists fearful. In a contemporary sense, the Lebanese economy is cornered between a war in Syria and a helpless public sector. The proximity to Syria, with the regular influx of refugees, ongoing cattle rustling and the tension of border conflicts, are concerns to any potential investor in the tourism sector. Rising political and sectarian tensions, a Lebanon travel ban by Gulf countries, and the impact of the war ravaging neighbouring Syria have combined to lead to a decline in tourism. For the first time since 1990 Lebanon's real GDP growth went down to an estimated 1.5%, according to the International Monetary Fund in 2011. The most notable outcome of the Syrian war is a political boycott of travel to Lebanon by Gulf Cooperation Council countries. Gulf Arabs traditionally flock to Lebanon every summer, swapping the searing heat of the desert for Lebanon's relatively cool mountain air and more permissive lifestyle. Mountain towns such as Aley and Bhamdoun are filled with restaurants and hotels that rely heavily on Gulf visitors, who normally comprise about 65% of all tourists (Blanford, 2013). Gulf Arabs are interesting tourists because they tend to stay longer than cultural tourists and are high spenders. Some of them are attracted by beach resorts, bars, nightclubs and prostitution while others seek out family-friendly cafés, restaurants and attractions that offer moral leisure spaces adhering to Islamic notions (Deeb and Harb, 2013).

However, disapproving of Lebanon's position towards the Syrian crisis, the United Arab Emirates (UAE), Saudi Arabia, Qatar, Oman and Kuwait all introduced travel warnings preventing their residents from travelling to Lebanon. Six Gulf countries slapped bans on their citizens travelling to Lebanon, citing the worsening security climate – although their hostility towards Lebanon's militant Shiite Hezbollah, which is fighting in Syria on behalf of the regime of Bashar al-Assad, is another factor. Although Lebanon officially follows a policy of neutrality towards the conflict in Syria, the country is split between supporters and opponents of the regime. These are the highest spenders in Lebanon's tourism industry, with many businesses relying on these tourists for profits and survival throughout the year.

Restaurants, nightclubs, hotels and other hospitality operators reported a drastic decline in business since the beginning of the year. Arrivals fell dramatically in 2011 and 2012, with slightly better prospects for 2013, although the looming threat of conflict may have caused renewed uncertainty prospects for 2013. In 2012, tourism in Lebanon was projected to generate $4.3 bn in direct revenue and provide roughly 9.5% of total employment in Lebanon.

The Ministry of Tourism launched campaigns early in 2013, aiming to attract Arabs back to Lebanon, but the result was not satisfactory, especially as the warnings remained in place. According to the *Daily Star* newspaper on 2 July 2013, restaurants and nightclubs were the hardest hit, with a whopping 50% decline in business since the start of 2013. Within Beirut, hotel occupancy rates were down to about 35%, while outside the capital rates dropped to as low as 5%. The report also pointed out how cities, historical sites and festivals with major visitor appeal had been affected, citing the example of the Baalbeck International Festival, which had to relocate to Beirut due to security concerns after the town of Baalbeck was struck by rockets (*Daily Star*, 2013).

The number of tourists to Lebanon dropped by 6.69% in 2013, compared with 2012, while European visitors topped the list ahead of Arabs for the first time in years. Regardless of all the gloom of the Lebanese crises, in 2012 Lebanon was ranked 20 among the world's top tourist destinations and in 2013, Lebanon was ranked among the top ten food destinations around the world according to Lonely Planet and it was named the world's best breakfast destination by CNN. Key findings of the report by Hodema (consulting services) which is published on an annual basis in *Le Commerce du Levant* 'Restos, bars, cafés edition' point out that the number of restaurants in Beirut has increased, while their total seating capacity has declined. The number of restaurants increased from 660 in 2012 to 708 in 2013 (*Le Commerce du Levant*, 2013), which is a very good indicator for the food and hospitality sector.

TOURISM GOVERNANCE IN LEBANON

After 1991, the overall policy in Lebanon was to promote Arab tourists from the Gulf as customers; in contrast there was little interest in Western tourists, whose contribution to income and expenditure was considerably lower and more limited. The Lebanese government, and in particular the Ministry of Tourism, tried to establish a strategy to plan for tourism governance, aimed at the development of the sector. In 1992, the transfer of the tasks of the CNT, including overseas promotion and execution of tourism projects, to the Ministry of Tourism will not be perceived by the stakeholders in the tourism field as a good strategy. The CNT was very active and contributed strongly to the promotion and the development of the tourism sector in Lebanon. A Tourism Master Plan was prepared in 1997 with the UNWTO, containing 72 priority actions and setting a national strategy for tourism development, focusing on eight priority areas including the Qadisha Valley and the mountain areas.

An Inter-Ministerial Council for Tourism Development exists, chaired by the Prime Minister, with ad hoc membership and irregular meetings to discuss various policy issues related to tourism. The Council for Development and Reconstruction (CDR) has not provided funding for tourist accommodation or attractions by the private sector, but it can undertake tourism-related activities under an 'Integrated Development and Environment' programme.

The government policy which wanted to move only towards a tourist elite currently recognizes the importance of also fostering Western tourism and tourism of young people, especially since 2000. This is demonstrated through editions of brochures and seminars. As a consequence, ecotourism, nature tourism and rural tourism are expanding in Lebanon and becoming very popular.

A UN project, the Cultural Heritage and Tourism Development Project (CHUD – US$31.5 million) has provided international grants and funds to support the public sector. The project was launched in 2002 and has been in its operational phase since 2004. The project is willing to finance site conservation, enhancement investments and associated urban infrastructure improvements in selected sites, and provide technical assistance to strengthen the capacity of the Directorate General of Antiquities, Ministry of Tourism, and targeted municipalities in cultural heritage preservation and tourism development. It aims to enhance the strategic links between Tourism, Antiquities, and key 'cultural heritage' municipalities and their local communities in the cities of Tripoli, Jbeil (Byblos), Saida, Baalbek and Sour (Tyre). The report (CDR, 2012) of this project indicated a major obstacle for the tourism development in Lebanon: the absence of a comprehensive vision for the future of the tourism sector. This has had a significant impact on the development of the industry. The tourism industry suffers from an absence of a national vision, an underdeveloped infrastructure, as well as a lack of a comprehensive strategy that would include the different stakeholders.

In May 2003, in response to the request of the government of Lebanon, the government of Japan decided to implement the study on the Integrated Tourism Development Plan. The Ministry of Tourism with the CDR co-sponsored the project and formed a steering committee to assist the Japan International Cooperation Agency (JICA) study team. The project was focusing on three of Lebanon's Qazas: Besharreh, Zahle and Baalbeck outside of Beirut. The study's primary goal was to create sustainable strategies and pilot projects that will protect and promote Lebanon's rich natural and cultural heritage and help Lebanon to return to its prominent role in regional and international tourism.

The budget of the Ministry of Tourism is very restricted especially compared with neighbouring countries such as Syria and Jordan. Its budget is less than 1% of the total Lebanese public budget at the time but the Ministry is trying to increase this budget. The marketing function takes up a major portion, about 36.5% of the Ministry of Tourism's budget (US$5.0 million in 2002), which is still small when compared with other tourist destination countries in the region. All stakeholders in the tourism field complain of the lack of public budget invested in the promotion of Lebanon as a major tourist destination in the area. With the means at its disposal the Ministry of Tourism takes part every year, along with several travel agencies and Lebanese hotels, in the main international travel fairs, including the ITB in Berlin (the world's largest tourism trade fair), the Arabian Travel Market (ATM) in Dubai, the leading international travel and tourism fair in the Middle East, and the World Travel Market (WTM) in London, trying to promote and relocate Lebanon on the international travel map. In 2012 Lebanon won the Award for Most Beautiful and Best Designed Stand at EMITT (East Mediterranean International Tourism and Travel Exhibition) in Istanbul.

In the public sector, very few municipalities and Union of Municipalities try to implement strategies to develop local tourism. Most notably there were a few initiatives, such as the ones of the Cazas[2] of the Chouf (2002), Batroun (2003) and Jezzine (2013), to edit guidebooks regrouping the touristic services in the Cazas with the collaboration of the Ministry of Tourism. The syndicates operating in the tourism field are the Association of Hotel Owners in Lebanon, the Syndicate of the Furnished Apartment Owners, the Syndicate of Touristic Seaside Resorts in Lebanon, the Syndicate of Restaurant Owners (includes owners of cafés, nightclubs and pastries) in Lebanon, the Association of Travel and Tourist Agents in Lebanon (ATTAL) and the Tourist Guide Syndicate. From these syndicates only two of them have a website, ATTAL (www.attallb.com) and the Tourist Guide Syndicate (www.tourguideslb.com). All these syndicates have tried to be involved in tourism governance and planning for tourism development in Lebanon but their actions are very limited and handicapped by long administrative procedures.

The stakeholders in the private sector and civil society are involved most of the time in the tourism governance through projects and actions planned by non-governmental organizations (NGOs) and associations supported by international funds. The US Agency for International Development (USAID) with SRI International[3] for example undertook an initiative to build capacity and growth in the tourism field (http://csted.sri.com/projects/catalyzing-tourism-rural-lebanon). SRI International worked with the Lebanese government, Lebanese tour operators and business owners, and other local partners to promote Lebanon as a touristic destination, especially the rural regions. The branding developed through the SRI project, including the website (a website adopted nowadays as the official website of the Ministry of Tourism – www.destinationlebanon.gov.lb), maps, brochures and other resources, have been adopted and sustained by the Ministry of Tourism and private stakeholders. SRI's approach aims to maximize participation and empower local stakeholders to absorb good practices, make new investments, build on momentum, and take on the responsibility of sustaining programmes and institutions to impact the industry and the country.

PRODES (an NGO – www.prodeslebanon.org) is implementing projects in the tourism sector with European Union (EU) funds especially to support women's entrepreneurship in the rural areas.

'Live your tour' is a project co-funded by the EU promoting sustainable and responsible tourism in Mediterranean countries involving Italy, Lebanon, Spain and Tunisia (www.liveyourtour.net). The goal of the project is to increase sustainable tourism, encouraging off-seasonal flows, the enhancement of their cultures, and reducing the non-sustainable pressures of tourism on coastal areas, in order to mitigate the impact on the environment. The project works with the authorities in terms of cross-border strategic planning on sustainable tourism but also involves large segments of the civil society through summer camps for young people, and design of games and awareness courses in schools. These are some examples, many led by local authorities or NGOs with international funds, involving stakeholders from the private sector and civil society. Some of these projects will contribute greatly to tourism development in Lebanon.

PLANNING APPROACH TO SUSTAINABLE TOURISM DEVELOPMENT IN LEBANON

This section explores paths to develop strategic planning for sustainable tourism in Lebanon, noting that there is little that has been written on the Lebanon in general planning literature and more specifically tourism planning literature.

The perceived and/or reality of insecurity in Lebanon is significant, impacting greatly on the potential of sustainable tourism products to support livelihood development in the region. The proximity to Syria, with the regular influx of refugees, ongoing cattle rustling and tension of border conflicts are concerns to any potential investor, let alone the international (or domestic) tourist.

This insecurity is the main obstacle to the stakeholders in the tourism industry but there are other challenges too, with a need to re-evaluate the nature of central government in relation to other stakeholders when it comes to decision and policy formulation for development strategies for sustainable development of tourism. The development of strategies to market Lebanon as a major touristic destination should be a priority and budgets invested for this issue should be raised significantly. As the report of the CHUD indicated, a comprehensive vision of this sector should be defined within a national vision, supported by all the stakeholders concerned. Little coordination has taken place between the different ministries that are directly or indirectly involved such as the Ministry of Tourism, the Ministry of Culture, the Ministry of Transport and Public Works, as well as the different syndicates such as the Association of Hotel Owners, the Syndicate of Restaurant Owners, the Syndicate of Furnished Apartment Owners and the Syndicate of Tour Operators, among others. The government needs to adopt a multi-pronged strategy that would include both the public and the private sector.

For sustainable tourism development through product diversification, the government and other stakeholders need also to develop an infrastructure which capitalizes on the rich potential for developing and promoting niche tourism products such as ecotourism, nature and rural tourism, health and religious tourism. In the context of insecurity in Lebanon and general weaknesses in tourism products they can contribute greatly to the development of the tourism economy. For the most part, tourism in Lebanon is founded on the consumption of goods, commercial leisure activities and services in the urban settings of the big cities and especially in and near Beirut. In the last decade interest in developing ecotourism, nature and rural tourism beyond Beirut is increasing. But the growth of this sector is linked to the need for greater attention to the country's infrastructure, accessibility within rural areas and other factors, with rural tourism having the capacity to rectify existing income disparities between regions, generating employment opportunities and alleviating poverty. At the heart of this, ecotourism professionals would like to see the government establish a legal framework for ecotourism organizations and activities in order to regulate the sector and promote it among the Lebanese. Ecotourism initiatives are spreading in the country even though the concept remains blurred and there is no legal framework for this type of tourism. The increasing number of NGOs and tour operators (about 25 including Lebanese Adventure, Cyclamen, Esprit Nomade and Baldat) are offering all sorts of activities under the label of ecotourism, including hiking, climbing, caving, paragliding, rafting, visiting farms, camping, watching birds or stars and skiing during the winter. Ecotouristic projects are also increasing in number, including Lehfed Country Club, Parc Arz Ehmej (www.ehmej-eco.org), Sharewood Camp (www.sharewoodcamp.com), Nature land (www.nature-land.com) and others (Farra-Haddad, 2013, p. 321–336).

Additionally, Lebanon has a long-standing heritage in being the 'Hospital of the Middle East', by offering high quality facilities for medical, spas and wellness services. This form of tourism is growing quickly in Lebanon, in recent years, because of the high costs of treatments in other countries and the long waiting lists. It has emerged from the rapid growth to become an industry where people from multiple countries travel to Lebanon to obtain medical, dental and surgical care. Such tourists are offered availability, accessibility, affordability and good quality. The Ministry of Tourism has accommodated the offices of the newly created National Council for Health Tourism (NCHT) since 2004. Lebanon has upgraded its technology, absorbed many medical protocols and emphasizes low cost and prompt attention. Trade in the Lebanese health services is expanding, becoming more competitive, and creating new dimensions of globalization, all elegantly packaged and functioning as the new niche of medical tourism.

Lebanon also has a huge potential for religious tourism as part of the 'Holy Land' and also as a 'land of Holiness', with thousands of religious sites spread all over its territory and a lot of vocations enabling the canonization of many saints. The sacred geography of Lebanon testifies to the importance of the 'cult of saints'. The so-called 'pilgrimage phenomenon' is deeply rooted in Lebanese culture, and involves a multitude of different religious communities. Countless places of worship scattered over all the Lebanese territory and vocations which have arisen and which allowed the canonization and beatification of many saints support these strategies.

The devotion to these saints, translated by the fervour of countless crowds is impressive. They became intercessors and defenders of all the Lebanese communities representing a national symbol. Religious tourism can grow drastically if the country is promoted as a part of the Holy Land. There are over 96 references to Lebanon in the Bible, and Jesus Christ himself is said to have walked on its soil. Christian communities were present from the apostolic period in Lebanon, and prior to his detention, Saint Paul made many visits to Lebanon, often travelling through the coastal city of Tyre. In May 1987, Pope Jean Paul II proclaimed Lebanon as a Holy Land for its privileged place in the Bible, its martyrs and its sacred places. Following this statement, many civil society initiatives, as well as the Ministry of Tourism, attempted to place Lebanon on the international tourism map of the Holy Land and pilgrimages. Examples of organizations that work towards this end include, 'The Association for the Development of the Pilgrimages and Religious Tourism in Lebanon',[4] 'Lebanon, Holy Land',[5] and 'In the Footsteps of Jesus Christ in Lebanon'.[6] In 2007, the Maronite Patriarchate created an association for the development of pilgrimages and religious tourism in Lebanon and this association in July 2010 organized the first meeting on religious tourism in Lebanon. Following this conference the Centre for Religious Tourism (CRT) was created by the former Prime Minister Saad El Hariri and the second conference on religious tourism in Lebanon was organized inside the Grand Serail in November 2010. Religious tourism has the capacity to grow quickly but unfortunately the situation in Lebanon and the Middle East will currently curb this potential.

Case Study: Promoting Domestic Rural and Nature Tourism as a New Niche for Sustainable Tourism Development in Lebanon

This case study explores the assets of domestic tourism in Lebanon and how the actors that operate in hospitality and travel, NGOs and the authorities are working to develop this niche tourism and how they can adopt proper measures for its development. Providing opportunities for authentic experiences of nature and rural tourism promotes sustainable tourism that benefits not only the tourist and the community but also the tourism industry in general, in Lebanon's contemporary crisis.

Since the early 20th century, affluent urban communities of all classes and backgrounds have pursued leisure in the country's rural and mountain hinterlands. They have participated in the summer ritual of leaving the city to spend time in a village. This form of domestic tourism is expanding and occupying an important place in the sector. Rural and nature tourism was formerly oriented only towards elites of intellectuals in Lebanon, but now is increasingly attracting clients from different backgrounds and origins. The National Seminar on the Development of Ecotourism in Lebanon, held as part of the Economic and Social Commission for Western Asia in January 2002, revealed the importance of ecotourism, nature and rural tourism in Lebanon that will support development. Since then there have been many other conferences and studies that have emphasized the importance

(*Continued*)

Case Study. Continued.

of developing sustainable tourism in Lebanon. Lebanon presents considerable touristic potential in rural and urban areas with unique offerings of sightseeing, attractions and sports carried out in natural surroundings. Due to the growing popularity of environment-centred adventure travel, in recent years there has been an increasing tendency towards nature and rural tourism locally. Nowadays, domestic tourism constitutes and embodies a real nationalizing practice. Lebanese people are struggling between political factions and social and geographical divisions and local tourism will lead to the creation of a shared ground for engaging with the concerns about the whole nation.

Different actors in the private sector and NGOs have been trying to promote sustainable tourism in Lebanon. Some international funds support local initiatives to develop tourism. For example INMA[7] with SRI International in collaboration with the Ministry of Tourism developed tourism and cultural programmes, funded by USAID, to edit brochures and maps and rehabilitate sites from 2002 until 2007. ECODIT[8] developed a project from 2005 until 2008, the Lebanon Mountain Trail (LMT) with grant funding from the American people through USAID, a national 440 km long-distance hiking trail stretching from the mountainous areas in the north of Lebanon to the foothills of Mount Hermon in the south. 'The LMT expands economic opportunities in rural trailside communities by promoting various forms of rural tourism and low-impact recreation, and at the same time encourages land conservation and natural resource management along the trail' (Ecodit, no date).

On 5 October 2011, the Beyond Beirut NGO held its first forum on 'Sustainable Tourism Development in Lebanon' at the UNESCO Palace, in partnership with USAID. The forum identified international trends and niche market options for Lebanon, examined lessons learned, and proposed ideas to exemplify sustainable tourism through the expansion of rural tourism products that enhance livelihoods. The event promoted dialogue and connections between local tourism stakeholders, entrepreneurs and municipalities from rural areas. Beyond Beirut organized the Second National Forum on Sustainable Tourism Development in collaboration with the School of Tourism and Hotel Management at the University of Balamand under the theme of 'Product Development' on the 14 December 2012.

On 23 and 24 May 2013, PRODES (an NGO) held a national forum on 'Sustainable development and local tourism: towards the creation of new job opportunities'. The forum was organized in Byblos (Jbeil) and focused on issues such as the role of women in socio-economic development, and sustainable tourism development projects initiated by municipalities and private initiatives. On the 24 May a workshop took place to discuss touristic guiding in Lebanon with the national Tourist Guide Syndicate in Lebanon. Former tourism minister Fady Abboud tried to focus on new trends in tourism in order to promote the different regions of Lebanon economically and socially all-year round.

The report of the JICA study, produced in 2004, noted:

> While the visitor arrivals to major tourism sites are decreasing in recent years, the number of Lebanese visitors surpasses the number of foreigner visitors at most tourism sites. The

(Continued)

Case Study. Continued.

> importance of domestic tourism would increase as income levels increase in Lebanon. Also with the established tourism sites of international class, the development of domestic tourism through regional tourism development would constitute tourism as an important part of the differentiation strategy for the international tourism as well in Lebanon as a whole.
>
> (JICA, 2004)

In 2013, the Ministry of Tourism supported an edition of a guidebook to boost local tourism: *Eco-Lebanon: Nature and Rural Tourism – A Guide to Unveil Lebanon* (Farra-Haddad, 2013). The book, a guided directory written in English, aims to promote internal tourism among the Lebanese because foreign visitors have dwindled alongside security. The idea of publishing the guidebook in this context is to develop domestic tourism since the international tourists are not coming to Lebanon from abroad because of the lack of security. This guidebook highlights the diverse attractions and unique opportunities in all regions of the country. Lebanon's rural regions – which contain most of the country's prominent historical, archaeological and natural attractions – are now a significant component of Lebanon's tourism marketing campaigns.

When Michel Pharaon was designated as the Tourism Minister, in February 2014, he decided with his team to adopt a strategy and a promotional campaign to develop domestic tourism and relocate Lebanon on the touristic map and to position Lebanon as a premier eco-destination. In this respect the Ministry of Tourism and other related government agencies, along with the private sector, are working hand in hand to develop and promote 'responsible tourism'. To lead to real sustainable development, promotional campaigns should target all levels of society and those from different geographical areas, raising the question as to whether the promotion of ecotourism, nature and rural tourism is for all Lebanese society or just the elite of the society. As stated earlier in the chapter, legislation is still missing to help sustainable development to emerge in this tourism sector. The general challenges described as 'ignorant' tourism on achieving sustainability include 'over consumption of natural resources, environmental degradation, exploitation of cultures and labour, displacement of people from their land, lack of consultation with local communities, poorly thought out tourism planning and high foreign exchange leakage which reduces local economic benefits' in the long run (World's Summit on Sustainable Development, 2002). Stakeholders in Lebanon should be aware that sustainability is an important issue and should work together, hand in hand, to avoid the negative impact of an 'ignorant' tourism. However, this remains a challenge for the private and the public sector to work together to elaborate and develop policies to encourage the conditions for sustainable tourism development. 'LiveLoveLebanon' (livelovelebanon.com) is the title of the large campaign of the Ministry of Tourism launched in Lebanon on the 18 May 2014, with a focus on niche tourism as an exemplar for sustainable development, but implementation of this aspiration remains a challenge for the tourism sector in the Lebanon.

CONCLUSION

This chapter investigated tourism planning and discussed how Lebanon can increase its potential for successful tourism development but the challenge ahead to realize sustainability will be to energize the tourism sector without compromising environmental quality and natural resources. Building on the UNWTO framework strategy, in 1996, the Lebanese government set a national strategy for tourism with priority actions such as: (i) rehabilitating and management of the historic sites in Tripoli; (ii) renovating the old city centre in Jounieh; (iii) managing the coastal plain of Damour; (iv) planning for the management of the Qadisha valley; (v) managing the Qaroun lake; (vi) elaborating a plan for the mountain areas; (vii) managing the coastal area in Batroun; and (viii) rehabilitating and upgrading public beaches (MOT/UNDP, 1996). This national strategy for the development of tourism is in urgent need of review to encourage low-impact tourism in an effort to reconcile tourism and development.

In this respect comprehensive planning should consider all of the tourism resources, organizations, markets and programmes within a region and consider the economic, environmental, social and institutional aspects of tourism development. The Lebanese government and the Ministry of Tourism are trying to adopt comprehensive tourism planning towards sustainability but without legislation there needs to be better coordination and consultation between the concerned stakeholders in order to identify objectives, and define and evaluate methods of achieving them all together. At the same time, this long-term planning process has to acknowledge the current major economic crisis.

Lebanon is a complex country of extraordinary promise; it is often thrust into crisis, including recent military assaults, political tensions, terrorist attacks and bombings. Before the war, Lebanon was a central tourist destination for the Arab world and beyond, sometimes referred to as the 'Switzerland of the Middle East', 'Las Vegas of the Middle East', the 'hospital of the Middle East' and Beirut as 'Paris of the Arab world'. Today, in the aftermath of a war and period of insecurity and political tensions, Lebanon is struggling to recover. Lebanon's tourism industry has suffered severe blows resulting from the war and political unrest that took place in the country. Even if the tourism statistics on Lebanon are not reliable and even with all the remaining concerns about insecurity, the tourism sector was one of the fastest growing sectors of the economy in the past several years up until the beginning of the Syrian crisis in 2011. During that period there has been a steady 10–14% increase in the number of tourists and as a result of this influx of tourists, several new hotels, guest houses, restaurants and nightclubs were established in Lebanon to host the potential increasing tourist flow. Also tourism development projects have taken place, supported sometimes by the international community.

Despite the increase in the number of arrivals to Lebanon, the industry is experiencing several weaknesses that are directly affecting its advancement. The tourism sector in Lebanon continuously proves to be resilient and robust. Born in a country accustomed to a great degree of uncertainty, entrepreneurs have become experts at gauging trends and reinventing business models. The Lebanese economy made a remarkable recovery from the war, supported by international aid, an open economic policy with a liberal banking system and overseas remittance among others. The macro economy of Lebanon is characterized by a large trade deficit that is compensated for by capital inflow and non-merchandise flow including overseas remittance and tourism receipts. With the economy slowing down in recent years, tourism revenue as well as the overseas remittance would be more important to fill the gap.

Stakeholders in the tourism field know very well that the country will not be successful in attracting tourists back or investment without achieving sustainable peace among its various secular factions and with its regional neighbours. But they know also that even with the political turbulence in the region and, at times, inside the country, tourism has always exhibited remarkable resilience and there are always crisis management solutions. The implications for tourism of political instability are significant. Possible solutions through the use and practice of crisis management are suggested, as this approach may help to prepare nations better for unexpected instability and the uncertainty of the future. Different tools can be used for future strategies leading to tourism advancement and improved satisfaction. The Lebanese government, helped by the private sector, is trying in different ways to reposition the country as a premier tourism destination, mainly orienting its strategy towards the promotion of niche tourisms and encouraging domestic tourism to boost the local touristic economy.

QUESTIONS

1. Can planning for domestic tourism alone lead to a sustainable development in the tourism sector?

2. For all the flavours of its storied past, holy sites, traditional villages and natural beauty, Lebanon is a well-kept tourist secret that begs exploration. How would you promote it in light of the current lack of security? What kind of tourism should be promoted in such times?

3. In regard to the current situation in the Middle East, can Lebanon be considered a tourism destination?

NOTES

[1] 'The Ministry of Tourism was composed of the Direction Générale des Affaires Touristiques. Decree no. 7142, issued on April 20, 1967, defined the specific tasks that fell under the separate responsibility of the National Tourism Council, including overseas promotion and execution of tourism projects. However, this decree was repealed by Decree no. 2829, issued on October 10, 1992, which nullified the tasks entrusted to the National Tourism Council and transferred them to the Ministry of Tourism.' (Ministry of Tourism, no date).

[2] In Lebanon there are six governorates (Mohafazat) (Beirut, North, South, Mount Lebanon, Bekaa, Nabatiyeh) and each one consists of several districts (Caza) combining a number of towns and villages.

[3] SRI International is a non-profit, independent research and innovation centre serving government and industry.

[4] At the request of the Maronite Patriarchate in Bkerke, The Association for the Development of Pilgrimages and Religious Tourism in Lebanon was created in 2007 regrouping rectors of shrines and travel agencies specializing in pilgrimages.

[5] 'Lebanon Holy Land' association chaired by Mrs Joyce Gemayel focuses its efforts on the file of

the Lebanese Cana seeking to have it officially recognized by the Vatican as the biblical Cana.

6 The project 'In the Footsteps of Jesus Christ in Lebanon' was initiated by the Maronite League in 2009. A team of specialists seeks to trace the pedestrian routes followed by Jesus Christ in South Lebanon.

7 The Social and Cultural Development Association (INMA), is a non-profit NGO, founded in 1997. INMA's scope of activities is based on servicing clusters of villages and specific neighbourhoods throughout Lebanon.

8 ECODIT is an international development firm working with governments, businesses and local communities to advance environmentally and socially responsible development around the world.

BIBLIOGRAPHY

Blanford, N. (2013) Shadow of Syria takes toll on Lebanon's tourist appeal. *The Christian Science Monitor*, 31 July. Available at: http://www.csmonitor.com/World/Middle-East/2013/0731/Shadow-of-Syria-takes-toll-on-Lebanon-s-tourist-appeal (accessed 5 February 2014).

Council for Development and Reconstruction (CDR) (2012) Cultural Heritage and Tourism Development Project (CHUD) Reports. Available at: http://www.cdr.gov.lb (accessed 20 March 2014).

Daily Star (2013) Baalbeck Festival Relocates to Metn. *Daily Star*, 2 July. Available at: http://www.dailystar.com.lb/Culture/Performance/2013/Jul-02/222204-baalbeck-festival-relocates-to-metn.ashx#axzz2t3BMqdUV (accessed 11 February 2014).

Deeb, L. and Harb, M. (2013) *Leisurely Islam: Negotiating Geography and Morality in Shi'ite South Beirut*. Princeton University Press, Princeton, New Jersey.

Ecodit (no date) Available at: http://www.ecodit.com (accessed 27 May 2014).

Fabricant, F. (2010) Restaurants help lead Beirut's revival. *New York Times*, 7 May. Available at: http://dinersjournal.blogs.nytimes.com/2010/05/07/restaurants-help-lead-beiruts-revival/ (accessed 5 February 2014).

Farra-Haddad, N. (2013) *Eco-Lebanon: Nature and Rural Tourism – A Guide to Unveil Lebanon*. Lebanon Traveler, Beirut, Lebanon.

International Council on Monuments and Sites (ICOMOS) (1999) Australia ICOMOS Burra Charter. Available at: http://australia.icomos.org/publications/charters (accessed 23 March 2015).

International Cultural Tourism Committee (ICOMOS) (2002) *ICOMOS International Cultural Tourism Charter Principles and Guidelines for Managing Tourism at Places of Cultural and Heritage Significance*. ICOMOS, Charenton-le-Pont, France.

Issa, I.A. and Levent, A. (2006) Impacts of political instability on tourism planning and development: the case of Lebanon. *Tourism Economics* 12(3), 361–381.

Japan International Cooperation Agency (JICA) (2004) JICA Reports. Available at: http://gwweb.jica.go.jp (accessed 21 April 2014).

Kanso, A. (2005) Reinvigorating the 'Switzerland of the Middle East': how the Lebanese Government can use public relations to reposition the country as a premier tourism destination. *Journal of Hospitality and Leisure Marketing* 12, 135–156.

Ladki, S.M., Zein, B. and Sadik, M.W. (2000) Effect of the Middle East peace process on pan Arab tourism industries. *Tourism Analysis* 6(3–4), 267–272.

Le Commerce du Levant (2013) Available at: www.hodema.net/pdf/MGLPDF137509711551.pdf (accessed 18 November 2014).

Ministry of Tourism (no date) *Official Site of the Ministry of Tourism*. Available at: www.mot.gov.lb (accessed 17 June 2014).

Ministry of Tourism (MOT)/United Nations Development Programme (UNDP) (1996) *MOT/UNDP Plan for the Reconstruction and Development of Tourism in Lebanon*. MOT/UNDP, Beirut, Lebanon.

Mitchell, S. (2009) Gordon Campbell Gray's New Beirut Hotel. *Travel and Leisure*, June. Available at: http://www.travelandleisure.com/articles/gordon-campbell-grays-new-beirut-hotel (accessed 5 February 2014).

United Nations Educational, Scientific and Cultural Organization (UNESCO) (1999) Tourism and culture: rethinking the mix. *The Courier* July/August. Paris. Available at: http://www.unesco.org/courier (accessed 6 December 2013).

United Nations Educational, Scientific and Cultural Organization (UNESCO) (2001) Urban development: finding a balance among land, sea and people. In: the Old City's Heritage and Archaeology: Urban Integration vis-à-vis the Challenge of Socio-cultural and Touristic Development workshop held at the University Institute of Technology, Saida, Lebanon, 28–31 May.

United Nations World Tourism Organization (UNWTO) (2006) Available at: http://unwto.org/ (accessed 21 April 2014).

World Bank (2003a) Revitalizing Lebanon's Historic Cities. Available at: http://go.worldbank.org/7TGJ36AUU0 (accessed 21 April 2014).

World Bank (2003b) World Bank Investment in Cultural Heritage and Development, Lebanon. Available at: http://www.unesco.org/most/saida_synthesis.pdf (accessed 24 September 2014).

World Bank (2008) Lebanon – Cultural Heritage and Urban Development Project. Available at: http://go.worldbank.org/FFV1HXTT70 (accessed 21 April 2014).

World's Summit on Sustainable Development (2002) Available at: https://www.un.org/jsummit/html/basic_info/basicinfo.html (accessed 7 June 2014).

chapter 11

Understanding the Balaton: A Hungarian Case Study in Planning for Tourism

Ágnes Raffay* and Alan Clarke

University of Pannonia, Veszprém, Hungary

LEARNING OBJECTIVES

- To understand the scale and complexities of governance around Lake Balaton.
- To be able to critically assess the challenge of working towards sustainable development within this destination.
- To evaluate challenges of planning multi-tourist activities in the Balaton area with competition between different Destination Management Organizations (DMOs).

INTRODUCTION

The central aim of this chapter is to introduce a truly complex tourism destination, the Balaton in Hungary, and examine the challenges raised in the destination when planning for sustainable development in the future. Not only is the Balaton the second most popular tourist destination in Hungary, it is a destination which consists of a wide variety of touristic assets. The assets include: (i) a sensitive cultural heritage; (ii) health and wellness sites; (iii) sites of scientific interest;

*raffay@turizmus.uni-pannon.hu

(iv) environmentally fragile environments including national parks; (v) wine and gastronomy; and (vi) leisure tourism and the associated pleasure periphery. Moreover, these complex destination components are planned and managed through formal governance arrangements which require agreement from a spatially, culturally and economically differentiated set of organizations and agencies. The analysis of the chapter begins by introducing the destination and then presents critical explorations of the issues underpinning sustainable development, including environmental, sectoral and political challenges.

INTRODUCING LAKE BALATON

Hungary's current borders were established following the end of the First World War with the signing, in 1920, of the Treaty of Peace between the Allied and Associated Powers and Hungary, known as the Trianon Agreement. The treaty regulated the independent Hungarian state, leaving it as a landlocked state containing only 28% of the pre-war territory and with only 36% of the pre-war kingdom's population. As five of the pre-war kingdom's ten largest cities were drawn into other countries, 3.3 million Hungarians were left outside the new country (Zeidler, 2003).

The most popular tourist destination in Hungary is the capital, Budapest, but Lake Balaton is the second most popular with almost 3 million guest nights each year. Moreover, given the destination is based around the largest freshwater lake in Europe, the challenges of sustainable development are ever present. It is important to note that the destination described in this chapter as the Balaton Development Region includes the lake shore and its hinterland. Lake Balaton is a freshwater lake in the Transdanubian region of Hungary. It is the largest lake in Central Europe, and one of its foremost tourist destinations. The Zala River provides the largest inflow of water to the lake, and the canalized Sió is the only outflow. The mountainous region of the northern shore is known for its historic character and as a major wine region, while the flat southern shore is known for its resort towns. Balatonfüred and Hévíz developed early as resorts for the wealthy, but it was not until the late 19th century when landowners, ruined by lice, began building summer homes to rent out to the burgeoning middle classes.

Figure 11.1 shows the location of Lake Balaton in Hungary.

LAKE BALATON RESORT AREA (LBRA) AND ITS INSTITUTIONAL CONTEXT

Located in the Transdanubian region of Western Hungary, the Lake Balaton catchment area, including the lake itself is 5775 km^2. The LBRA is an area of 3780 km^2 that lies almost entirely in the watershed of the lake and comprises 164 municipalities. The area has a permanent population of 250,000; however, the summer peak population in July and August swells to 1–2 million, exerting substantial stress on the environment and infrastructure (Balaton Development Agency, 2013). Lake Balaton proper is shared by three counties and its entire watershed by four, which are second level administrative units in Hungary. Regional resource allocation is complicated by the LBRA falling into three

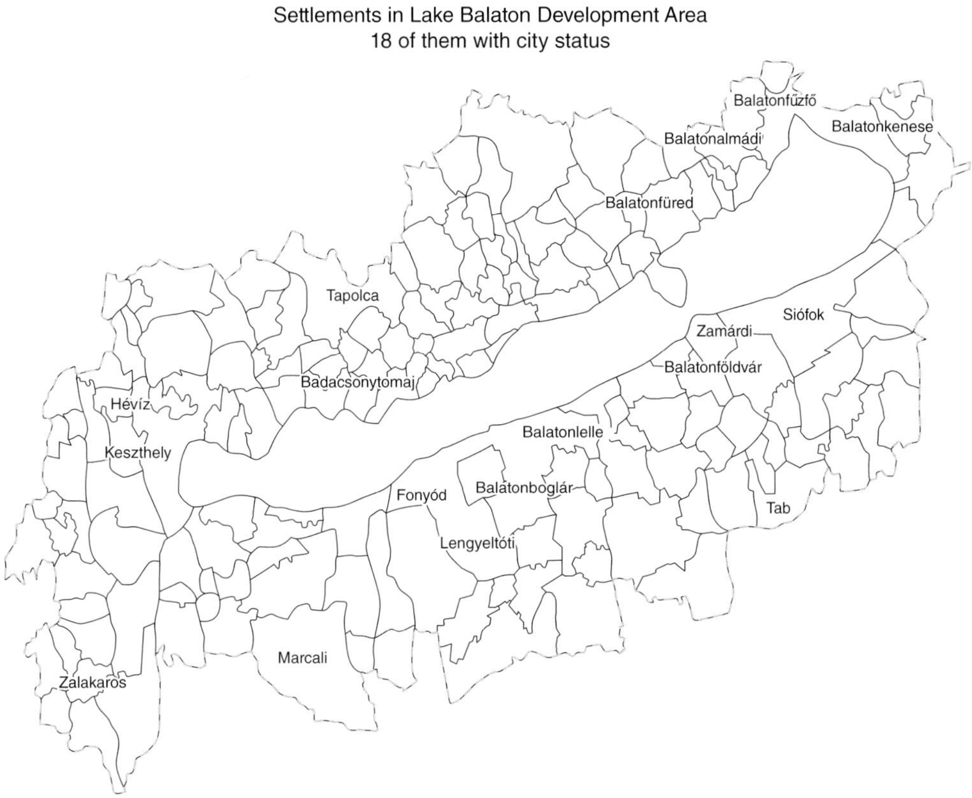

Fig. 11.1. Map of Hungary showing Lake Balaton, the largest freshwater lake in Europe. (Courtesy of the Veszprém Tourism Association.)

different Development Planning Regions, and the division of governance responsibilities are currently shared by three Environment and Water Management Directorates, three Environmental Inspectorates, three Water Inspectorates – whose areas of operation do not overlap with those of the counties or regions – and three Public Health Authorities. Following recent restructuring, nature conservation in the LBRA became the sole responsibility of the Balaton Uplands National Park (Martonné Erdős, 2001). The complications of management, planning and development will be considered throughout the rest of the chapter.

Land use

The Lake Balaton catchment area is predominantly agricultural land and the elevation of the mildly rolling hilly landscape varies between 100 m and 400 m above sea level. The catchment can be divided into three characteristic regions. The northern catchment follows the shoreline of the lake in a wide band and represents about one-fifth of the total area. The streams (both permanent and seasonal) draining this area are relatively short and with steep gradients. Due to the topography and to the soil conditions this area is least suitable for plough land (row crops) and the forest land fraction is the highest here.

The ratio of low-value grassland is also high here. Wine-grape plantations form the dominant agricultural use and they have a special importance to the water quality of the lake (Kovács, 2007).

The area of the southern catchment is 1450 km^2, about 1.5 times larger than the northern catchment. Streams flow perpendicularly to the shoreline and drain this area. There are marshes with groves along the near-lake area, where grassland and forestry are the dominant land uses. Forest soil, underlain by loess, clay and silt deposited by the wind, form the area of the Marcali table, land prone to erosion, along with the indented area of Külsô-Somogy. The western catchment area is nearly as large as the above two combined. The river Zala drains this catchment and carries about half of the total inflow and total sediment load of the lake. These loads impact directly on the Keszthely Basin, which represents hardly one-twentieth of the total volume of the lake. The area is hilly with the exception of the Kis-Balaton and the surrounding marshy area. Soils of the area are prone to erosion, a characteristic accentuated because this area receives the highest precipitation (Kaiser and Dlusztus, 2005).

The historical origins of tourism in the Balaton

The changes in Hungarian approaches to tourism planning have moved from the centralized controls of the Soviet Period to a free market approach currently in place with a neo-liberal government. This change is also explored by Panyik and Ratz (2014). They identify that the roots of tourism development around the lake lie in voluntary civic associations taking responsibility at local destination level. These were disbanded by the new regime installed by the Soviet Union at the end of the Second World War. There was also a Land Reform Bill in 1945, which redistributed land ownership from large estate owners to peasants and changed the ownership of land surrounding the lake. Industry was nationalized and agriculture collectivized. This centralized economy also directed the development of tourism as the industries and trade unions were responsible for providing accommodation and amenities for their workers. The international profile of Hungary was raised in 1956 when attention was focused on the short-lived and ultimately unsuccessful uprising against the Soviet Union. The valiant defeat attracted a great deal of attention and sympathy, creating a basis for future visitor motivations. The 1960s saw a series of liberalization initiatives, freeing political prisoners and church leaders and increasing the rights of farmers and industrial workers. In 1968 the 'New Economic Mechanism' was introduced, bringing elements of the market to the Hungarian economy. This was the first time tourism investment could be brought in from international developers. One of the characteristics of tourism during this period was tourist visitation from the country's neighbours – Slovakia, Ukraine, Romania, Serbia, Croatia, Slovenia and Austria – which is counted as international tourism but also includes a high proportion of tourists who would identify themselves as Hungarian. These are the territories which were lost during the Trianon Agreement and therefore it could also be seen as a form of homecoming tourism.

Panyik and Ratz (2014) highlight that in May 1989, the border with Austria was officially opened, making the first breach in the 'Iron Curtain' and the Communist State was dismantled in the transition to multiparty democracy, seeing the Soviet forces leaving Hungary in 1991 and the dissolution of the Warsaw Pact. The 1990s and the new regime witnessed a wide ranging change in consumption and investment possibilities, leaving Hungary as a member of NATO (North Atlantic Treaty Organization) (in 1999) and the European Union (in 2004). Hungary also became a member of the Schengen Agreement and this was implemented on the 21 December 2007. Another feature of the new economy was the advent of low-cost and charter flights into Hungary and the Balaton particularly benefitted from the opening of a former military airfield which now operates Hévíz-Balaton as a 100% locally owned airport. The local authority justify its continued support of this airport on the basis of the bed nights generated from the passenger numbers, especially highlighting the numbers from Russia.

The concept of social tourism is a concept which the authors define as a form of tourism with an added moral value, of which the primary aim is to benefit either the host or the visitor in the tourism exchange. Although a definition such as this one does not, in itself, do justice to the different manifestations of social tourism and their individual characteristics, it has the benefit of positioning social tourism against regeneration projects that can be seen as mainly economic (Minnaert *et al.*, 2011, p. 414). As noted, this is not a new concept for Hungary. In the socialist era, social tourism took place throughout the country, as a part of the social services supported by (state-owned) companies and trade unions (Puczkó and Rátz, 2011).

The facilities, such as the accommodation establishments and some spa services, were often of very low quality (Deák, 1979). Employees, union members and their families were eligible to use these premises on trips which were either heavily subsidized or totally free. Following the regime changes around 1990, the old support system collapsed with the workers and their families buying many of the accommodation units. This system was replaced by the so-called supported holiday system. Social tourism establishments were either privatized or closed down. The role and power of trade unions diminished, the central government's budgets were cut and/or significantly reorganized. Organizations and companies allocated lower amounts for social benefits and gradually removed inherited obligations.

The planning of social tourism has undergone significant changes in the last decade, leading to a wider scale of supported services as well as increased demand for such services. A system of tourism cheques was introduced which gave citizens money that could only be spent on Hungarian tourism. This has been developed into the current SZÉPCARTYA (www.szecheny-ipihenokartya.hu/index.php) which again targets resources on Hungarian tourism destinations and is particularly important for the accommodation providers around the lake. The central government has introduced the framework but it is operationalized locally and specifically.

The origin of the international tourists is interesting, as the official figures (see Box 11.1) do not reveal the full story. The presence of Germany at the top of the list partly reflects a different post-war settlement. When Germany was divided at the end of the Second World War, Hungary and particularly the Balaton found itself in an unusual position. The West Germans could travel because they were sharing in the freedoms of the capitalist West, but the East Germans were living under stricter controls. Hungary as a member of the Eastern bloc was deemed to be a safe site for socialist tourism and therefore the Balaton became a place where East and West Germans could both travel and meet up, reuniting families and friends split by the division of their country. The tourism choice became available for the tourists from the single Germany. The details shown in Box 11.2 now provides evidence of the 'opening up' of the Balaton to a wider tourism visitation post-1990 and the emergence of a new post-cold war Europe.

Box 11.1. Countries of origins for tourists at the Balaton 2011.

Main countries of origin of international arrivals 2011

Germany (708,401)
Russia (212,087)
Austria (209,042)
Netherlands (101,316)
Czech Republic (86,445)
Denmark (55,384)
Poland (52,188)
Slovakia (43,753)
Ukraine (31,839)

(From Balaton Development Agency, 2013)

Box 11.2. Statistics of Balaton tourism.

The Essential Balaton

- Number of domestic guests: 946,137
- Number of domestic guest nights: 2,700,589
- Number of foreign guests: 350,216
- Number of foreign guest nights: 1,726,984

(From Balaton Development Agency, 2013)

The popularity of this destination with tourists provides a range of challenges highlighted in the next section.

HIGHLIGHTED PLANNING CHALLENGES AT THE BALATON

Seasonality

- There is a spatial challenge in planning for tourism as the high season is very compressed, lasting a mere 2 months in the middle of summer with large numbers of tourists settling by the lake to enjoy the 'Hungarian Sea', as the Balaton is known in Hungary, utilizing a range of accommodation from the camping grounds through to the self-catering houses and apartments to the top of the range hotels.

Matching and reconciling competing usages of the lake and its environment

- The history and development of tourism within this area becomes complicated as it contains a number of locations which could be presented as different destinations. Heviz, at the eastern end of the lake, is a spa town (http://www.lakeheviz. com/), but other parts of the Balaton offer simple lake-side tourism, heritage tourism, active and ecotourism (biking, hiking, water sports), health and wellness tourism, and wine tourism, gastronomy tourism and cultural tourism. (Tourism partnerships have grown to support six wine regions in the Balaton region which are regulated not only by the public authorities but also by the wine guilds. The guilds have slowly recognized the need to establish wider linkages within the tourism sector.) Catering for these different tourism activities present different planning challenges.

- Permission has been granted for sailing but not for crafts dependent upon engines except for pleasure boats and the ferries. Jet skis and water skiing are not permitted although there is one electric cable which makes skiing possible. The planning structures have allowed a rapid growth in the number of marinas around the lake and therefore an increase in the number of boats on the lake. There have been several controversies about extending the usages of the lake, for instance with motor boats and water skiing but no agency has ever pushed these claims and the priority of the environmental claims has remained unchallenged.

Primacy of the natural environment in decision making

- A critical exploration of sustainability in Lake Balaton draws us almost inevitably to the lake and the water. Following many years of water quality problems a negative water balance induced water shortages starting in 2000 and lasting for 4 years (Vizugy Igazgatóság, 2010). This raised and continues to raise serious sustainability concerns in the Lake Balaton area. Due to these trends, sensitivity of Lake Balaton to climate change and its impacts came to the fore both for policy and for science. Because of Lake Balaton's high profile and the relative immaturity of the vulnerability and adaptation policy agenda, there is a strategic opportunity to influence the way this agenda unfolds in Hungary and other countries of the region.
- The constant concerns with the quality of the water in the lake have been of great interest. However, there is a touristic issue involved which concerns the quantity of the water in the lake rather than the quality. The levels of water in the lake are problematic as it fills in winter and empties in summer – in fact it overfills in winter and drains too quickly in the summer. This therefore requires management, but there are contesting interests in determining what the water levels should be, especially between those planning authorities based on the northern shore as compared to those on the southern shore. For those directly involved in tourism, high water levels are good for business (people prefer the beaches of the northern shore as the water is deep enough for swimming just a short distance from the shore whereas on the southern shore they have to walk out for some distance to find deeper water). However, high levels can lead to flooding and damage to property, therefore the planning and monitoring has to be carefully balanced.

- Planning the Balaton has raised issues over carrying capacity because of the sites of special scientific interest, National Park and select nature reserves, not to mention the water quality. This has not prevented there being over 170 hotels being built, with 40 of these offering four- and five-star accommodation. At the present time, no coordinated response has emerged to the control and limiting of capacity as proposals for development are considered locally and individually.

Zonal planning

- Access to the lake is a planning challenge that has to be considered. For example, Balaton Széplak refers to an area in the resort town of Siófok, which is a highly touristified city on the southern shore of the lake famous for its summer holidays and nightlife. This area of the city was designated and largely developed for tourism by the central government and contains 53 hotels. This zonal planning was a direct expression of the recognition of the importance of tourism but also demonstrated the lack of awareness of the needs of the environment and even of tourists. The hotels were allowed to build right up to the lake's edge and then deemed to be private property, thus making it impossible to freely walk by the lake.
- The Balaton has planned for a number of activities, including 210 km of bicycle routes, which connect the different public authorities around the lake. This is recognized as a positive approach to planning for sustainable tourism in the region. However, studies of cyclists show that the provision is not maintained to an appropriate standard and reflects the different priorities placed on the significance of the cycling route in the different jurisdictions. There are even more hiking trails which also cross local boundaries and raise similar problems about routing, maintenance and signage.
- The lake offers a range of child-friendly beaches and on the north shore there is an aqua park set back from the lake shore. Adventure parks have proliferated as agricultural land has been re-designated or allowed to be redeveloped for leisure purposes. There have been two golf courses developed, which may not seem too significant until it is put against the six in total in the whole country. The pressure on land use is intensifying as the demands of the pleasure periphery continue to grow.

A MICROCOSM OF PLANNING CHALLENGES FOR BALATON TOURISM: TIHANY

Tihany is a settlement situated on the peninsular jutting out from the middle of the north shore of the lake, and it serves as a microcosm of the analysis of tourism by the lake. Tihany's 1358 inhabitants (2013 estimate) have the highest per capita income and the village has the highest housing prices in the whole of Hungary (Tihany TDM, 2014). Defining Tihany denies any simple classification. It is a lakeside, or beach as they are called in Hungary, resort with a small marina (non-motorized), water sports facilities, fishing opportunities, tourism shops and entertainments.

However, the planning of the tourism development adds another dimension to the work of local councils, as it is also the home to one of the Balaton Highlands (Balaton Felvidéki) National Park's Visitor Centres, the Levendula Ház or the 'Lavender House' in English (Balaton Highland National Park, 2014; Levendula Ház, 2014). Despite its name Lavendula Haz does far more than tell the story of lavender (which has been cultivated in the area for centuries) but also has an erupting volcano, to explain the geological history of the area, the extinct volcanoes, the tunnels and the flora and fauna of the region.

Planners have had to deal with the sensitivity of allowing tourism around the Benedictine abbey. This is one of the outstanding Benedictine abbeys in the country and was founded in AD 1055 by András I, who is buried in the crypt in the only original grave of a Hungarian king which remains untouched. The founding charter of the abbey, although written in Latin, contains one sentence in Hungarian and is the first extant record of a document written, in part, in the Hungarian language. The abbey also features in the history of the Habsburg empire as the last Habsburg Emperor of Austria, Charles I, was briefly held prisoner here following his second attempt to regain the throne of Hungary. In 2000 the abbey gained permission for a conversion of the buildings for use as a café and subsequently for the abbey's associated Visitor Centre, completed in 2011, adds an interpretative dimension to the visitor experience.

Tihany also offers festivals and events throughout the year, including classical music concerts in the abbey and an open-air theatre festival. Traffic management becomes a challenge in the high season and often sees complications with the overspill from other Balaton attractions, such as the Harley Davidson Festival (over 1000 bikes in Balatonalmadi over a single weekend) causing congestion, noise and nuisance in the narrow streets round the abbey.

THE DEVELOPMENT OF DESTINATION MANAGEMENT ORGANIZATIONS (DMOS)

In order to understand the complexity of the proposal to establish DMOs by the Hungarian national government, this section first explores the existing institutional arrangements for tourism in Hungary and then moves further into the consideration of the details. There are still very strong remnants of centralized state structures from the previous Communist regime. Hungary has been governed by a centralized system, where local discretion has flourished within the confines of clearly defined central policies. The recognition of numerous problems arising in tourism development due to the lack of coordination and cohesion among the vast number of players in the tourism industry has brought with it the need for finding ways of bringing the 'interested parties' together to address the issues of complexity, coordination and adaptation (Bramwell and Pomfret, 2007).

Formal planning powers have not been transferred to the DMOs but reside with the local authorities. There has been no attempt to officially involve the DMOs or the Regional Development Agencies (RDAs) in the planning system, despite the concentration of expertise in these newer bodies. Traditional powers are difficult to remove and relocate but there has been a growth in the linkages between those planning powers and the DMOs. This leads to a complex and often bureaucratic system, where the 'centre' establishes strategic frameworks and targets which the regions and destinations have to comply with, and this has implications for the governance and planning of the Balaton.

Much has been made of the tourism planning organization needs for a unified approach to the promotion of the destination. This raises serious questions when the situation around the lake is considered. The details of tourism governance around Lake Balaton reveal a complex situation with:

- 179 settlements (43 shoreline);
- three county councils;
- three RDAs; and
- nearly 20 turisztikai desztinációs menedzsment (TDMs; local DMOs) including the only regional TDM in Hungary and three sub-regional ones.

Figure 11.2 shows key relationships between the various bodies and agencies with a direct role in tourism. This figure, while perhaps simplistically, focuses on the major direction of the flows of deliberative practices, power, influence, resource allocation and decision making within the governmental system. The bottom right of Fig. 11.2 demonstrates where the potential DMOs fit into the

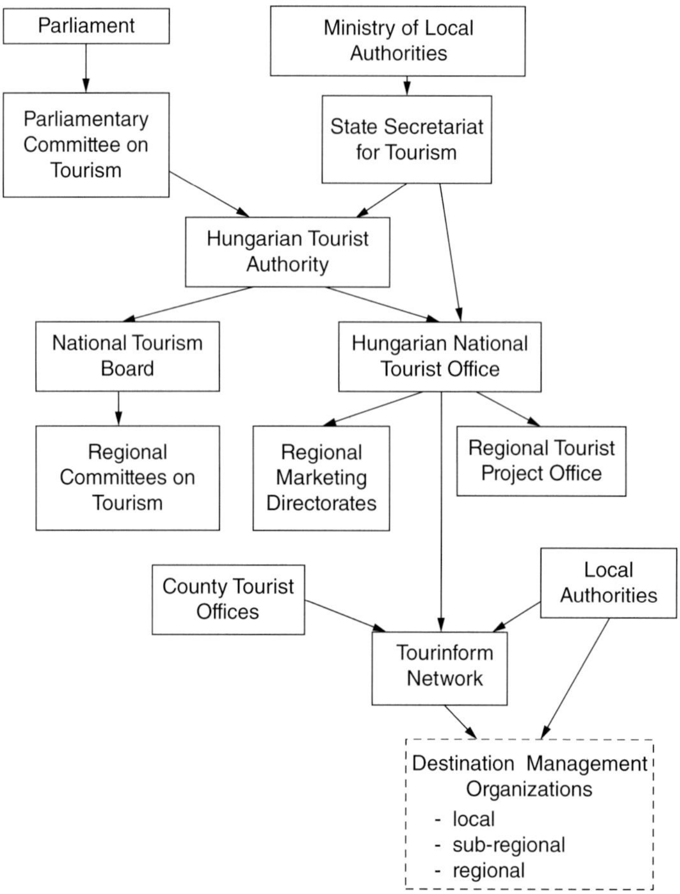

Fig. 11.2. Public-sector tourism bodies in Hungary. (Adapted from Clarke and Raffay, 2011.)

organization of tourism in Hungary. At the destination level both the local authority and the Tourinform offices will have a key role in the future DMOs, as there is no project funding available for any DMOs that are set up without the participation of the local authority or tourist information centres. Equally, to secure funding, private sector businesses and at least one civil organization have to be members of the new tourism DMO.

The diagram in Fig. 11.2 only records the management which is dedicated to tourism and does not include the RDAs, primarily conceived as economic agencies, or the National Park Authorities which were designed for conservation. They operate across boundaries which are not co-terminous with those of the tourism agencies. This poses the question of how all these bodies are coordinated, which in theory should happen but in practice is lacking. In practice we see an emerging consensus on the importance of tourism around the Balaton and for sustainable tourism.

This section examines the processes through which tourism TDMs (DMOs) were established in Hungary and how they impact on the Balaton. DMOs became an integral part of the national tourism strategy and are seen by the Secretariat for Tourism within the Ministry of Local Authorities as the vehicle for sustainable tourism development throughout the country (Lőrincz *et al.*, 2007). This development was the first time in Hungary that funds had been allocated for a major development of DMOs, where destination areas had to compete for the recognition to be allowed to operate and to secure funding. DMOs have been widely promoted as an effective strategy for strengthening tourism. One of the crucial research issues identified by Svensson *et al.* (2005, p. 33) concerned the inclusiveness of partnership arrangements: who participates, in what roles and who controls access to the partnership. Furthermore, coherence is necessary within DMO partnerships for them to be successful; coherence increases credibility by projecting a common outlook and helps to mobilize resources to implement the agreed agenda.

In Gyenesdias, one of the first DMOs set up in 2003, the local council closed its Tourism Committee in 2010, arguing that this role was now fulfilled by the DMO in which it had representation. The decisions made in the DMO still need to be taken through the planning committee of the local council but a direct link has been established. In Balatonfüred the lake front has been redeveloped in line with recommendations from the DMO but again this was processed through the planning committee (Kamondi, 2014). For the Balaton this becomes a crucial question when the competition between the 20 TDMs within the one destination region is considered. There are also areas that fall into the gaps between this plethora of DMOs and different financing regimes as well. Concerns about transferring power and responsibility to other agencies can prove very difficult to overcome. In these examples, there was a transfer of power and authority from a previously established free-for-all towards a structured approach of development through statutory DMOs, which involve both formal and informal negotiations of planning powers.

There are no formal linkages between these different agencies at the planning level. There are reporting regimes that could impose a sense of order from above but in practice this has not happened and a sense of the 'Balaton' has emerged from the informal needs of having to work together on a day-to-day basis. These informal connections require further research in order to elaborate how they work.

GOVERNANCE OF TOURISM IN THE BALATON

On a general level, governance is not only about institutions, processes and legal frameworks, but it increasingly concerns leadership and trust (Laesser and Beritelli, 2013). There are many advantages of stakeholder participation in progressing the transition towards sustainability; however, it is not often fully coordinated in the sustainable management of tourism. Each stakeholder group has a different set of needs and expectations relating to a destination's performance and its sustainability goals. A destination adopting a sustainable management approach to tourism should attempt to design one development strategy that achieves the objectives of various stakeholders. Furthermore, stakeholder perceptions are accepted as crucial for evaluating participatory processes and devising effective strategies for implementing sustainable tourism (Swarbrooke, 1999).

On a destination level, 'governance thinking' requires an increase in professionalism to achieve successful and sustainable development. The Balaton Development Agency has been working to bring stakeholders together to promote collaborative best practice. For the Balaton it is important to recognize that governance can only be produced through processes of collaboration, including cooperation and leadership, and through these collaborative moments, finding a common understanding. This sense of governance is a constant political negotiation, a learning process and requires a business model with high tolerance for trial and error. Conceptually, governance occurs through raising levels of community trust. From a practical perspective, destination governance production involved the act of creating a physical forum and formalized proceedings. In the case of the Balaton the authors feel that the entrepreneurial, cultural and political actors fulfil more decisive roles than those of the operational strategists/planners. In the governance process, the roles are a reflection of the perspective they represent. The political players focus on building coalitions, negotiating, and using power and influence in the network. Strategic individuals drive progress and operations, design strategy, define the goals and tasks, and drive the implementation.

It is argued that any destination management involves network management, so there is a need to develop means and ways to deal with different stakeholder logics/rationalities in thinking and acting. (Laesser and Beritelli, 2013) This becomes readily apparent in any examination of the Balaton as we briefly saw in the microcosm of Tihany. There are many issues involved in the development of planning contracts and development approvals around the lake. Partnership programmes can be designed to increase the number of actors that can be seen to be capable and legitimate players in processes of tourism development. The emergence of private-sector-led partnerships has been a major feature of the new Hungarian regime that has emerged since the shaking off of communism. Attempts to marry market forces with state provision have proved less than successful in a number of areas, from health to power supplies, but in tourism there are possibilities to bring the public and the private sector together to work more effectively to advance the interests of tourist destinations and the country more generally. However, in around the Balaton it is not even clear that information is routinely shared across the boundaries of the DMOs, local authorities and RDAs.

The first step in the creation of the DMOs was the 'registration stage', where the potential partnerships could demonstrate that they fulfilled all the conditions. These conditions included criteria such as: (i) the minimum number of guest nights spent in the destination; (ii) the proportion of the tourism businesses that had to be part of the DMO; (iii) all three sectors (public, private, voluntary) had to be members of the DMO; and (iv) the financial contribution of the members, etc. (It must be noted that the necessary minimum number of guest nights per year was defined by the relevant RDAs based on statistical data from the previous years, as a way of recognizing the destinations with real potential for tourism, i.e. they did not wish to fund partnerships that they did not deem viable in the long run.) In this case, fewer areas submitted bids than had been anticipated because the processes enforced were seen as too daunting. The documents were so complicated that many were given to consultants. For those still in the running, this meant that the rewards were higher as the pot of funding remained the same, but was shared between fewer DMOs.

One rather interesting issue must be mentioned as an example of using top-down tourism policy in a process where the bottom-up approach is necessary. During the bidding process, in every DMO-related document and at every event the emphasis was on explaining the key features of successful DMOs – including the need for the initiatives to come from the potential members of the partnership. However, at two of the information and support-providing events, the RDA allocating the fund was almost urging two destinations that were not eligible to apply to form a DMO together to be able to submit a bid. The destinations were in relative geographic proximity to each other but had rather different characteristics and a significantly different tourism offer, so neither of the two parties saw any potential in collaboration other than being able to qualify for government money. Another example of a similar approach suggested that two settlements of a larger lake tourism destination formed separate DMOs and submitted two applications in order to receive twice the sum they would be eligible for as one DMO. This was the point where we thought that the split in the destination recommended by the RDA posed a real threat in terms of dividing the (tourism) community of an otherwise coherent destination.

In practice the complexity of the organizational arrangements around the lake have rendered the regional DMO largely impotent. The funding made available by the central government has been restricted to local DMOs, which has left the regional Balaton DMO entirely dependent on income from subscriptions and meant it has lacked the resources to drive new developments. If we look for the positive from this situation, it can be argued that it places greater emphasis on the need for partnership working and highlights the formal and informal stakeholder interactions. Stakeholder involvement and stakeholder engagement studies allow for a demonstration of the dynamics of managing sustainable tourism development around the lake, and demonstrate that it is fraught with challenges and yet also presents opportunities, as demonstrated below in the case study on the critical exploration of working towards sustainable development in the Balaton.

Case Study: Critical Exploration of Sustainable Development Issues – Sustainabilty and Stakeholder Involvement in the Balaton

In 2013–2014, the Balaton Development Agency was charged with undertaking a stakeholder review of the sustainable tourism development of the region within the framework of the INTOURACT project. The INTOURACT project was designed to develop integrated tourism action plans in functional regions of South-Eastern Europe, one of which was the Lake Balaton Resort Area (LBRA).

The important tasks of the project are to foster the expansion of the tourism season and to ensure synergies in tourism in the rural, urban and peri-urban areas in accordance with sustainable development goals, in order to balance the different tourism concentration levels between the popular tourism destinations and rarely visited background areas. Furthermore, the project aims to increase knowledge and disseminate the highest economical standards of environmental protection, which provide protection for residents in the rural areas, and also guarantee safe rural development. This was undertaken by a series of meetings held around a series of themes:

- environmental protection and ecotourism;
- mobility;
- culture and cultural heritage; and
- innovation and technical competencies.

Stakeholder representation was structured by invitation and the 'invitees' were defined as those affected stakeholders in the LBRA who are competent in the four themes of the project workshops, as well as other interested groups committed to the development of the area. This was a significant attempt to reach out to stakeholders and increase the sense of involvement in the future of sustainable tourism in the LBRA. The four meetings generated audiences of around 40 people, which in Keszthely, where the meetings were held, is a good turnout. However, it is on the eastern end of the north shore of the lake and it is likely that the spatial as well as the local politics ruled out some people participating. You are looking at a 2 h drive to reach a meeting in which you may not be recognized, either personally or professionally, as someone with a valid contribution to make. However, the complications of the Balaton argue that questions must be asked about the representativeness of the audience. By stating that attendance was for those who were competent in all of the four themes, it was difficult to see where these stakeholders were going to come from. The earlier discussions on governance and the growth of collaboration pointed to a need to build a sense of competence and to strengthen the capabilities of individuals to ensure that people feel enabled to take part in such consultations. Sectoral and vested interests were challenged by the four areas of competence required but there was still room for tensions to emerge in the perceptions within the consultative group of what developments could be said to be sustainable and how different components

(Continued)

Case Study. Continued.

of the tourism development could be held together. This could be seen in a discussion of the role of spa tourism and health tourism in the offers of the Balaton. One town, very close to Keszthely, is renowned for its spa waters and medical treatments but the voice of the stakeholders – or at least one of them – was heard saying 'Hévíz is full of old, ill Hungarians'. Whether this is an accurate account can be – and was – discussed but it does point to the difficulty of offering medical tourism in the context of a destination strategy looking to promote a range of interests and activities. The voice was legitimate, it came from an invitee, but the observation does not lead to a simple conclusion. The spas, with their associated health and medical tourism facilities, are a significant feature of tourism in certain parts of the Balaton and in the wider context of Hungary. The planning issue is to accommodate both the health tourists and other market segments in such ways as to maximize the benefits of all segments using the spa and the waters.

While sustainability and particularly sustainable tourism is becoming increasingly popular, Jenkins and Schröder (2013) indicated that doubts have been raised about whether the promised harmonization of ecological, social and economic goals associated with sustainability is actually achievable. These doubts are emphasized by the complexities of the formal and informal relationships involved in the planning of the Balaton. We have shown how the formal actors in the governance of the lake region are many and varied and the sectoral interests are equally diverse. This has led to uneven development within the destination area, despite the significance ascribed to the Balaton as an environmental resource and the recognition nationally and locally of tourism as a major economic force. Consensus has been difficult to achieve and hence the developments within the region continue to vary in both quality and sustainability. The case study demonstrates the need for greater collaboration among stakeholders from around the lake, spatial distances need to be closed organizationally to enhance the tourist experience.

CONCLUSION

One of the challenges highlighted is whether the LBRA should be treated as a planning unit or not. The planning of destinations and the governance of destinations have focused on a lower level unit of analysis and it is possible to see the homes of the 20 TDMs can be seen as 20 different and distinct destinations, with their own specific needs, lacks, specialisms and senses of future directions. In sustainable planning terms, subsidiarity would argue that these micro-level destinations are the correct level of agency to direct, monitor and evaluate the sustainable development of tourism. The DMOs and the work undertaken with stakeholders have great significance locally. However, the Balaton is not divisible in one important sense – the lake always connects and decisions made in one part of the lake have direct and indirect impacts in all the other parts of what is always the same lake. The sustainable planning model argues that the socio-cultural, economic and environmental elements must be held in balance but the lake suggests that one element is more important than the others. If the planning regime, and more likely if it is divided the planning regimes, allow the environment to be eroded and the future of the ecosystems threatened then all of the lake area will be affected and none of the stakeholders can afford to neglect this. However, there are significant socio-cultural systems to be found in the settlements around the lake and although these are not as easily monitored, they need to be protected and promoted just as much as the other parts of the natural environment. The lesson to be taken from the case study is that complexity does not have to weaken our ability to plan for sustainable tourism. The central values of sustainability have to be enshrined in the planning frameworks which allow local plurality to prosper. This is difficult and Hungary has failed to legislate for tourism so far, although the debate has been held in the last two parliaments. However, it does have an Office of the Ombudsman for Future Generations which can examine cases where developments have taken place that are not seen as being sustainable. This obviously only works in retrospect and damage has already been done but the principle of intervention for sustainability does exist in Hungarian regulation. The Balaton calls out for this to be enshrined from the 'top down' and then for the recognition of the local to be allowed to prosper within this framework. Subsidiarity, with a collective conscience, regional structures and local stakeholder participation, points the way forward for a sustainable Balaton.

QUESTIONS

1. Is the Balaton too complex an area to plan as a single tourism destination?

2. How can stakeholder involvement be promoted throughout the Balaton Development Region?

3. What are the principle drivers for sustainable tourism development in the Balaton?

BIBLIOGRAPHY

Aas, C., Ladkin, A. and Fletcher, J. (2005) Stakeholder collaboration and heritage management. *Annals of Tourism Research* 32(1), 28–48.

Balaton Development Agency (2013) Sustainable Tourism for the Balaton. Presentation to the INTOURACT Project. Balaton Development Agency, Siofók, Hungary.

Balaton Highland National Park (2014) Available at: www.bfnp.hu/english/pages/tihany_lavender_house/1/1 (accessed 9 June 2014).

Bornhorst, T., Brent Ritchie, J.R. and Sheehan, L. (2010) Determinants of tourism success for DMOs and destinations: an empirical examination of stakeholders' perspectives. *Tourism Management* 31(3), 572–589.

Bramwell, B. and Lane, B. (1993) Sustainable tourism: an evolving global approach. *Journal of Sustainable Tourism* 1(1), 1–5.

Bramwell, B. and Lane, B. (1999) Collaboration and partnerships for sustainable tourism. *Journal of Sustainable Tourism* 7(2), 179–181.

Bramwell, B. and Meyer, D. (2007) Power and tourism policy relations in transition. *Annals of Tourism Research* 34(3), 766–788.

Bramwell, B. and Pomfret, G. (2007) Planning for lake and lake shore tourism: complexity, coordination and adaptation. *Anatolia* 18(1), 43–66.

Bramwell, B. and Sharman, A. (1999) Collaboration in local tourism policymaking. *Annals of Tourism Research* 26(2), 392–415.

Cheong, S. and Miller, M.L. (2000) Power and tourism – a Foucauldian observation. *Annals of Tourism Research* 27(2), 371–390.

Clarke, A. and Raffay, Á. (2007) Understanding stakeholder relations in urban tourism management. In: Long, P. (ed.) *Researching Destination Management, Policy and Planning: Linking Culture, Heritage and Tourism*. Leeds Metropolitan University Press, Leeds, UK.

Clarke, A. and Raffay, Á. (2008) Stakeholder relationships and sustainable tourism partnerships: power, identity and sustainability. In: Long, P. (ed.) *Proceedings of the EUTO Conference 2008 Attractions and Events as Catalysts for Regeneration and Social Change*. Christel DeHaan Tourism and Travel Research Institute, University of Nottingham and The Centre for Tourism and Cultural Change, Leeds Metropolitan University, September 2008. The University of Nottingham, Nottingham, UK, pp. 49–67.

Clarke, A. and Raffay, Á. (2011) The introduction of tourism destination management organisations in Hungary: top down meets bottom up. In: Dredge, D. and Jenkins, J. (eds) *Stories of Practice: Tourism Policy and Planning*. Ashgate, Farnham, UK, pp. 311–334.

Currie, R.R., Seaton, S. and Wesley, F. (2008) Determining stakeholders for feasibility analysis. *Annals of Tourism Research* 36(1), 41–63.

d'Angella, F. and Go, F.M. (2009) Tale of two cities' collaborative tourism marketing: towards a theory of destination stakeholder assessment. *Tourism Management* 30(3), 429–440.

de Araujo, L.M. and Bramwell, B. (2000) Stakeholder assessment and collaborative tourism planning: the case of Brazil's Costa Dorada project. In: Bramwell, B. and Lane, B. (eds) *Tourism, Collaboration and Partnerships: Politics, Practice and Sustainability*. Channel View Publications, Clevedon, UK, pp. 272–294.

Deák, É.V. (1979) *Felfedezem a Balatont*. Móra Könyvkiadó, Budapest.

Dredge, D. (2006a) Policy networks and the local organisation of tourism. *Tourism Management* 27(2), 269–280.

Dredge, D. (2006b) Networks, conflict and collaborative communities. *Journal of Sustainable Tourism* 14(6), 562–581.

Dredge, D. (2010) Place change and tourism development conflict: evaluating public interest. *Tourism Management* 31(1), 104–112.

Glaser, B. and Strauss, A. (1968) *The Discovery of Grounded Theory: Strategies for Qualitative Research*. Weidenfield and Nicholson, London.

Gray, B. (1985) Conditions facilitating interorganizational collaboration. *Human Relations* 38(10), 911–1036.

Gray, B. (1989) *Collaborating: Finding Common Grounds for Multiparty Problems*. Jossey-Bass, San Francisco, California.

Hall, C.M. (2000) Rethinking collaboration and partnership: a public policy perspective. In: Bramwell, B. and Lane, B. (eds) *Tourism Collaboration and Partnerships: Politics, Practice and Sustainability*. Channel View Publications, Clevedon, UK, pp. 143–158.

Healey, P. (1997) *Collaborative Planning: Shaping Places in Fragmented Societies*. Macmillan, Basingstoke, UK.

Healey, P. (2005) Place, identity and governance: transforming discourses and practices. In: Hillier, J. and Rooksby, E. (eds) *Habitas: A Sense of Place*. Ashgate, Aldershot, UK, pp. 189–218.

Hillier, J. and Rooksby, E. (2005) *Habitas: A Sense of Place*. Ashgate, Aldershot, UK.

Huxham, C. (1993) Collaborative capability: an intra-organisational perspective on collaborative advantage. *Public Money and Management* (July–September), 21–28.

Jamal, T.B. and Getz, D. (1995) Collaboration theory and community tourism planning. *Annals of Tourism Research* 22(1), 186–204.

Jenkins, I. and Schröder, R. (eds) (2013) *Sustainability in Tourism: A Multidisciplinary Approach*. Springer Grabler, Wiesbaden, Germany.

Kaiser, O. and Dlusztus, I. (2005) *Balaton*. Pécsi Direkt Kft, Alexandra Kiadója, Pécs, Hungary.

Kamondi, Z.Á. (2014) The state of Hungarian Destination Management Organisations through the examples of the DMOs of Balatonfured and Siofok. MA dissertation, University of Pannonia, Italy.

Kovács, E. (2007) A 19. Századi Balaton Világa. Széphalom Könyvműhely, Budapest.

Ladkin, A. and Bertramini, A.M. (2002) Collaborative tourism planning: a case study of Cusco Peru. *Current Issues in Tourism* 5(2), 71–93.

Laesser, C. and Beritelli, P. (2013) St. Gallen consensus on destination management. *Journal of Destination Marketing & Management* 2(1), 46–49.

Levendula Ház (2014) Available at: http://www.budapest.com/hungary/lake_balaton/lavender_house (accessed 09 June 2014).

Lőrincz, K., Raffay, Á. and Clarke, A. (2007) Best practice in rural destination management – exploring Hungarian excellence. In: Long, P. (ed.) *Researching Destination Management, Policy and Planning: Linking Culture, Heritage and Tourism*. Leeds Metropolitan University Press, Leeds, UK.

March, R. and Wilkinson, I. (2009) Conceptual tools for evaluating tourism partnerships. *Tourism Management* 30(3), 455–462.

Marien, C. and Pizam, A. (1997) Implementing sustainable tourism development through citizen participation in the planning process. In: Wahab, S. and Pigram, J.J. (eds) *Tourism, Development and Growth. The Challenge of Sustainability*. Routledge, London, pp. 164–178.

Martonné Erdős, K. (2001) *Magyarország tájföldrajza*. Kossuth Egyetemi Kiadó (Debreceni Egyetem), Debrecen, Hungary.

Mason, P. (2008) *Tourism Impacts, Planning and Management*. Butterworth Heinemann, Oxford.

Michael, E. (2007) *Micro-Clusters and Networks: The Growth of Tourism*. Elsevier, Oxford.

Minnaert, L., Maitland, R. and Miller, G. (2011) What is social tourism? *Current Issues in Tourism* 14(5), 403–415.

Okumus, F., Altinay, L. and Roper, A. (2007) Gaining access into organisations for qualitative research. *Annals of Tourism Research* 34(1), 4–26.

Panyik, E. and Ratz, T. (2014) Tourism policy in Hungary: from central planning to multi-level destination management. In: Costa, C., Panyik, E. and Buhalis, D. (eds) *European Tourism Planning and Organisation Systems: The EU Member States*. Channel View Publications, Clevedon, UK, pp. 225–242.

Pforr, C. (2005) Three lenses of analysis for the study of tourism public policy: a case from Northern Australia. *Current Issues in Tourism* 8(4), 323–343.

Porter, M.E. (1995) *Competitive Advantage of Nations*. Palgrave Macmillan, Basingstoke, UK.

Puczkó, L. and Rátz, T. (2011) Social tourism in Hungary: from trade unions to cinema tickets. *Current Issues in Tourism* 14(5), 459–473.

Reed, M.G. (1997) Power relations and community-based tourism planning. *Annals of Tourism Research* 24(3), 566–591.

Ritchie, J.R.B. and Crouch, G.I. (2000) The competitive destination: a sustainable perspective. *Tourism Management* 21(1), 1–7.

Sainaghi, R. (2006) From contents to processes: versus a dynamic destination management model (DDMM). *Tourism Management* 27(5), 1053–1063.

Scott, N., Baggio, R. and Cooper, C. (2008) *Network Analysis and Tourism: From Theory to Practice*. Channel View, Clevedon, UK.

Selin, S. and Chavez, D. (1995) Developing an evolutionary tourism partnership model. *Annals of Tourism Research* 22(4), 844–856.

Simpson, K. (2001) Strategic planning and community involvement as contributors to sustainable tourism development. *Current Issues in Tourism* 4(1), 3–39.

Spradley, J.P. (1979) *The Ethnographic Interview*. Holt, Rinehart and Winston, New York.

Stevenson, N., Airey, D. and Miller, G. (2008) Tourism policy making: the policymakers' perspectives. *Annals of Tourism Research* 35(3), 732–750.

Svensson, B., Nordin, S. and Flagestad, A. (2005) A governance perspective on destination development – exploring partnerships, clusters and innovation systems. *Tourism Review* 60(2), 32–37.

Swarbrooke, J. (1999) *Sustainable Tourism Management*. CAB International, Wallingford, UK.

Tihany TDM (2014) Available at: www.tihanyinfo.com/en/ (accessed 9 June 2014).

Tosun, C. (2000) Limits to community participation in the tourism development process in developing countries. *Tourism Management* 21(6), 613–633.
United Nations Environment Programme/United Nations World Tourism Organization (UNEP/UNWTO) (2005) *Making Tourism More Sustainable: A Guide for Policy Makers*. UNEP Division of Technology, Industry and Economics (DTIE), Tourism Programme, Paris.
Veszprémi Egyetem Turizmus Tanszék (2004) *Üdülőhelyi igazgatóságok a Balaton régióban – Tanulmány*. Veszprémi Egyetem Turizmus Tanszék, Veszprém, Hungary.
Vizugy Igazgatóság (2010) Available at: https://www.vizugy.hu/uploads/csatolmanyok/44/balaton2.pdf (accessed 9 June 2014).
Wahab, S. and Pigram, J.J. (1998) Tourism and sustainability. Policy considerations. In: Wahab, S. and Pigram, J.J. (eds) *Tourism, Development and Growth. The Challenge of Sustainability*. Routledge, London, pp. 277–290.
World Travel and Tourism Council (WTTC) (1996) *Agenda 21 for the Travel and Tourism Industry: Towards Environmentally Sustainable Development*. WTTC, London.
Yasarata, M., Altinay, L., Burns, P. and Okumus, F. (2010) Politics and sustainable tourism development: can they co-exist? Voices from North Cyprus. *Tourism Management* 31(3), 345–356.
Yuksel, F., Bramwell, B. and Yuksel, A. (1999) Stakeholder interviews and tourism planning at Pamukkale, Turkey. *Tourism Management* 20(3), 351–360.
Zeidler, M. (ed.) (2003) *Trianon*. Osiris Kiadó, Budapest.

chapter 12

National Approaches to Planning and Tourism in Latvia

Dzintra Atstāja,[1*] Ivars Brīvers[1] and Agita Līviņa[2]

[1]*School of Business and Finance, Riga, Latvia;* [2]*Vidzeme University of Applied Sciences, Valmiera, Latvia*

LEARNING OBJECTIVES

- To understand the physical and human geography and historical development of tourism in Latvia.
- To review information about the present policy in tourism governance in Latvia and the linkages with the implementation of that policy in tourism planning and management.
- To be able to discuss the sustainability of tourism policy and planning in Latvia and other Baltic States.

INTRODUCTION

The main focus of the chapter is to provide an evaluation of sustainability of tourism in relation to planning processes in Latvia, initially providing an overview of the natural geographical and cultural heritage and the history of tourism development in Latvia. The chapter culminates in a case study of dark tourism, which is a key product that is being developed within contemporary Latvia, representing aspects of its recent past. The first section is an overview of the physical and

*dzintra.atstaja@ba.lv

human geography of Latvia and also within the context of the Baltic region. A short historical review of development of Latvia and the Latvian nation is included. The second section deals with the historical development of tourism in Latvia split into four distinct time periods: (i) the historical phase up until the First World War; (ii) the period of independence 1918–1940; (iii) the Soviet time period 1945–1990; and (iv) the post-Soviet period after 1990. This is followed by the third section which deals with the contemporary structure and processes of tourism governance including strategic development planning initiatives, tourism planning initiatives, niche tourism development and sustainability and pedagogy. The fourth section includes the case study about dark tourism in Latvia.

OVERVIEW OF THE PHYSICAL AND HUMAN GEOGRAPHY OF LATVIA WITHIN THE CONTEXT OF THE BALTIC REGION

Latvia is a comparatively small country in Europe, on the East coast of the Baltic Sea. The territory of Latvia is 64,600 km^2 (CSB, 2014) and has borders with Estonia in the north, Russia in the east, Belarus in the south-east and Lithuania in the south. On the west Latvia's closest neighbour is Sweden. Between the two countries lies the Baltic Sea – the distance between the Latvian and the Swedish coast is approximately 300 km. The capital of Latvia is Riga with approximately 650 thousand inhabitants (CSB, 2014). Riga was founded in 1201 near the place where the Daugava River – the largest river in Latvia – flows into the Baltic Sea. Riga may be considered the centre not only of Latvia, but the whole Baltic region which consists of three countries – Latvia, Lithuania and Estonia – as Riga is the largest of the three capitals, and is situated in the centre of the region.

The distance from Riga to Vilnius (the capital of Lithuania) and Tallinn (the capital of Estonia) is approximately the same – 300 km. The distance as the crow flies from Riga to some other cities in Europe is: London – 2250 km, Paris – 2320 km, Stockholm – 390 km, Helsinki – 370 km, Moscow 920 km. The other large towns are: (i) Daugavpils in the south-east; (ii) Liepāja in the west; (iii) Jelgava in the centre, 40 km south of Riga; (iv) Jūrmala, a seaside resort in the neighbourhood of Riga; (v) Ventspils in the west; (vi) Rēzekne in the east; (vii) Valmiera in the north; and (viii) Ogre and Jēkabpils in the centre, to the east of Riga. Historically, up to the 20th century the right shore of the Daugava River developed as an agricultural area, while the left shore, since the times of Courland Duchy in the 17th century, was more industrialized. The events of the 20th century has meant there is now less of a difference, and Kurzeme, being one of the most densely populated regions of Latvia before the First World War, now has a lower population density, compared with other regions, except Latgale, which has been rapidly losing its population in recent years.

The population of Latvia was approximately just under 2 million people in 2014 (CSB, 2014). Two-thirds of them live in towns, one-third in rural areas. More than half the population lives in the Riga megalopolis area. Latvians make up approximately 60% of the whole population; other ethnical groups are Russians, Byelorussians, Ukrainians, Poles and Lithuanians. In all the seven largest cities Latvians are the minority or close to the minority, not exceeding 50% of the population. The rural area is mainly populated by Latvians and it has the external European Union (EU) border with Russia and Belorussia. Most incoming international tourists come from Lithuania, Estonia, Russia, Finland and Germany. Lithuania and Estonia are our closest neighbours with a common tourism policy and marketing activities.

The Latvian language belongs to the small group of Indo-European languages – the Baltic language group. At present there are only two languages belonging to this group – Latvian and Lithuanian. Although there are some similarities both with Slavonic languages, particularly Russian, and Germanic languages, particularly German in the Latvian language, the Latvian language does not belong to any of these groups. The three Baltic nations – Latvians, Lithuanians and Estonians – have a lot of similarities in their history, but also a lot of differences. Latvians and Lithuanians have the same roots, but for 700 years they had lived separately. The border between Latvia and Lithuania is perhaps the oldest border in Europe. Estonians belong to quite a different language group – Finno-Ugrians, but for 700 years these two nations have lived closely together without any border between them. That explains why the Latvian mentality and traditions are closer to Estonians, than to Lithuanians. The three nations historically have lived together harmoniously and there have not been any serious conflicts between them (except some rifts concerning football or basketball).

The three capitals – Riga, Vilnius and Tallinn – look different, as visitors usually expect. Latvia and Estonia for centuries have been under the influence of German culture; Riga and Tallinn (Reval) have been members of the Hanseatic League. Lithuania, on the other hand, has been under the influence of Polish culture. Catholic and Orthodox religions constitute approximately equal groups.

Latvia's strategic location has made it into an international crossroad for trade, commerce and cultural exchange since ancient times. The modern history of Latvia starts from the end of the 11th century, when German merchants arrived and established a settlement near the mouth of the Daugava River and the present territory of Latvia consists of four historical regions: (i) Vidzeme (Livonia) in the centre; (ii) Latgale (Inflantia) in the east; (iii) Kurzeme (Courland) in the west; and (iv) Zemgale (Semigallia) in the south. In the 18th century the present territory of the Baltic was gradually included in the Russian Empire – first Livonia and Estonia as a result of the Northern War between Russia and Sweden, then Inflantia as the result of the Second division of Poland between Russia and Prussia, and finally Courland and Lithuania as the result of the Third division of Poland, when the Polish–Lithuanian Commonwealth ceased to exist. The Baltic States declared their independence in 1918, when the First World War led to the collapse of the Russian Empire. In 1940 all the Baltic States were occupied and included in the Soviet Union as formally independent states. In 1991 Latvia, Lithuania and Estonia restored independence.

Latvia has one international airport – Riga Airport. It is the largest airport in the Baltic, and is comparatively well connected with other parts of Latvia and even the largest towns of Lithuania and Estonia by the national airline's (airBaltic) transfer coaches.

The territory of Latvia is mainly flat with lakes and hills, the highest being situated in the eastern part of the territory with the highest point a little more than 300 m above sea level. Latvia's advantage in the field of tourism is the long, sandy beach extending for 500 km in overall length, as well as the comparatively mild climate and serene environment with some historical health resorts and places of recreation, such as Baldone, Ķemeri, Jūrmala, Liepāja, Saulkrasti and Ogre. The climate in Latvia is not extreme, with rather mild winters, when the air temperature seldom falls below 10°C, and not very hot summers, when the air temperature seldom exceeds 30°C. The water temperature in the Baltic Sea in summer is on average about 20°C, so tourists who are accustomed to the warm waters of the southern seas at first feel uncomfortable trying to bathe in the Baltic Sea, but later enjoy it, finding it refreshing. The environment of Latvia creates a competitive advantage in tourism for that segment of tourists who are not seeking excitement and extreme adventures, but rather prefer a peaceful rest and mild environment.

THE HISTORY AND DEVELOPMENT OF TOURISM WITHIN LATVIA

The concept of tourism was established only at the beginning of the 19th century, and the travelling activities until then cannot be considered as tourism in a modern sense. The most significant specialist from the historical perspective of tourism in Latvia was Kārlis Vanags, who considered the origins of tourism in Latvia in 1862, when different organizations started to organize regular excursions. This section highlights the historical development of tourism in Latvia split into four distinct time periods: (i) up until the First World War and the foundation of the independent Latvia state; (ii) the period of independence from 1918 until 1940; (iii) the Soviet period from 1945 until 1990; and (iv) the post-Soviet period after 1990.

Tourism before the 19th century

In the 17th century there was only one hotel in Riga, which was situated on the *Lielā Smilšu iela* and it was recognized as unprofitable and was closed in 1653. In the 18th century new hotels opened, among which the most popular was the Petersburg hotel, opened in 1763. In 1864 there were 13 hotels in Riga and by 1901, 25 hotels (Rozīte, 1999). The first guidebook about Riga was published in German in 1880 (Rozīte, 1999). Besides Riga the seaside resort was opened in Liepāja in 1810, and after 1875, when the well-house was opened, Liepāja became a significant place for visitors (Rozīte, 1999). In 1795 medicinal baths were opened in Baldone, 30 km from Riga. A year later in 1796 Ķemeri (40 km from Riga) became a health resort owing to the sulfur springs (Rozīte, 1999).

The opening of the Riga–Tukums railway in 1877 made visits to the seaside common for Riga citizens. By the end of the 19th century there was a passenger train that ran from Moscow to Ķemeri, which made this resort available for a wider range of inhabitants of the Russian Empire. In the beginning of the 20th century, there was the electric tram in Ķemeri, which took visitors from the railway station to the recreation places at the seaside. Besides that, other places of rest and recreation were developed at the seaside, such as Neubaden (now Saulkrasti) and Vecāķi, and nice quiet resort places such as Ogre and Sigulda about 40–50 km distance from Riga.

Tourism in the period of independence

The Ministry of the Interior, Ministry of Foreign Affairs and the Railway Central Board established the Latvian Central Tourism Society (*Latvijas Centrālā Tūristu biedrība*) in 1929 which oversaw the transformation of a mass movement in tourism by foreign standards. This society joined the International Congress of Official Tourist Traffic Associations at The Hague. In 1931 the Bureau of Tourism at the Foreign Passports Department of the Ministry of the Interior was established and it was transferred to the Ministry of Social Affairs as the Tourism Department in 1937. The administrator of both the Bureau of Tourism and the Tourism Department was lawyer Kārlis Vanags (born 1907), who was the author of the guidebook *Ceļvedis pa dzimto zemi* in 1937, as well as other books about tourism (Švābe, 1955).

Literature about tourism and guidebooks were regularly published from the mid-1920s onwards, such as the guidebook *Riga* by Mārtiņš Celms in 1926, and many others. There was also the Latvian Excursion Society (*Latvijas ekskursiju biedrība*), which published a monthly paper *Ekskursants* since 1933 (Švābe, 1955). The tourism movement in Latvia had two objectives: (i) to involve a wide range of the indigenous population to 'Travel and get to know your Native land'; and (ii) to attract foreign tourists by developing the facilities and amenities in the most beautiful places. Since then Latvia became a serious competitor for the other Baltic and Scandinavian countries. For instance, the number of foreign tourists in 1931 was 85,719 in Latvia, compared with approximately 40,000 in Lithuania, 37,470 in Finland and 71,975 in Norway. Each year approximately 50,000 Latvian citizens went on foreign tours. There was a special fund for the popularization of tourism, which received financing from the fee from foreign passports. This fund supported tourist centres, built observation towers (in the castle ruins in Bauska and Cēsis), established booths and sheds for tourists, information stands and road signs, stairs and benches, etc. (Švābe, 1955).

Postage stamps, maps and posters were widely used for the popularization of tourism. The Tourism Department maintained boat traffic on the Daugava River from Pļaviņas to Koknese, on Usma Lake, and carried passengers between the ports of Kurzeme and Vidzeme Seaside. The State Travelling Bureau Ceļtrans organized cheap trains for tourists and ski-runners each Sunday. Groups of tourists with 10 or more persons would pay only half of the price for railway tickets. In 1937 there were 185 hotels, 68 boarding houses and 441 tourist houses in Latvia, where 34,363 travellers stayed for 57,279 nights (Švābe, 1955). Further development of tourism was obstructed by the shortage of modern long-distance buses and the poor quality of unpaved roads. Modern hotels were built mainly for foreign tourists in Jelgava, Cēsis, Valmiera and Rēzekne. For implementing a common tourism policy of the Baltic States and facilitating the travelling processes the Baltic Tourism Council was established (Švābe, 1955).

Tourism in Soviet Latvia

After the Second World War, when Latvia was incorporated in the Soviet Union, the Latvian Republican Tourism and Excursion Board (*Latvijas Republikāniskā Tūrisma un ekskursiju pārvalde*) was established in 1946, under the subjection of the Latvian Republican

Trade Union Council. The Tourism section at the Committee of Physical Culture and Sports was established in 1954 and since 1962 the development of tourism was managed by the Latvian Republican Tourism and Excursion Council. The tourism activities among schoolchildren were administered by the Central Children Excursion and Tourism Station (*Centrālā bērnu ekskursiju un tūrisma stacija*) (Samsons, 1970). Since 1963, the Autostop system emerged, where tourists who were hitch-hiking received special coupons with which they could pay the car drivers, who later received remuneration for this. Foreign tourist affairs in the whole territory of the Soviet Union, as well as travels of soviet citizens abroad, were administered by the All-Union agency 'Intourist'. In 1966, the Foreign Tourism Board at the Council of Ministers of the Latvian Soviet Socialist Republic (SSR) was in operation, and until the collapse of the Union of Soviet Socialist Republics (USSR) the internal tourism policy was implemented by the Central Board of Tourism and Excursions. Foreign tourism policy issues were the responsibility of the USSR State Foreign Tourism Committee and the International Youth Tourism bureau 'Sputnik' (Samsons, 1970).

In 1973, the USSR started recreation planning, which was updated in 1980, with the whole USSR territory being divided into four recreation zones and into 20 districts, on the basis of staged tourism development. Latvian SSR was included in the first and second zones, with a high 'stage' of tourism development. In the late 1980s, serious research of tourism resources and tourism potential was conducted, particularly in Latvia by Trušiņš and Buka (Trušiņš and Buka, 1986; Klepers, 2012). This methodology assessed the capacity and appropriateness of places and facilities to domestic and international visitor flows. This included a panel of experts making an assessment of perception of each place and facility.

Tourism since 1991

In 1991, Latvia declared the restoration of its independence and the governance structure of sectors were changed. From 1993 until 2003, the responsibility of the tourism sector was in the hands of the Ministry of Environmental Protection and Regional Development. In 1998, the Law on Tourism was accepted by the Parliament of the Republic of Latvia. The purpose of this law was to create a legal basis for the development of the tourism industry in Latvia and to specify the procedure in which State administrative institutions, local governments and merchants operate in the area of tourism and protect the interests of tourists. This law set objectives for a state tourism policy, in order to ensure the rational use and protection of tourism resources and to specially promote the development of local and international tourism (Cabinet of Ministers of the Republic of Latvia, 1998). During this period the first tourism planning documents were created but the implementation stage of planning documents was weak because it was not connected with an allocation of financial resources.

Local scholars turned their attention to tourism development too. The first doctoral thesis on the subject was M. Rozīte's 'Urban tourism, its development and spatial expressions in Rīga' in 1999 (cited in Rozīte, 1999). This research naturally focused on urban tourism in Riga, as the capital is still the most visited place by foreigners. Since 2004, the Ministry of Economics was responsible for tourism policy. In 2002, the Government of the Republic of Latvia, the Government of the Republic of Estonia and the Government of the Republic of Lithuania have cooperated in the field of tourism, establishing a Joint Tourism Committee. In recent years, responsible ministers of tourism affairs took decisions to strengthen Baltic tourism activities. The Latvian Tourism Development Agency is a tourism policy executive institution established in 1997, and reorganized in 2002, and this agency operated under the Ministry of Economics. The agency is taking part in organizing a marathon, 'The Great Baltic Travel' promotion on domestic tourism, 'Discover Latvia again' and other marketing and promotional activities for tourism entrepreneurs and visitors. Latvia was a full member of the United Nations World Tourism Organization (UNWTO) from 2005 to 2012. In fact, Latvia interrupted its membership of UNWTO from 2013, based on an analysis of the benefits of potential opportunities set against the cost of paying the membership fee. In 2014, the Government of Latvia has been starting negotiations to enter the Organisation for Economic Co-operation and Development (OECD). Tourism officials from all three Baltic States have prepared common tourism activities for visitors, for instance the campaign for popularizing tourism products of Baltic States 'The Great Baltic Travel', particularly focuses on travellers from Estonia, Latvia and Lithuania.

THE STRUCTURE AND PROCESSES OF TOURISM GOVERNANCE

As stated above, from an historical point of view tourism in Latvia can be divided in four stages: (i) up to the First World War; (ii) the first Republic of Latvia 1918–1940; (iii) under Soviet rule 1945–1991; and (iv) the current period in a free market economy since 1991. The tourism balance was positive in the first Republic of Latvia, nowadays the tourism balance is negative, but it is slowly improving.

Figure 12.1 demonstrates tourism governance in Latvia from the international level to the local level and highlighted are the main organizations and stakeholders of tourism planning and policy in Latvia.

In the next section the main institutions involved in the tourism industry in Latvia at different policy-making levels are discussed.

Strategic development planning initiatives

In 2002, the Cabinet of Ministers adopted the *Guidelines of Sustainable Development of Latvia* (Cabinet of Ministers of the Republic of Latvia, 2002). The document included an overview of the current situation on sustainable development and provided a description of sustainable development principles and goals, as well as policy outcomes and performance indicators in 16 fields (e.g. water protection, regional development, education, poverty, industry, transport and tourism). A key priority for achieving tourism policy goals was to increase the intensity of tourism flows and to regulate tourism flows within environmentally sensitive areas. Other aims included: (i) to improve the quality of tourism services; (ii) to develop quality standards; (iii) to promote availability of tourism services for every target group (seniors, children, physically challenged people, unemployed); and (iv) to continue cooperation with other countries around the Baltic Sea.

These 'Guidelines' identified six indicators for policy achievement performance,

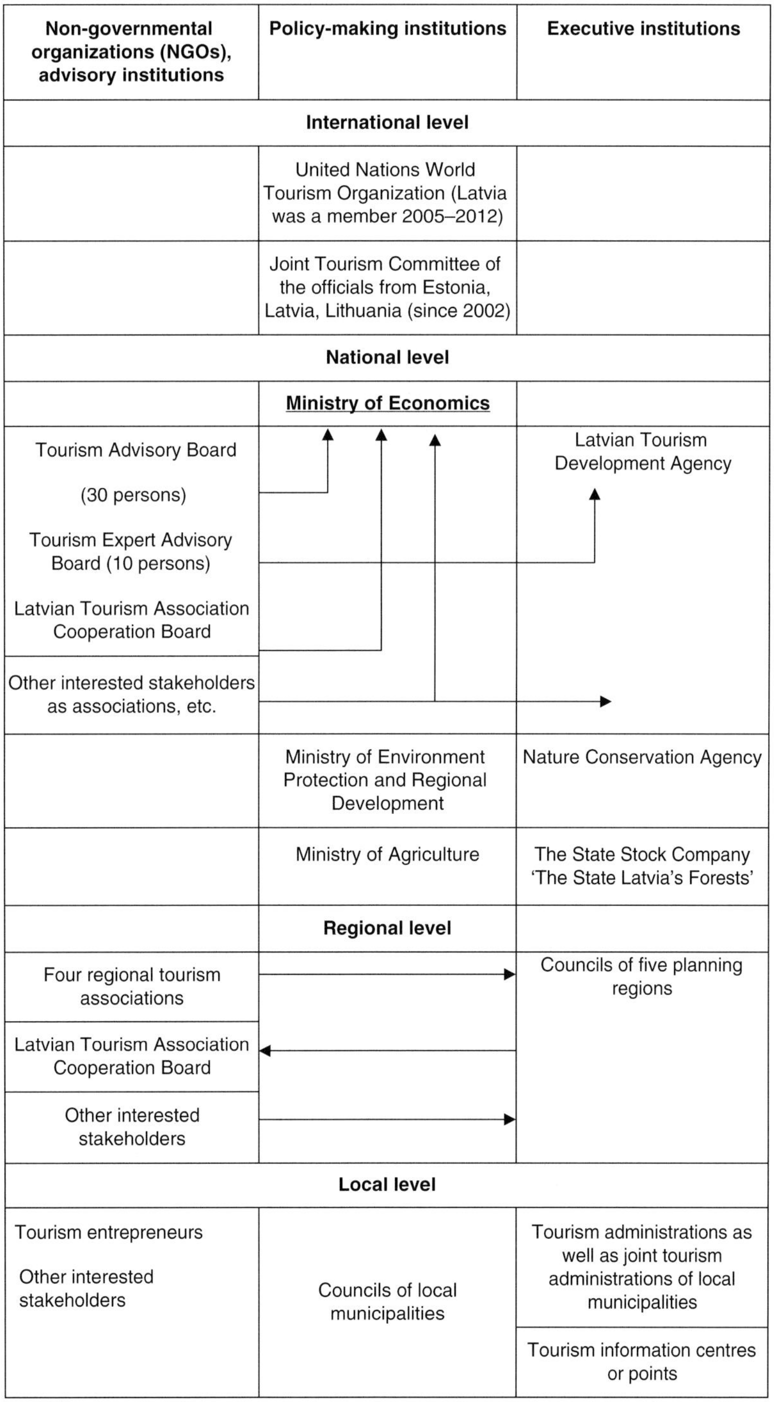

Fig. 12.1. Overview of the tourism governance in Latvia. (Adapted from Druva-Druvaskalne and Līviņa, 2014.)

for example, tourism's contribution to gross domestic product (GDP), tourism balance, an increase of the ecotourism proportion, and tourism companies with quality certificates or environmental management certificates. The 'Guidelines' (Cabinet of Ministers of the Republic of Latvia, 2002) were in force until 2010, when the government adopted a new planning document. The most crucial difference between the 2002 document and the latest planning documents, particularly in relation to sustainability, is that the number of international tourists is not the most crucial indicator, rather the most significant indicator for sustainable development, including regional development, is to extend the length of stay of tourists.

The umbrella long-term planning document in Latvia is the *Sustainable Development Strategy of Latvia 2030* (referred to as *Latvia 2030*) (Ministry of Regional Development and Local Government of the Republic of Latvia, 2010), approved by the Latvian Parliament in 2010. It emphasizes the need for united and coordinated strategic planning across different policy sectors and result-oriented approaches to using policy planning indicators – from macro-level indicators to policy and action indicators (ESPON, 2013).

This planning document identifies that almost each sector can find a main long-term goal which has commonalities with the sustainable development strategy of Latvia. Tourism is indirectly connected with four priorities: (i) the development of cultural space; (ii) an innovative and eco-efficient economy; (iii) nature as future capital; and (iv) perspectives in spatial development. Nature as future capital focuses on the management of natural capital and on the facilitation of a sustainable lifestyle. There is also direct interaction between tourism and the promotion to spend leisure time in a qualitative way, encouraging the local population to engage in domestic tourism. This correlates with the aims of sustainable tourism development throughout the country.

Latvia 2030 is an initiative which includes several indicators for measuring performance of achieving priorities and goals. One indicator directly connected with the tourism industry focuses on a sustainable understanding of tourism linked to the number of foreign tourists (who are staying for 4 days and more). This approach continues in tourism planning documents at a national and local level. One of the six perspective spatial development areas is the coast of the Baltic Sea, which is identified in *Latvia 2030*. This demonstrates continuity in the planning principles in practice because in 2011 the Cabinet of Ministers adopted the *Strategy of Development of Coastal Areas for 2011–2017* (Ministry of Environmental Protection and Regional Development of the Republic of Latvia, 2011). Coastal areas have a crucial role for recreation and tourism, incorporating both cultural heritage and nature conservation goals to sustain high levels of biodiversity.

Consequently, there is a coordinated effort by state and municipal governments to plan coastal infrastructure to increase economic activity. In 2014, there was an open tender to launch an environmental impact assessment of the 'State long-term thematic plan for Baltic Sea coast', which anticipates the implementation of a public infrastructure network concept, incorporating development plans for coastal territories. The main problem of the Baltic Sea coastal planning is that local municipalities and the private sector are being encouraged to act locally, in advance of the emergence of national guidelines for coastal area development. The consequence of this is that several places on the coast have restricted accessibility for society.

Tourism planning initiatives

The National Development Plan of Latvia 2014–2020 (NDP 2014–2020) was approved by the Latvian Parliament in 2012. In the last 20 years, two important tourism planning documents have been elaborated at a national level: (i) the *National Programme of Latvian Tourism Development 2000–2010* (Cabinet of Ministers of the Republic of Latvia, 1999); and (ii) the *Guidelines of Latvian Tourism Development 2014–2020* (Ministry of Economics, 2014). The most successful planning document in the implementation stage was the *Latvian Tourism Marketing Strategy 2010–2015* (TAVA, 2010). This document is the only tourism strategic planning document at national level published since 2010, the previous tourism policy planning document *Guidelines of Latvian Tourism Development* was in force from 2004 to 2008. An upward trend of foreign visitors staying in tourism accommodation shows that the Latvian Tourism Development Agency (TAVA) has been working effectively by participating in segmented tourism marketing activities. In the first quarter of 2014, there was an increase of 15% in the numbers of foreign visitors in tourism accommodation. In addition the rural tourism association *Lauku ceļotājs* mentioned that last year the number of tourists from Germany reduced but in 2014 it increased sharply. In April and May 2014 occupation rates in hotels increased by 2–3% in comparison with the same period a year ago (TAVA, 2014).

The national tourism destination management organization, as stated above, is the TAVA. In 2014, the Cabinet of Ministers adopted the *Guidelines of Latvian Tourism Development 2014–2020* (Ministry of Economics, 2014), with other stakeholders including regional tourism associations, entrepreneurs and municipalities establishing the main priorities and supported activities in the next 6 years in Latvia. The broader goal of the guidelines is the development of sustainable tourism in Latvia by promoting an increase of competitive tourism products in foreign markets. There are four goals with a consistent aim of increasing income from tourists in a sustainable way: (i) increasing the length of stay; (ii) increasing the capacity of accommodation out of season; (iii) increasing profit from tourism products; and (iv) increasing annual tourism product exports. The development of tourism products must be based on seven key values:

- quality;
- sustainability;
- personalization;
- high added value;
- cooperation;
- tourist involvement/experience; and
- cooperation for competitiveness.

The tourism development guidelines support the slogan 'Latvia. Best Enjoyed Slowly' and promote the value of Latvia's beautiful nature supported by enjoyable activities provided by high quality services. Since 2007, TAVA has managed a foreign visitor survey in Latvia. The main goal of the survey is to evaluate the current situation in tourism development and to use the information for tourism planning, development and marketing for the next stage of the evolution of tourism. Since 2009, TAVA has not had a regional office or employees in the regions but in four regions of Latvia, non-governmental organizations (NGOs) operate as regional destination management organizations called regional tourism associations. The working performance of each regional tourism association is different but there are common functions and activities. These include: (i) supervision of cooperation among private companies, public organizations, local municipalities and education institutions; (ii) initiating and implementing international projects related to tourism development; (iii) organizing annual conferences and specific training seminars for members of associations; (iv) marketing activities including printed and online materials; (v) specific activities for increasing domestic tourism into the region based partly on the development of new tourism products such as bike routes around the Kurzeme lakes; (vi) initiatives in social media; and (vii) promoting family-friendly accommodation and catering companies.

These initiatives are indicative of sustainable tourism integration in practice with support for domestic tourism by seeking new ways to use local resources for leisure activities of individuals and small groups. Regional tourism associations have a membership and allocate financial resources from different external resources with the main aim to improve quality of hospitality and tourism services, accessibility of infrastructure and to attract visitors to regions by increasing the length of stay. Representatives of regional tourism associations are members of the Tourism Expert Advisory Board of TAVA and the Tourism Advisory Board of the Ministry of Economics. The councils of the planning regions are responsible for spatial development plans and the strategies of the regions involved in different types of tourism projects, for instance the Ancient Cult Sites for Common Identity on the Baltic Sea coast in Vidzeme, Kurzeme and in the Riga region, and Riverways in Kurzeme and Vidzeme regions.

Niche tourism development

Under the auspices of the marketing strapline 'Latvia. Best Enjoyed Slowly', the planning and development of tourism in contemporary Latvia focuses on a range of niche forms of tourism which can be viewed as exemplars of sustainable tourism.

The capital Riga has a destination management organization 'Riga Tourism Development' known as the brand 'Live Riga' since 2009. In 2014, the city of Riga was a European Capital of Culture and the city council, in cooperation with other organizations, developed a new visitor attraction with objects and activities which will continue beyond 2014. The companion town of the cultural capital of Riga is Sigulda, and it is one of the ways of how to distribute and diversify activities in different places by increasing visits out of the capital. The brand 'Live Riga' is different from the slogan 'Latvia. Best Enjoyed Slowly' because the life and offer of urban places is more dynamic than that of the countryside.

In 2012, the Health Tourism Cluster was established, with the main location in the city of Jūrmala. This followed on from 2008 when the association of Latvian resort towns was established. The Health Tourism Cluster consists of nine towns which are coordinating activities with the main aim of promoting common marketing activities to attract visitors for medicine and well-being tourism.

The rural tourism association *Lauku ceļotājs*, established in 1993, is still operating in the rural tourism field with a strong focus on sustainable development. The main functions of the association are: (i) to responsibly contribute to the development of tourism policy; (ii) to sell rural tourism services to customers (domestic and international); (iii) to provide regular training courses and field trips to other places to get new experiences; and (iv) to provide quality services for customers. This last is achieved by, for example, implementing 'Butterfly', a quality grade system introduced in 2005 which is the environmental quality label 'Green certificate'. There is also a Latvian 'Heritage mark', established in 2013. For a service to be awarded a certificate or mark they must meet certain criteria and indicators confirming the level of quality. The association also produces handbooks of places of botanical interest and forests and guidelines of traditional building rules of the Baltic Sea coastline. The association uses the latest technologies and methods in serving rural tourism services.

The Nature Conservation Agency of Latvia contributes to the development of tourism infrastructure by helping to organize flows of visitors in a sustainable way. For example in 2014, four new environment awareness centres were created with an emphasis on environmental education not exclusively directed at tourism. The Nature Conservation Agency of Latvia provides visitor centres in national parks in Latvia, and in 2012 they organized a travelling activity during the summer season called 'Discover protected areas in Latvia'.

In recognizing the importance of the relationship between local communities and tourists in the adoption of principles of sustainability, there are several cases when local municipalities in the context of tourism development are focusing also on local inhabitants. Tourism information offices offer services for local inhabitants including distributing tickets for different events and providing specific events for locals, for example the tradition in the tourist city of Sigulda of organizing a 'Musical breakfast' outside in the park every Sunday morning in July. The main target group for this event is local people; many visitors come by bicycle to the park. Also Sigulda municipality provides discounts for local people to use tourism services inside the municipality. This is how to create a bridge between local inhabitants and visitors. This is one of the most significant issues which requires responsible action from both parties. The main challenge for sustainable development of tourism in protected areas is to balance the flow and behaviour of visitors with the protection goals set up for the area at different political levels.

Sustainability and pedagogy

Higher education institutions contribute to sustainable tourism development by: (i) providing tourism and hospitality education in several universities; (ii) providing advice and applied research for the industry; and (iii) organizing scientific conferences in higher education institutions.

In the 1990s, the International School of Tourism was founded in Latvia. The institution provided higher education programmes that were based on the experience of the Swiss partners and used adapted study programmes and materials. Later on, the International School of Tourism was included in the University Turiba as a separate faculty. In 2000 an international, scientific and practical conference 'Tourism practice, education and research integration: experience and evaluation' was held in Riga. The conference materials were published in conference proceedings. In 2011 the Association for Tourism and Leisure Education and Research (ATLAS) 20th anniversary conference 'Landscape and tourism: the dualistic relationship' was held in the Valmiera Vidzeme, organized jointly by ATLAS and Vidzeme University of Applied Sciences. Vidzeme University of Applied Sciences offers four higher education study programmes in the field of tourism for local and international students.

Case Study: Dark Tourism in Latvia

While it is important to acknowledge that national tourism planning documents do not directly focus on planning for dark tourism, within the context of the national tourism brand 'Best Enjoyed Slowly' dark tourism can be viewed as an example of the development of niche forms of tourism. Furthermore, in most dark tourism cases in Latvia, these dark tourism destinations could be integrated in different tourism offers and also in nature tourism products, for example birdwatching tourism, and do feature as part of more local and regional plans for tourism.

Before looking specifically at examples of dark tourism destinations in Latvia, more generally dark tourism is defined by Tarlow (2005, p. 48) who stated that it involves 'Visitations to places where tragedies or historically noteworthy death has occurred and that continue to impact our lives'. Examples of such places can be found in each country with different micro-tourism types including disaster tourism, prison tourism, ghost tourism, slavery-heritage tourism, cemetery tourism, battlefield tourism, Holocaust tourism and military tourism. Dark tourism places stimulate different reasons for visitation, such as curiosity, respect/empathy for the victims, personal connection to the tragedy place, intellectual and cultural interest, remembrance, something different, understanding how such terrible things could have happened, risk seeking, iconic sites, political discourse, and how horrific war is.

In Latvia there are several cases of dark tourism, which are already known on a global scale. Over the course of the centuries, Latvia and the Baltic region have been devastated by wars, occupations and foreign armies, leaving the landscape scarred by military installations, redoubts, fortresses, shoreline batteries, bunkers, dugouts, airfields and ghostly battle sites. The rural tourism association *Lauku ceļotājs* has developed a map of military heritage with 200 different objects in Latvia on the basis of evaluation using nine indicators: (i) safety; (ii) ease of finding; (iii) physical accessibility; (iv) utilities; (v) local guides; (vi) uniqueness; (vii) historical value; (viii) landscape; and (ix) availability of infrastructure (LLTA *Lauku ceļotājs*, 2011). Military tourism objects in the Vidzeme region, for example the Zeltiņi Nuclear Base, have the potential to be integrated as part of tourism routes (Baltiņš, 2014). The Zeltiņi Nuclear Base is one of many Soviet missile bases shut down in 1989, and on this site missile hangars, firing systems, bunkers, tunnels and residential buildings survive to this day. A huge bust of Lenin on one of the missile sites used to stand in the town of Alūksne in the Vidzeme region.

(Continued)

Case Study. Continued.

Prison tourism: the Liepāja Naval Port Prison

Liepāja city is located on the Baltic Sea coast. The *Tourism Development Strategy of Liepāja City 2003–2010 and Actions 2003–2006* (Liepāja City Council, 2004), developed on strategic planning principles, identified cultural and historical tourism based on the military and industrial heritage of the city. Planners of Liepāja city recognized the significance of the city's unique space in developing dark tourism products (Liepāja City Council, 2004). The Liepāja Naval Port Prison (Fig. 12.2) was a military prison until 1997 and is the only military prison in Europe open to visitors.

The Liepāja Naval Port Prison has been opened for visitors since 2002 and it received the EU Award of European Destinations of Excellence in 2011. This tourism attraction is operated by an NGO and since 2003 has staged the show *Behind Bars* and from 2004, a new tourism attraction 'Extreme night'. *Behind Bars* is a historical interactive reality show involving the audience, created on the basis of true stories from life in the naval port prison. 'Extreme night' includes a night's lodging in the prison as a unique adventure experience which could be even more impressive than Alcatraz in San Francisco, USA. Tourists can sleep on a prison bunk or an iron bed and have a prison meal and can experience what it is like to be a prisoner without being convicted by a judge.

The graph in Fig. 12.3 shows an upward trend in the total number of visitors to the Liepāja Naval Port Prison. The number of visitors reached a peak in 2006 of 28,585 visitors which is a sharp increase on previous numbers. Numbers declined due to the global

Fig. 12.2. The Liepāja Naval Port Prison. (Picture courtesy of Dzintra Atstāja)

(Continued)

Case Study. Continued.

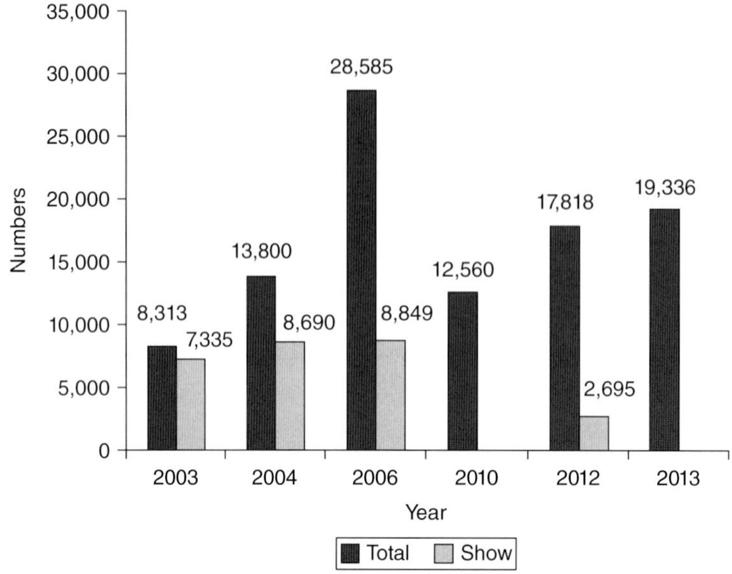

Fig. 12.3. Visitors to the Liepāja Naval Port Prison from 2003 to 2013. 'Show' indicates the number of visitors who took part in a show, see text for details. (From data supplied by Liepāja Naval Port Prison in 2013.)

recession in 2007/8 but there has been a moderate growth of visitors since 2009 annually. Visitors can combine a visit to the prison with leisure pursuits and visiting the adjacent beautiful sandy beach. The main target groups of the prison visitors are groups of domestic pupils and foreign tourists. In 2013 the total number of visitors was 19,336 and shows gradual development. Guided tours including interactive elements such as uniforms and commands for prisoners are available if they are booked in advance. The Liepāja Naval Port Prison Visitor Centre is open all year round offering bicycle hire and guided bicycle tours of the naval port and it is developing new tourism products such as 'Naval Port Brunch' on Sundays and camps for photographers.

Soviet/military tourism: the Secret Soviet Bunker

Based in Līgatne, the bunker is located in the Gauja National Park and its surrounding area is rich in scenic landscapes, cultural heritage sites and creative tourism offers. Tourism development was set out as one of four municipality specializations in the *Development Programme of Līgatne Municipality 2012–2018* (Līgatne Municipality Council, 2012) with the main priority of this strategic planning document being the sustainability of the natural environment. At the same time the Development Programme indicates that the Līgatne Municipality is marketed in the Baltic States and Europe as it has a non-traditional tourism product, the 'Secret Soviet Bunker' (Līgatne Municipality Council, 2012). This is an

(Continued)

Case Study. Continued.

underground well-equipped 2000 m^2 facility, whose secrecy stamp was removed in 2003. The bunker was built only for the top politicians and government persons in 1980, in case of a nuclear war. It was built as a self-sustaining structure with all the necessary equipment for survival, including a telephone exchange facility. The furniture has been retained as well as different maps, books, materials, communication equipment, gas masks and a Soviet-style canteen, with a typical menu from this time. The local population did not know about the existence of the bunker and it was for this reason that local people wanted to visit this place. It is important to note that this area is a popular place for short leisure trips in Latvia because it is inside the Gauja National Park. The Līgatne Municipality mentioned in the Development Programme has included this dark tourism site as it is an added specialization to the surrounding area. This aspect is significant from a sustainable planning point of view as it is important to find strengthening and collaboration opportunities instead of competitive elements. The bunker has been open for visitors since 2000, and is beginning to attract a large number of visitors. In 2012 alone the bunker was visited by 19,000 visitors. The bunker is increasing the services offered for visitors with guided tours in Latvian, Russian and English and reality games, catering and corporate events.

It is important to note that the Liepāja Naval Port Prison and the Secret Soviet Bunker in Līgatne were included in the Lonely Planet year book *Best in Travel 2011* offers, in a list of the top ten exceptional destinations regarded as exciting Soviet heritage tourism attractions. Within the Baltic States there are a range of dark tourism offers including: (i) the Patarei sea fortress-prison in Tallinn, Estonia, which is a unique and wild place; and (ii) the Grutas Park in Lithuania which is a private offer with a collection of Soviet monuments (from the time of Lenin).

CONCLUSIONS

The small country of Latvia as an independent state and member of the EU has proved to be an attractive travel destination especially because of the friendly offers and convenient location.

The main challenge for sustainable development of tourism in protected areas is to balance the visitor flow, provide accessibility to sea, lakes and other unique natural places, and objects for society and suitable infrastructure elements and behaviour of visitors with the protection goals set up for the area at different political levels. In the relatively short period of time of about 20 years, Latvia has become an internationally equal partner in the promotion of sustainable tourism experiences. The main challenge of the tourism industry is to implement in practice the latest strategic tourism planning documents and to strengthen partnership and cooperation with stakeholders in sustainable tourism planning and development. The dark tourism offer fits with the slogan 'Latvia. Best Enjoyed Slowly' and dark tourism has a future potential to diversify and to develop new tourism places and objects.

QUESTIONS

1. Name the priority tourism types in Latvia since 1991. Are there significant changes in achieving sustainable tourism development?

2. What kind of institutions lead tourism planning and development in the regions of Latvia?

3. Is dark tourism a national priority of tourism development or is it a priority of specific local municipalities?

BIBLIOGRAPHY

Baltiņš, K. (2014) Latvian northeastern Soviet military heritage facilities and their potential to be used in tourism. Bachelor thesis, Vidzeme University of Applied Sciences, Valmiera, Latvia.

Cabinet of Ministers of the Republic of Latvia (1998) *Law on Tourism*. Saeima of the Republic of Latvia, Riga, Latvia.

Cabinet of Ministers of the Republic of Latvia (1999) *National Programme of Latvian Tourism Development 2000–2010*. Saeima of the Republic of Latvia, Riga, Latvia (in Latvian).

Cabinet of Ministers of the Republic of Latvia (2002) *Guidelines of Sustainable Development of Latvia*. Saeima of the Republic of Latvia, Riga, Latvia (in Latvian).

Central Statistical Bureau of Latvia (CSB) (2014) Available at: http://www.csb.gov.lv (accessed 30 June 2014).

Druva-Druvaskalne, I. and Līviņa, A. (2014) Tourism in Latvia: from fragmented resorts of Russian empire to a national brand on international level. In: Costa, C., Panyik, E. and Buhalis, D. (eds) *European Tourism Planning and Organisation Systems*, Vol II. Channel View Publications, Clevedon, UK, pp. 118–130.

ESPON (2013) *Key Indicators for Territorial Cohesion and Spatial Planning*. Main report. ESPON (the European Observation Network for Territorial Development and Cohesion) and the National University of Ireland, Maynooth, Ireland.

Klepers, A. (2012) Tourism spatial structures in Latvia, their formation, expression and imporatance in promotion of destinations on the market. Doctoral thesis, University of Latvia, Riga, Latvia.

Liepāja City Council (2004) *Tourism Development Strategy of Liepāja City 2003–2010 and Actions 2003–2006*. Liepāja City Council, Liepāja, Latvia, p. 18.

Līgatne Municipality Council (2012) *Development Programme of Līgatne Municipality 2012–2018*. Līgatne Municipality Council, Līgatne, Latvia, p. 180.

LLTA *Lauku ceļotājs* (Latvian Rural Tourism Association) (2011) *Military Heritage Management Guidelines*. Available at: http://www.celotajs.lv/cont/prof/proj/GreenBelt/dokumenti/vadlinijas/Militaro_objektu_apsaimniekosanas_vadlinijas_eng_kopija.pdf (accessed 30 June 2014).

Lonely Planet (2011) *Best in Travel 2011*. Lonely Planet, London.

Ministry of Economics (2014) *Guidelines of Latvian Tourism Development 2014–2020*. Ministry of Economics, Riga, Latvia, 80 pp. (in Latvian).

Ministry of Environmental Protection and Regional Development of the Republic of Latvia (2011) *Strategy of Development of Coastal Areas for 2011–2017*. Order no. 169. Approved by the Cabinet of Ministers on 20 April 2011. Cabinet of Ministers of the Republic of Latvia, Riga, Latvia.

Ministry of Regional Development and Local Government of the Republic of Latvia (2010) *Sustainable Development Strategy of Latvia 2030 (Latvia 2030)*. Approved by the Latvian Parliament in 2010. Saeima of the Republic of Latvia, Riga, Latvia.

Rozīte, M. (1999) Tūrisma attīstība Latvijā. In: Rižovs, N. and Pētersone, V. (eds) *Latvijas Arhīvi*. Latvijas Nacionālais Arhīvs, Riga, Latvia, pp. 96–110. Available at: http://www.visitdaugavpils.lv/sites/default/files/96-110-vesture-rozite-turisms.pdf (accessed 30 June, 2014).

Samsons, V. (ed.) (1970) *Latvijas PSR mazā enciklopēdija*, Vol. 3. Zinātne, Rīga.

Švābe, A. (ed.) (1955) *Latvju enciklopēdija*, Vol. 3. Trīs Zvaigznes, Stockholm.

Tarlow, P.E. (2005) Dark tourism. The appealing 'dark' side of tourism and more. In: Novelli, M. (ed.) *Niche Tourism Contemporary Issues, Trends and Cases*. Elsevier Butterworth-Heinemann, Oxford.

TAVA (Latvian Tourism Development Agency) (2010) *Latvian Tourism Marketing Strategy 2010–2015*. Approved by A.Slokenbergs, Director of TAVA. TAVA, Riga, Latvia.

TAVA (Latvian Tourism Development Agency) (2014) *Latvijas tūrisma nozares pārstāvji gandarīti par ārvalstu viesu straujo pieplūdumu*. TAVA, Riga, Latvia. (in Latvian)

Trušiņš, J. and Buka, E. (1986) *Latvijas rekreācijas sistēma*. LPSR Celtniecības Zinātniskās pētniecības institūts, Rīga.

United Nations World Tourism Organization (UNWTO) (2013) Available at: http://www2.unwto.org/ (accessed 30 June 2014).

chapter 13

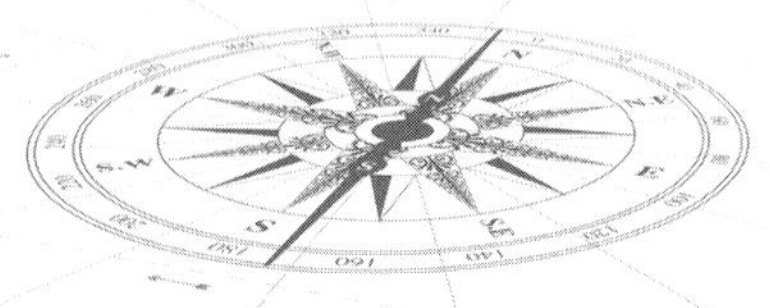

Beyond Enclave Tourism: The Road to Sustainability of Mauritius

Girish Prayag*

University of Canterbury, Christchurch, New Zealand

LEARNING OBJECTIVES

- To understand the nature of tourism governance in Mauritius.
- To recognize the consequences of enclave tourism on sustainable tourism development.
- To understand the challenges of Mauritius in achieving sustainability within the existing tourism governance structure.

INTRODUCTION

Small island developing states (SIDS) are attractive tourist destinations. Enclave tourism is a popular form of tourism development in many SIDS (Nunkoo and Ramkissoon, 2013). Enclave tourism is tourism that promotes all-inclusive facilities and services centred on controlling the cultural and physical environment that tourists experience as part of their stay (Freitag, 1994), leading to limited economic links to the local community (Wall, 1997). Accordingly, enclave tourism is controversial (Mbaiwa, 2005) and raises a number of challenges for sustainable tourism development (Sharpley, 2000). The purpose of this chapter is to evaluate the opportunities and constraints that tourism planning and the governance structure in Mauritius have on sustainable tourism development. The chapter starts with a brief overview of the physical

*girish.prayag@canterbury.ac.nz

and human geography of Mauritius, followed by an outline of the key mandates of public and private sector bodies involved in tourism as well as the role of community participation in tourism planning and development. In effect, the effectiveness of tourism governance in Mauritius is assessed. Thereafter, tourism development in Mauritius from a sustainability perspective is critically examined with a focus on the challenges in achieving sustainability. The case study on Grand Baie at the end of the chapter exemplifies the challenges that enclave tourism and the existing tourism governance impose on sustainable tourism development. In particular, the consequences of non-participatory tourism planning and development processes on residents are highlighted.

THE PHYSICAL AND HUMAN GEOGRAPHY OF MAURITIUS

The Republic of Mauritius, located off the east coast of Africa in the Indian Ocean, has a multi-racial population estimated at 1.23 million in 2011 (CSO, 2011) with a lineage of immigrants from India, Europe, China and Africa. This island nation consists of four main islands (Mauritius, Rodrigues, the islands of Agalega, and the archipelago of Saint Brandon). Considered part of Africa geographically, Mauritius is a member state of the Southern African Development Community (SADC) and the African Union. Among the many objectives of the African Union is the promotion of sustainable development at the economic, social and cultural levels as well as the integration of African economies (African Union, no date). Mauritius has enjoyed unprecedented economic success driven by industries such as agriculture (sugarcane), manufacturing (textiles) and, more recently, services (tourism, information technology and financial services). Tourism contributes approximately 28% to gross domestic product (GDP) and employs (directly and indirectly) 26.4% of the country's formal workforce (WTTC, 2013). According to the world *Travel and Tourism Competitiveness Index* (TTCI) of 2013, Mauritius is ranked the second best performing country in sub-Saharan Africa ahead of South Africa (ranked third) and behind the Seychelles (ranked first) (World Economic Forum, 2013). On a global scale, Mauritius performance is laudable, ranking 58th out of the 140 countries assessed, with the travel and tourism regulatory framework ranking first in sub-Saharan Africa but 32nd on a global scale (World Economic Forum, 2013). With a land mass of 1864 km^2, Mauritius has, to date, been essentially a resort-based destination with its sun, sand and sea (3S) offering (Prayag, 2011).

THE HISTORY AND DEVELOPMENT OF TOURISM IN MAURITIUS

The beginning of a formal tourism industry can be traced back to the 1950s, driven by the recognized potential of the island as an attractive '3S' (sun, sand and sea) destination, the need for economic diversification, and employment creation for locals (Prayag, 2011). By the 1980s, tourism enclaves became the norm. Private investors seeking new investment opportunities beyond the agriculture industry and government planners favoured the development of enclave resorts in coastal areas of the island. Concentrated in remote areas, enclave tourism is all inclusive and often fails to take into account the needs of local communities (Freitag, 1994). Enclave tourism has been heavily criticized for its negative economic, socio-cultural and environmental impacts on host communities (Sharpley, 2000; Mbaiwa, 2005). These negative impacts are often the result of a top-down planning approach for tourism development (Nunkoo and Ramkissoon, 2013).

In the early years, tourism development centred on restricting tourist numbers, air accessibility and the types of holiday activities undertaken by tourists. The government, through the Hotel Development Incentive Act of 1974, provided tax incentives and loans at reduced interest rates to fuel hotel development. Enclave tourism persisted for almost 20 years (1980–2000) with the number of hotels increasing dramatically from 43 in the 1980s to 95 in the 2000s. The period 2000–2005 saw growth in tourism arrivals plummeting to its lowest levels in comparison with the previous decade (Prayag, 2011). While tourism receipts increased from Rs (rupees) 14.2 to 25.7 bn, the number of hotels stagnated at 95 in 2002, increasing to 103 in 2004 and dropping to 99 in 2005 (Ministry of Tourism, Leisure, and External Communications, 2008a). Economic, social and environmental problems, a heavy reliance on repeat tourists, and a destination image that was no longer fashionable were order of the day. The tourism product and positioning of the island were inadequate in sustaining tourist numbers and receipts (Prayag, 2011). The *National Tourism Development Plan* (NTDP), funded by the European Community in 2000, was commissioned to help the diversification and rejuvenation of the tourism industry (Deloitte and Touche, 2002). As a result, the destination upgraded its existing hotel facilities and increased its overall tourist capacity by building new hotels. The NTDP suggested among others, that tourism growth can be achieved by diversifying the tourism product and targeting new markets with the intention of positioning the island in the luxury segment of tourist destinations. Since then, various policies such as the *Hotel Development Strategy* (HDS) of 2008, focusing on hotel development of only four- and five-star rated hotels, have sought to reinforce this positioning (Ministry of Tourism, Leisure, and External Communications, 2008b). As of 2012, 107 hotels were in operation with an average occupancy of 62%, tourism receipts amounted to Rs 44.4 bn, and tourist arrivals reached 965, 441 (CSO, 2013).

TOURISM GOVERNANCE IN MAURITIUS

Governance refers to 'systems of governing and ways in which societies are governed, ruled or steered' (Bramwell, 2011, p. 459). Mauritius has 'a strong track record of good governance, fiscal management, social protection and competitiveness reforms. None the less, Mauritius faces a number of challenges to strengthen the foundations of its economic growth and ensure sustainability over the long term' (ADB, 2009, p. 18). There are five main players (see Table 13.1) in the public sector involved in tourism planning and development in Mauritius. The Ministry of Tourism and Leisure (MTL) remains the key player in providing the legal and institutional framework for planning, monitoring and regulation of tourism activities. The Tourism Authority (TA), which falls under the aegis of the MTL, has among its key mandates tourism planning, policy formulation and implementation.

Table 13.1. Key public sector entities in the tourism industry of Mauritius.

Main players	**Key mandates**
Ministry of Tourism and Leisure (MTL)	• To promote Mauritius worldwide as a prime holiday island destination • To pursue the upmarket and selective tourism policy • To plan, manage and control tourism development in Mauritius • To define, adjust and implement sustainable tourism strategies which are environmentally sound, socially acceptable and economically viable • To institute a legal and operational framework to plan, monitor and regulate tourism activities in line with the defined tourism strategies and policies
Tourism Authority (TA)	• To optimize the social, economic and environmental benefits to Mauritius from tourism • To promote a sustainable tourism industry in Mauritius • To formulate such policy and nationally integrated planning for the tourism industry • To coordinate the activities of the various organizations concerned with the tourism industry, in both the private and the public sectors • To issue guidelines for the development of the tourism industry
Mauritius Tourism Promotion Authority (MTPA)	• To promote Mauritius abroad as a tourist destination • To provide information to tourists on facilities, infrastructures and services available to them in Mauritius • To conduct research into market trends and market opportunities and disseminate such information and other relevant statistical data on Mauritius • To initiate such action as may be necessary to promote cooperation with other tourism agencies
Tourism Employees Welfare Fund (TEWF)	• The Fund shall be to set up schemes and projects and, generally, to carry out such other activities as may be considered desirable for promoting the welfare of employees of tourism enterprises and their families

(*Continued*)

Table 13.1. Continued.

Main players	Key mandates
Board of Investment (BOI)	• To stimulate the development, expansion and growth of the economy by promoting Mauritius as an international investment, business and service centre • To promote and facilitate the development of all forms of investment and business activities • To formulate investment promotion policies and plans and marketing strategies and undertake promotion to attract foreign and local investments • To advise government on strategies for investment policies, national investment marketing and investment aftercare, economic and industrial planning and country image building

The TA has a strong focus on tourism policy that will encourage and support sustainable tourism development. The Mauritius Tourism Promotion Authority (MTPA) focuses its activities around destination management, including a strong marketing component. Both the TA and MTPA have an advisory role in decision-making processes of the MTL. The Tourism Employees Welfare Fund (TEWF) is solely responsible for the general well-being of tourism industry employees and their families. The Board of Investment (BOI) has an advisory role in MTL and facilitates international investments in Mauritius, including the tourism industry.

The governance structure of the key players suggests a top-down approach where decision making is concentrated in the hands of a few people and the MTL substantially dictates tourism planning and development. In effect, the existing structure reflects Hall's (2011) hierarchical governance structure, characterized by hierarchical relations between different levels of the state, whereby tourism policy is mainly the responsibility of policy makers, legislators and government, with a strong focus on whether policy goals are met. Unsurprisingly, local communities are not always represented in this type of governance structure (Hall, 2011). Traditionally, local communities have had little input in tourism planning and development in Mauritius (Nunkoo and Ramkissoon, 2013). The hierarchical governance structure is often 'elitist' with policy made by the top and implemented by the bottom, with policy instruments geared towards licensing, permits, consents and meeting standards (Hall, 2011). This is very much the role of the TA in Mauritius.

The private sector is represented in the decision making of MTL and MTPA through various sectorial groups and associations. The Association des Hôteliers et Restaurateurs de l'ile Maurice (AHRIM) represents and promotes the interest of hotels and restaurants in Mauritius. AHRIM's vision is to 'make the tourism industry a viable prospect for the prosperity of the economy, leading to a rise in the quality of living of all local citizens' (AHRIM, no date).

Founded in 1973, AHRIM members represent 66 of the 109 hotels in operation in 2011. The Association of Inbound Operators Mauritius (AIOM), founded in 1992, represents the interests of professional inbound tour operators and car rental companies. Among its key objectives include: (i) helping members provide the highest service standards to tourists; (ii) representing the interests of its members to government and/or other statutory bodies to ensure a suitable framework and environment that allows their business to grow and prosper; (iii) assisting members in preserving the image of Mauritius as a destination while working in an environmentally conscious framework; and (iv) promoting cooperation among members and other associated organizations in their effort to increase visitor numbers to Mauritius (AIOM, no date). Some key features of governance such as allocation of resources, exercising control, application of rules, and promotion of coordination and cooperation among numerous actors (Rhodes, 1996; Bramwell, 2011) are evident among the key mandates and objectives of the public- and private-sector tourism bodies outlined above. Also, there seems to be a strong reliance on external consultants and organizations to develop tourism plans. Whether this system of governance promotes sustainable tourism is discussed later in the chapter. Next, the importance of community participation in tourism planning is discussed.

COMMUNITY PARTICIPATION IN TOURISM PLANNING

Community participation in the tourism planning process has been advocated as a way to implement sustainable tourism (Okazaki, 2008) that can eventually increase a community's carrying capacity through reducing the negative impacts and maximizing the positive impacts of tourism (Jamal and Getz, 1995). Community participation in tourism planning is not without implementation barriers such as conflicting vested interests, business inexperience of local communities, and insufficient financial assistance from national and local government among many others. However, participatory tourism planning is considered the preferred course of action for developing sustainable tourism in local communities due to its ability to: (i) increase collaboration and community engagement; (ii) redistribute power; and (iii) contribute positively to the building of social capital (Okazaki, 2008). The use of collaborative and cooperative approaches in tourism planning can bring together stakeholders that would otherwise not participate in tourism development (Reed, 1997). The lack of coordination and cohesion among stakeholders in tourism development is a well-known problem for destination planners and managers (Jamal and Getz, 1995). Existing models of community participation in tourism planning suggest that when community participation is promoted, power redistribution will be facilitated and social capital will gradually emerge from the participatory planning process. In this way community participation in tourism development can contribute to improve destination sustainability through the creation of synergy both within the community and between the community and other stakeholders (Okazaki, 2008). Hence, the involvement of residents in tourism planning is critical for generating support both for the tourism industry as a viable economic alternative and for community tourism development (Nunkoo and Ramkissoon, 2011b). Generally tourism planning at community level evolves through successive phases of non-participation, degrees of tokenism, and degrees of citizen power (Okazaki, 2008). In the early stages, the local community is not involved in the tourism planning process which is driven mostly by other stakeholders such as local government and external agents. In the late stages, the local community is considered a partner in tourism planning and development and has considerable power and control on policy making and management (Okazaki, 2008). In this stage, sustainable tourism development can build social capital through addressing issues of social equity.

The enclave tourism model is notoriously known for a lack of community involvement in tourism planning (Andriotis, 2003; Nunkoo and Ramkissoon, 2013) and heavy reliance on this form of tourism development in Mauritius has impeded community-based tourism development initiatives. Tourism policy has a strong compliance component that undermines the ability of local communities to define their own sustainable futures. The enclave tourism model in Mauritius relies extensively on stakeholders such as external investors, consultants and the national government to chart the path of tourism development for local communities. This model of tourism development still persists with variations such as the integrated resort scheme (IRS) and the real estate scheme (RES) being popular since 2005. These variations are not without controversy in terms of their impacts on local communities. However, one noticeable change in implementing this 'new' form of enclave tourism has been local community consultation. For example, in 2010 community involvement in tourism planning was seen as part of the Local Economic Development (LED) plan for Le Morne Cultural Landscape (LMCL). The plan was commissioned to build up the economic capacity of a local area in such a way that it improves the quality of life of residents. LMCL is a World Heritage Site (WHS) since 2008 and is located on the south-west tip of Mauritius. It is one of the least developed coastal areas in Mauritius (Le Morne Heritage Trust Fund, 2011). A participatory planning process involving residents, government, and other local stakeholders such as youth and women's associations were used to identify community needs, economic development opportunities and challenges, and a vision statement for the community. Through local stakeholder consultations, community workshops, socio-economic survey, local government input and the use of consultancy services, the *Local Development Plan for Le Morne* (LDPLM) emerged to create better conditions for economic growth and employment generation in the local community of 'Le Morne' (Le Morne Heritage Trust Fund, 2011). Yet, this approach is not the norm in Mauritius and this particular instance of participatory planning process is merely a requirement for maintaining the site's WHS status.

SUSTAINABLE TOURISM DEVELOPMENT IN MAURITIUS

The concept of sustainable development is hinged on three broad approaches and concerns, namely: economic, social and ecological sustainability (Mbaiwa, 2005). Central to the concept is the notion of 'balance' between economic, social and environmental concerns (Bramwell and Lane, 2011). In reality, the three approaches often require trade-offs and the idea of balance is often used to mask that economic growth is often the ultimate concern of countries (Hunter, 2002). United Nations Educational, Scientific and Cultural Organization (UNESCO) recommendations for SIDS such as Mauritius include the monitoring of tourism impacts so that tourism development as well as social and environmental priorities are mutually supportive at all levels (UNESCO, 2010). The 'Ile Maurice Durable' project was launched by the Mauritian government in 2008 to promote sustainable development of the island. Principles of sustainable tourism for Mauritius were first highlighted in the NTDP in the year 2000 and subsequently reinforced through various policy documents and reports such as the HDS in the year 2008 and more recently, the Mauritius Sector Strategic Plan for Tourism (MSSPT) (2008–2015) and the National Assessment Report of 2010 by the MTL and Ministry of Environment and Sustainable Development (MESD), respectively. Sustainable tourism development is certainly a priority of the current government with several initiatives in place to promote sustainable tourism. This section reviews only key strategies and initiatives.

Economic sustainability

Economic sustainability generally refers to the economic needs of a population being met through the use of natural resources in a sustainable way that maximizes economic efficiency (Mbaiwa, 2005). Tourism policy has, over the years, emphasized low-impact and high-spending tourism (Nunkoo and Ramkissoon, 2013). This is not surprising of Mauritius, given that most SIDS have a limited natural resource base that generates a dependency on the services sector for economic stability (Hall and Page, 1996). Despite the recognition that sustainability is of concern to tourism stakeholders (see Table 13.1), including the government, the primary focus of tourism development in Mauritius has been the maximization of tourist numbers within the constraints of the existing natural and cultural resources (Prayag, 2011). Watt (2007) notes that the political discourse on tourism in Mauritius suggests that 2 million tourists/year by 2020 is desirable but no scientific evidence confirms this number as the ultimate carrying capacity of the island. Both the government (e.g. the TA) and the private sector (e.g. the AIOM) embrace the discourse on sustainable tourism by emphasizing its importance in their mandates but are implicitly supporting strategies of economic growth by increasing tourist numbers and tourism receipts.

In this path to economic sustainability, the MTL has used several strategies. First, product-led tourism through a strong focus on packaged tours and enclave tourism drives tourism development to this date. Hunter (1997) suggests that product-led tourism is a weak interpretation of sustainable development, given its emphasis on resource optimization as opposed to resource conservation, but often is the most viable and cost-effective option in the short term. Second, through product diversification (e.g. adventure tourism, golf tourism and second-home tourism), the government has attempted to lower the risk associated with attracting only '3S' tourists. While '3S' tourism is not inherently unsustainable (Aguilo *et al.*, 2005), Mauritius faces substantial difficulties in retaining its competitive advantage vis-à-vis its main competitors (the Seychelles and the Maldives) in this type of tourism. The island has seen its price competitiveness decline considerably (ranked 75th, down from 18th in the last competitive assessment) – primarily the result of increasing hotel and fuel prices and high ticket taxes and airport charges (World Economic Forum, 2013). Third, market diversification strategies have been pursued to lower the risk of dwindling tourist numbers by tapping into new inbound markets such as India and China as well as Eastern Europe along with the more traditional inbound markets of France, the UK and Germany (Prayag, 2011). Fourth, there is a strong focus on retaining existing visitors as part of the recommendations made to SIDS for sustainable tourism development (UNESCO, 2010).

In summary, these strategies have embarked Mauritius on its journey to economic sustainability but some contradictions are evident between the existing tourism policies and the sustainable tourism discourse. For example, as part of product-led tourism development the government now allows foreign ownership of properties on the island through the IRS. Known as 'second-home' tourism in other countries, this form of tourism development has been controversial for small island communities (Marjavaara, 2007). Despite leading to tourism receipts and job creation opportunities, second-home tourism can result in resident–host conflicts, displacement of permanent residents, changing local culture, and poses a serious threat to community identity and physical environments (Hall and Muller, 2004; Marjavaara, 2007). Likewise, adventure tourism (e.g. windsurfing, big-game fishing and diving) is attracting new tourists to Mauritius but needs to be regulated sufficiently to control environmental impacts and ensure the safety and security of such tourists (Prayag and Jankee, 2009).

Ecological sustainability

Geographic isolation of island destinations often leads to the creation of unique ecosystems that shape the physical attributes of the place and these often have special appeal to travellers (Reid and Reid, 1994). There is no agreed definition of ecological sustainability (Aarts and Nienhuis, 1999), but the concept generally refers to the limits set by the carrying capacity of the natural environment (physically, chemically and biologically), so that human use does not irreversibly impair the integrity and proper functioning of its natural processes and components (De Groot, 1996). Compared with other countries in Africa, the policy environment in Mauritius is supportive of tourism development (World Economic Forum, 2013). Mauritius typically benefits from high levels of safety and security for visitors but the country's environmental sustainability has received a weakened assessment by the TTCI. Of particular concern is the natural environment given its importance for the country's tourism industry (World Economic Forum, 2013). Tourism in Mauritius is provoking ecological stress (Daby, 2003) through the high influx of tourist arrivals, substantial hotel construction and over-exploitation of coastal and marine resources (Sobhee, 2008). Several studies (Daby, 2003; Sobhee, 2004, 2006) have highlighted the over-exploitation or destruction of marine resources that have adverse effects on coastal communities. Resort hotels, in particular, have placed severe pressure on existing natural resources and several cases of hotels destroying the marine environment have been reported in the media (Nunkoo and Ramkissoon, 2013). For years, resort hotels have been removing sea grasses in the belief that they are unsightly or harbour organisms causing injury to bathers. 'Such disruptions draw down the resilience of marine habitats and make the island more vulnerable to environmental change and extreme events, with higher risks of chaos and ecological collapse' (Daby, 2003, p. 313).

Tourism development in Mauritius has been geared towards exploitation of natural assets for economic gains for several decades rather than environmental conservation for future generations. This is not surprising given that enclave tourism and integrated resorts are highly resource consumptive, can degrade the natural environment (Andriotis, 2008) and may reduce the economic value of natural resources (Lee, 1997). Integrated resorts in particular are thought to be an attempt at careful design, to plan and control a well-defined tourism product that ensures complementary and compatible land uses (Stiles and See-Tho, 1991; Wall, 1997) but the reality suggests that their 'development can be undertaken in an unplanned manner and ad hoc basis outside of their borders' (Smith, 1992, p. 215). Integrated resorts can strain local resources by increasing demand for electricity and water as well as generating waste that may be harmful to communities and the local ecosystem (Stonich, 1998; Andriotis, 2008). Such developments demand large stretches of land for construction of recreational, sports, residential and infrastructural facilities (Andriotis, 2008). Sobhee (2008) argues the need for an ecological tax on foreign visitors to Mauritius that can be used towards various forms of conservation policies such as awareness campaigns to sensitize locals and visitors to the fragility of coastal and marine ecosystems, public infrastructure to protect the coastline, and co-financing of hotel waste water treatment facilities, for example, by the government.

Only recently, tourism policy has caught up with existing environmental laws to address issues of ecological sustainability of the island and the tourism industry. A prime example is the HDS that emphasizes the need for eco-friendly practices by hotels such as renewable energy supply, energy saving appliances, water conservation and treatment (Ministry of Tourism, Leisure, and External Communications, 2008b). Both the MTL and the TA have included environmental issues associated with tourism development as part of their mandates (see Table 13.1). Specifically, with the increase in the number of coastal hotels and associated activities and given the sensitive nature of coastal zones, Environmental Impact Assessment (EIA) is required of developers under the Environment Protection Act of 2002 for coastal hotels (Department of Environment, 2002). Interestingly, an evaluation of the EIA report of nine hotels in Mauritius reveals several weaknesses, for example: (i) inadequacy of the screening mechanism; (ii) access to information and public participation in the EIA process needs improvement; and (iii) no environmental management plans included (Ramjeawon and Beedassy, 2004). The authors note that the legal and institutional framework for environmental management in Mauritius 'is fragmented, with the relevant provisions being dispersed across a number of different acts and regulations, often giving rise to jurisdictional overlaps' (Ramjeawon and Beedassy, 2004, p. 541). Daby (2003) notes that despite EIAs being approved officially by government, activities such as shore modification and landscaping, lagoon seabed clearing and dredging, and even coral reef blasting are not uncommon of hotel developers. Activities such as organic waste disposal by pleasure crafts, hydrocarbon spills, and coral damage by speed boats, snorkelling and fishing remain largely uncontrolled. The pace and scale of tourism development in Mauritius has been described by environmentalists and residents as 'coastal zone high-jacking' (Daby, 2003, p. 314).

In an attempt to reconcile hotel development and environmental sustainability, several policies have been enacted such as the 'Speed Limit and Prohibited Activity Zone for Pleasure Crafts' of 2008 and the Planning Policy Guidance for coastal development that puts emphasis on the need to preserve the natural, exquisite and pristine characteristics of the seascapes. Based on the recommendations of the MSSPT, Mauritius has also embarked on a Blue Flag Programme with objectives including: (i) to promote the sustainable use of the coastal resources; and (ii) to introduce national policies on lagoon water quality, reefs, protection of the beaches and safety. Spatial planning of the lagoons has also become of prime importance, which has prompted the need for the preparation of a master plan for the zoning and sustainable management of the lagoon. To move towards the 'greening' of the tourism industry, the government is in the process of introducing an eco-label scheme for the environmental sustainability of the sector (UNDP, 2013).

While these initiatives are laudable, there is an urgent need to set up, for example, Strategic Environmental Assessments for coastal development rather than EIA, given that the latter gives no consideration to cumulative effects of a proposed development (Ramjeawon and Beedassy, 2004). Environmental impacts of a proposed hotel should not be considered in isolation, but rather should be viewed as the additional and cumulative impacts that it has on the existing ecosystem. The government should also clarify its stance on environmental quality of the destination and environmental tourism policy considerations should be integrated within an overall sustainable development framework (Hall, 2008). Perhaps the fragmented and sectorial approach to environmental sustainability is the result of governance structures, including tourism governance, that are rigid, non-participatory and detached from local communities' expectations. This is not surprising given the recognized difficulties associated with governance for sustainable tourism. Often, sustainable tourism concerns can cut across several sectors and span diverse policy domains that make it difficult to secure a coordinated approach within the tourism sector and across sectors (Bramwell, 2011). Typically, hierarchical governance structures are often ineffective in relation to environmental management, which has implications for the sustainability of tourism activities (Meadowcroft, 2007; Hall, 2011).

Social sustainability

For SIDS such as Mauritius, small geographic size and often specific cultural identity assist in the promotion of a sense of community and personal ties, which make it easier to preserve local culture and identity. These are important elements in giving island destinations distinctiveness and enable tourism planners to be selective in their choice of tourism development approaches and tourism markets for promotion (Reid and Reid, 1994). Social sustainability is a wide-ranging multidimensional concept that hinges on issues of social inclusion, social equity, social cohesion and building of social capital through sustainable communities (Dempsey *et al.*, 2011). Ensuring participatory approaches and the involvement of women and major groups in decision making should be considered important by SIDS. In particular, culture is an integral element in the promotion of sustainable development and:

> investment in education and training to broaden the skills base in all aspects of cultural activities and entrepreneurship, technical assistance for collecting data on cultural industries, developing better market intelligence and stronger distribution networks for cultural products and services in SIDS are challenges that must be addressed.
> (UNCTAD, 2005, p. 14)

Unfortunately, enclave tourism does not always lead to better guest–host relationships. In fact, enclave tourism can exacerbate problems of social inequality and disparity (Sharpley, 2000). Specifically, enclave resorts tend to widen the social and cultural gap between hosts and guests, while dampening local entrepreneurship opportunities (Shaw and Shaw, 1999). Through their exclusive nature, enclave resorts often ignore cultural linkages with local communities. Frequently, the only linkage is through employment of workers for construction and subsequent staffing of the resort (Freitag, 1994). In the case of the Canary Islands, Gil (2003) highlighted that this form of tourism development is conducive to issues of poorly paid jobs for hospitality and tourism employees and a lack of future career paths for locals. Negative social impacts such as commercialization and commodification of culture, criminality, change in social values and negative demonstration effects are often associated with integrated resorts in particular (Freitag, 1994; Andriotis, 2003; Mbaiwa, 2005). In this way, communities often experience a sense of 'lesser' control over their natural resources and identity. In summary, enclave tourism allows limited social benefits to accrue to host communities (Freitag, 1994; Nunkoo and Ramkissoon, 2013).

The sustainable tourism discourse in Mauritius attempts to create social justice through tourism development. However, social equity issues have been more of a concern for the private sector than the public sector (Prayag, 2011). Public sector initiatives have been slow and uncoordinated but include the Tourism Employees Welfare Fund Act of 2002 that attempts to safeguard the social and economic welfare of employees involved in the tourism industry. This was followed a few years later by the Tourism Fund Act in 2006; the government forced hotel developers to contribute to this fund for the purpose of financing infrastructural development that benefits local communities. Also, the government has enacted a corporate social responsibility (CSR) policy in July 2009 whereby registered companies have to pay 2% of their book profit towards programmes that contribute to the social and environmental development of the country. Typical programmes can include: (i) promotion of civic education and responsibility; (ii) development of community-based projects; (iii) promoting business start-ups; (iv) gender issues and women empowerment; (v) youth and children development; and (vi) development of sports, arts and crafts, environment conservation projects and education/training initiatives (National Empowerment Foundation, 2012). This has prompted several hotel management companies (e.g. Naiade Resorts, Beachcomber and New Mauritius Hotel) to strengthen their existing CSR initiatives. For example Le Meridien Hotel, part of the Starwood group, funds several initiatives that are linked to environmental conservation and education of previously disadvantaged children (Le Meridien Hotel, no date). However, it remains unclear whether hotels are engaged with CSR for compliance reasons or altruistic motives. Traditionally, hotel management companies, the TA and MTL, and foreign investors substantially influenced tourism policy with little input from residents. In fact residents were often marginalized and alienated from the planning and development process (Nunkoo and Ramkissoon, 2013). While this is slowly changing, resident involvement and consultation in tourism development matters remain a rare occurrence. A non-participatory tourism planning regime persists as evidenced by tourism policies failing to give voice to communities on the way tourism should be developed. As a result, Mauritius is still in the early stages of its journey to social sustainability.

Case Study: Grand Baie – When Enclave Tourism Goes Bad

Located on the north-west coastline of Mauritius, Grand Baie has transformed itself from a traditional fishing and agricultural village to the major tourist hub of the island (Nunkoo and Ramkissoon, 2011a). As of 2012, 19 of the 107 hotels in Mauritius are located in the village or its immediate surroundings. For tourists visiting Mauritius, Grand Baie offers substantial luxury shopping, nightlife and entertainment, and fine-dining restaurants, making the village a 'must' for international visitors. Grand Baie also owes its popularity to the high quality of its beaches, safe swimming facilities, and several other water-based activities such as sailing, windsurfing and water skiing. It is also the departure point for deep-sea fishing trips and boat excursions to the islands in the north of Mauritius (ABZ Global Solutions, no date). In the 17th century the Dutch used to call Grand Baie 'De Bogt Zonder Eynt', which meant the 'Bay Without End' but ironically, today, it appears as though it is tourism development and not 'the bay' that is without end (Lonely Planet, 2014).

In the 1970s tourism became the preferred economic development option for transforming this traditional fishing village into a tourist destination that would create economic opportunities for locals. The enclave tourism model was applied, resulting over the years in what has been described as 'concrete tourism' – bungalows and hotels line up along the coastline. Poorly controlled tourism development has led to several economic and environmental issues in the area (UNEP, 2006). The tourism industry in particular has over the years intensified the impacts on existing infrastructure and the environment, resulting in many negative perceptions of the tourism industry by residents (Nunkoo and Ramkissoon, 2011a). Resort developers have exacerbated this situation by restricting access of locals to public beaches and construction of jetties and marinas. Tourism development in this village has been mostly driven by the government through the MTL, BOI and resort

(*Continued*)

Case Study. Continued.

developers. Residents' lack of tourism knowledge and experience of tourists have been used to support arguments that effective tourism development should be in the hands of government departments, consultants and foreign investors. In this way the level of community participation in tourism development has been low and power redistribution to local community has been non-existent.

Tourism development in Grand Baie has resulted in both positive and negative impacts on residents. Positive impacts include significant infrastructural developments such as road networks, waste water management, recreational sites and amenities (e.g. showers and parking) at public beaches. Several hotel groups in the area are also engaging in CSR to improve the quality of life of residents (Nunkoo and Ramkissoon, 2011b). With tourism development several negative impacts have materialized over the years such as traffic congestion, pollution, crime, prostitution and lack of public and green spaces (Nunkoo and Ramkissoon, 2011a). Compounding the existing issues, the BOI and MTL have approved several second-home projects as part of the government's IRS scheme in recent years. The problems facing the community are commonly experienced when a real-estate-driven resort growth model is used (Keller and Bieger, 2008), which is characterized by limited public input in tourism decision making. Recently, non-governmental organizations and residents have teamed up to protest against further encroachment by hotel developers on existing public beaches (*L'express*, 2014). Parts of public beaches are being earmarked for hotel development and/or IRS schemes. This is not an isolated incident given that local media often reports on residents' opposition to hotel development and expansion, construction of recreational second homes and private bungalows (Nunkoo and Ramkissoon, 2011b). This perhaps is indicative of tourism failing to fulfil resident expectations as a viable form of community development. Economic benefits from tourism development in the region tend to go to resort developers and local elites while local residents' participation in tourism is limited to wage employment and small businesses (Nunkoo and Ramkissoon, 2011b).

The benefits and consequences of tourism on the village of Grand Baie are the result of the enclave tourism model under a hierarchical tourism governance structure. This structure relies heavily on external consultants, private investors and national/local government for community tourism development. A non-participatory tourism planning regime over the years has led to residents feeling detached from the benefits of tourism while experiencing the negative impacts of it on a day-to-day basis (e.g. traffic congestion). In the enclave tourism model, residents are perceived more as elements of destination attractiveness, a labour pool and/or a resource to be exploited (Hall, 2005). No overt role for residents in the tourism planning process is evident at community, local and national levels. There seems to be a strong belief in the public sector that successful tourism development requires the support of tourism planning specialists, consultants and other external

(*Continued*)

Case Study. Continued.

agents, and a top-down policy implementation. In this form of governance, community tourism development fails to be integrated into other community development plans and residents are generally unaware of who is responsible for tourism planning issues in their community (Moscardo, 2011). Tourism governance, thus, appears to be driven by the assumptions that local communities have limited knowledge of tourism or tourists. Local residents are considered limited in their capacity to contribute to tourism and in the conservation of the natural environment.

Nunkoo and Ramkissoon (2013) found that, unsurprisingly, residents associate enclave tourism in Mauritius with environmental degradation and denial of freedom to enjoy public goods such as beaches, and they think it has negatively impacted local culture. Residents of Grand Baie are acutely aware of the negative socio-cultural and environmental impacts of tourism on their community. This difference in knowledge of and perspectives on tourism impacts between residents and the government suggests that community tourism development for Grand Baie is not necessarily based on a consultative process. Decision making is centralized and may reflect path dependency, whereby decisions made today for tourism development are constrained by decisions made in the past (e.g. enclave tourism and real-estate-driven growth models) by the government and other external agents. This implies that tourism institutions are self-reinforcing (Pierson, 2000). As tourism grows further, the community may become less attractive as a place of residence for both second-home buyers and potential residents. Grand Baie as a tourist destination is certainly at the beginning of its path to environmental and social sustainability and the current governance structure, as well as the lack of community participation in tourism planning, may be problematic in implementing the sustainable tourism agenda for SIDS.

CONCLUSION

Although Mauritius is engaged at several levels in tourism sustainability strategies, the governance structure in the tourism industry will be increasingly problematic in achieving ecological and social sustainability. As Bramwell (2011, p. 461) observed: 'Destinations wanting to promote sustainable tourism are more likely to be successful when there is effective governance'. The hierarchical tourism governance structure and enclave tourism model used in the past limits the development of low-impact tourism centred on maintaining community and environmental integrity. Bramwell and Lane (2011, p. 412) noted: 'Tailored and effective governance is a key requirement for furthering the objectives of sustainable tourism'. Non-participatory tourism planning, top-down tourism policy implementation, heavy reliance on external agents for community tourism development, lack of legislation and at times confusing regulations, and lack of a holistic view of sustainability whereby government departments work together to achieve sustainability are the major challenges that tourism governance needs to address.

To improve the competitiveness of the tourism industry, the governance system should be viewed as the tool for destination change management. On the road to tourism sustainability, several recommendations can be made to improve tourism governance. First, giving more power to resident communities is a necessary prerequisite for sustainable approaches to tourism and this can be achieved through more inclusive governance structures (Moscardo, 2011). Including residents in tourism decision making can potentially increase community ownership of tourism as a viable option for economic development. Second, within an enclave tourism model, it is possible to set up community monitoring programmes that would measure economic, social and environmental indicators of change. Such monitoring would also include a feedback system to the community based on community reactions to changes in the indicators (Gill and Williams, 2011).

QUESTIONS

1. What are the limitations of the enclave tourism model in achieving sustainability of island destinations?

2. What measures do you think should be put in place to improve community participation in tourism decision making in Mauritius?

3. What features of a governance structure for tourism would make sustainability more achievable for Grand Baie?

BIBLIOGRAPHY

Aarts, B.G.W. and Nienhuis, P.H. (1999) Ecological sustainability and biodiversity. *International Journal of Sustainable Development and World Ecology* 6(2), 89–102.

ABZ Global Solutions (no date) Available at: www.mauritiusattractions.com (accessed 7 May 2014).

African Development Bank (ADB) (2009) *Country Strategy Paper Draft 2009–2013, Mauritius*. ADB Report. ADB, Abidjan, Ivory Coast.

African Union (no date) Available at: www.au.int (accessed 7 May 2014).

Aguilo, E., Alegre, J. and Sard, M. (2005) The persistence of the sun and sand tourism model. *Tourism Management* 26(2), 219–231.

Andriotis, K. (2003) Tourism in Crete: a form of modernisation. *Current Issues in Tourism* 6(1), 23–53.

Andriotis, K. (2008) Integrated resort development: the case of Cavo Sidero, Crete. *Journal of Sustainable Tourism* 16(4), 428–444.

Association des Hoteliers et Restaurateurs de l'ile Maurice (AHRIM) (no date) Available at: www.mauritius.net (accessed 7 May 2014).

Association of Inbound Operators Mauritius (AIOM) (no date) Available at: www.aiom.mu (accessed 7 May 2014).

Bramwell, B. (2011) Governance, the state and sustainable tourism: a political economy approach. *Journal of Sustainable Tourism* 19(4/5), 459–477.

Bramwell, B. and Lane, B. (2011) Critical research on the governance of tourism and sustainability. *Journal of Sustainable Tourism* 19(4/5), 411–421.

Central Statistics Office (CSO) (2011) *Digest of International Travel and Tourism Statistics*. Ministry of Finance and Economic Development, Port-Louis, Mauritius.

Central Statistics Office (CSO) (2013) *Digest of International Travel and Tourism Statistics*. Ministry of Finance and Economic Development, Port-Louis, Mauritius.

Daby, D. (2003) Effects of seagrass bed removal for tourism purposes in a Mauritian bay. *Environmental Pollution* 125, 313–324.

De Groot, R.S. (1996) Towards a conceptual framework for measuring ecological sustainability of ecosystems. In: Sprengers, S.A., Nienhuis, P.H. and Elias, P. (eds) *Report of the Workshop Sustainability of Ecosystems: Ecological and Economic Factors*, 27–29 March 1993, Bratislava, Slovac Republic. Royal Netherlands Academy of Arts and Sciences, Amsterdam, pp. 17–49.

Deloitte and Touche (2002) *National Tourism Development Plan for Mauritius and Rodrigues*. Deloitte and Touche Consulting Report, Port-Louis, Mauritius.

Dempsey, N., Bramley, G., Power, S. and Brown, C. (2011) The social dimension of sustainable development: defining urban social sustainability. *Sustainable Development* 19(5), 289–300.

Department of Environment (2002) Environment Protection Act 2002. Department of Environment, Government of the Republic of Mauritius, Port-Louis, Mauritius. Available at: http://www.law.pace.edu/sites/default/files/IJIEA/primary_sources/Mauritius_Environment_Protection_Act_2002.pdf (accessed 20 November 2014).

Freitag, T.G. (1994) Enclave tourism development: for whom the benefits roll? *Annals of Tourism Research* 21, 538–554.

Gil, S.M. (2003) Tourism development in the Canary Islands. *Annals of Tourism Research* 30(3), 744–747.

Gill, A.M. and Williams, P.W. (2011) Rethinking resort growth: understanding evolving governance strategies in Whistler, British Columbia. *Journal of Sustainable Tourism* 19(4/5), 629–648.

Hall, C.M. (2005) *Tourism: Rethinking the Social Science of Mobility*. Prentice-Hall, Harlow, UK.

Hall, C.M. (2008) *Tourism Planning. Policies, Processes and Relationships*. Pearson, Harlow, UK.

Hall, C.M. (2011) A typology of governance and its implications for tourism policy analysis. *Journal of Sustainable Tourism* 19(4/5), 437–457.

Hall, C.M. and Muller, D.K. (eds) (2004) *Tourism, Mobility and Second Homes: Between Elite Landscape and Common Ground*. Channel View Publications, Bristol, UK.

Hall, C.M. and Page, S.J. (1996) *Tourism in the Pacific: Issues and Cases*. International Thomson Business Press, London.

Hunter, C. (1997) Sustainable tourism as an adaptive paradigm. *Annals of Tourism Research* 24(4), 850–867.

Hunter, C. (2002) Aspects of the sustainable tourism debate from a natural resources perspective. In: Harris, R., Griffin, T. and Williams, P. (eds) *Sustainable Tourism: A Global Perspective*. Butterworth-Heinemann, Oxford, pp. 3–23.

Jamal, T.B. and Getz, D. (1995) Collaboration theory and community tourism planning. *Annals of Tourism Research* 22(1), 186–204.

Keller, P. and Bieger, T. (2008) *Real Estate and Destination Development in Tourism*. Erich Schmidt Verlag, Berlin.

Lee, C.-K. (1997) Valuation of nature-based tourism resources using dichotomous choice contingent valuation method. *Tourism Management* 18(8), 587–591.

Le Meridien Hotel (no date) Available at: www.lemeridien-mauritius.com (accessed 7 May 2014).

Le Morne Heritage Trust Fund (2011) *Local Development Plan for Le Morne*. Le Morne Heritage Trust Fund, Reduit, Mauritius.

L'express (2014) Accaparement des plages: des habitants de Bel-Ombre et Baie-du-cap montent au creneau. *L'express*, 30 March 2014.

Lonely Planet (2014) Introducing Grand Baie. Available at: http://www.lonelyplanet.com/mauritius/north-mauritius/grand-baie (accessed 7 May 2014).

Marjavaara, R. (2007) Route to destruction? Second home tourism in small island communities. *Island Studies Journal* 2(1), 27–46.

Mbaiwa, J.E. (2005) Enclave tourism and its socio-economic impacts in the Okavango Delta, Botswana. *Tourism Management* 26, 157–172.

Meadowcroft, J. (2007) Who is in charge here? Governance for sustainable development in a complex world. *Journal of Environmental Policy and Planning* 9, 299–314.

Ministry of Tourism, Leisure, and External Communications (2008a) *Handbook of Statistical Data on Tourism*. Ministry of Tourism, Leisure, and External Communications Report. Ministry of Tourism, Leisure, and External Communications, Port-Louis, Mauritius.

Ministry of Tourism, Leisure, and External Communications (2008b) *Hotel Development Strategy*. Ministry of Tourism, Leisure, and External Communications Report. Ministry of Tourism, Leisure, and External Communications, Port-Louis, Mauritius.

Moscardo, G. (2011) Exploring social representations of tourism planning: issues of governance. *Journal of Sustainable Tourism* 19(4/5), 423–436.

National Empowerment Foundation (2012) *Corporate Social Responsibility: Approved Programmes under CSR*. Available at: http://www.csr.mu (accessed 7 May 2014).

Nunkoo, R. and Ramkissoon, H. (2011a) Residents' satisfaction with community attributes and support for tourism. *Journal of Hospitality and Tourism Research* 35(2), 171–190.

Nunkoo, R. and Ramkissoon, H. (2011b) Developing a community support model for tourism. *Annals of Tourism Research* 38(3), 964–988.

Nunkoo, R. and Ramkissoon, H. (2013) Stakeholders' views of enclave tourism: a grounded theory approach. *Journal of Hospitality and Tourism Research*, in press. Available at: jht.sagepub.com/content/early/2013/09/29/1096348013503997.full.pdf (accessed 21 November 2014).

Okazaki, E. (2008) A community-based tourism model: its conception and use. *Journal of Sustainable Tourism* 16(5), 511–529.

Pierson, P. (2000) Increasing returns, path dependence, and the study of politics. *The American Political Science Review* 94(2), 251–267.

Prayag, G. (2011) Rejuvenating paradise: changing products, changing markets and changing visitor behaviour in Mauritius. In: Carlsen, J. and Butler, R. (eds) *Island Tourism: Sustainable Perspectives*. CAB International, Wallingford, UK, pp. 157–170.

Prayag, G. and Jankee, M. (2009) Exploring the relationship between motives and perceived risks associated with windsurfing as an adventure activity – the case of 'Le Morne', Mauritius. *Tourism Recreation Research* 34(2), 169–180.

Ramjeawon, T. and Beedassy, R. (2004) Evaluation of the EIA system on the island of Mauritius and development of an environmental monitoring plan framework. *Environmental Impact Assessment Review* 24, 537–549.

Reed, M.G. (1997) Power relations and community based tourism planning. *Annals of Tourism Research* 24(3), 566–591.

Reid, S.D. and Reid, L.J. (1994) Tourism marketing management in small island nations: a tale of micro-destinations. *Journal of International Consumer Marketing* 6(3/4), 39–60.

Rhodes, R. (1996) The new governance: governing without government. *Political Studies* 44(4), 652–667.

Sharpley, R. (2000) Tourism and sustainable development: exploring the theoretical divide. *Journal of Sustainable Tourism* 8, 1–19.

Shaw, B.J. and Shaw, G. (1999) Sun, sand and sales: enclave tourism and local entrepreneurship in Indonesia. *Current Issues in Tourism* 2(1), 68–81.

Smith, R.A. (1992) Review of integrated beach resort development in Southeast Asia. *Land Use Policy* 9, 209–217.

Sobhee, S.K. (2004) Economic development, income inequality and environmental degradation of fisheries resources in Mauritius. *Environmental Management* 34, 150–157.

Sobhee, S.K. (2006) Fisheries biodiversity conservation and sustainable tourism in Mauritius. *Ocean & Coastal Management* 49, 413–420.

Sobhee, S.K. (2008) Tourists' willingness to pay and sustainable tourism policies in Mauritius. *International Review of Business Research Papers* 4(3), 178–191.

Stiles, R.B. and See-Tho, W. (1991) Integrated resort development in the Asia Pacific region. *Travel and Tourism Analyst* 3, 22–37.

Stonich, S.C. (1998) Political ecology of tourism. *Annals of Tourism Research* 25(1), 25–54.

United Nations Conference on Trade and Development (UNCTAD) (2005) *Mauritius Strategy for the Further Implementation of the Programme of Action for the Sustainable Development of Small Island Developing States*. Report by UNCTAD. UNCTAD, Geneva, Switzerland.

United Nations Development Programme (UNDP) (2013) *National Report of the Republic of Mauritius*. Third International Conference on Small Island Developing States. UNDP, Apia, Samoa.

United Nations Educational, Scientific and Cultural Organization (UNESCO) (2010) *Tourism Resources: Extract from the Mauritius Strategy*. Report by UNESCO. UNESCO, Paris.

United Nations Environment Programme (UNEP) (2006) Africa Environment Outlook 2: Our Environment, Our Wealth. Available at: http://new.unep.org/dewa/africa/docs (accessed 19 November 2010).

Wall, G. (1997) Integrating integrated resorts. *Annals of Tourism Research* 23(3), 713–717.

Watt, I. (2007) Sustaining Beauty in Mauritius. *BBC News, 25* September 2007.

World Economic Forum (2013) *Travel and Tourism Competitiveness Index*. The Travel and Tourism Competitiveness Report 2013. World Economic Forum, Geneva, Switzerland.

World Travel and Tourism Council (WTTC) (2013) *Mauritius Travel and Tourism Report*. WTTC, London.

chapter 14

Cooperation, Local Communities and Planning of a Religious Heritage Site: Valle del Colca, Peru

Dolors Vidal-Casellas* and Neus Crous-Costa

University of Girona, Girona, Spain

LEARNING OBJECTIVES

- To understand the role of public planning nationally and in an emerging tourism destination: the Colca Valley, Peru.
- To explore the role of external bodies and the local population in the development of the tourism industry in the Colca Valley.
- To understand the use of heritage resources and discourse as a sustainable tool for local benefits and improving the tourist experience.

INTRODUCTION

This chapter describes the tourist setting in Peru, followed by relevant policies, planning and stakeholders involved in the tourism industry from a social perspective (from the decision-making process to the delivery of the service to customers). It uses a combination of insights from local scholars and uses a range of public planning documents, such as the 2012–2021 National Strategic Tourism Plan

*dolors.vidal@udg.edu

(*Plan Estratégico Nacional de Turismo*, PENTUR) in the analysis of the Colca Valley. The Colca Valley is located in the most southern region of the country, relatively close to the Bolivian and Chilean borders. It is one of the deepest valleys in the world and the region is popular among tourists mainly as a place for sighting condors, but also because of its natural beauty. This chapter discusses the use of heritage resources in the development of a sustainable tourism industry and empowering local communities by involving them in the hospitality and tourism decision-making processes, through training and related jobs. The case study within the chapter demonstrates how the use of endogenous resources (here, Catholic churches from the Hispanic period) are viewed as a potential source of sustainable development, by both academic researchers and the Peruvian section of the Spanish Agency for International Development and Cooperation (*Agencia Española de Cooperación Internacional para el Desarrollo*, AECID).

TOURISM, CULTURE AND DEVELOPMENT

Tourism has traditionally been considered a sector that can facilitate economic, social and cultural development in areas where it is located. Recent years have witnessed an increase in projects related to responsible, sustainable, community and rural tourism, among others, and especially among indigenous peoples, who have not achieved major industrial and economic development and therefore find themselves at risk of social exclusion (Figueroa, 2014). Economically, tourism represents a catalyst for the depressed economy in certain regions, not least throughout Latin America (those containing a 'sight' from a tourist perspective or which award value to some element of their heritage, or less ideally, develop leisure services), generating revenue and employment for indigenous people. Both national governments and international organizations, led by the United Nations World Tourism Organization (UNWTO), propose the tourism industry as a vehicle for sustainable development and eradicating poverty. It must be borne in mind, however, that this stance is being criticized by some authors, such as Figueroa (2014). As Sen points out (1999, 2000 and 2007, cited in Figueroa, 2014), however, it is not only economic aspects that define the lives of individuals, but other factors must also be taken into account such as society, freedom and the environment (both natural and cultural) in which they live. On a socio-cultural level, by connecting people from different parts of the world and with different backgrounds, tourism becomes a facilitator of social change.

AN OVERVIEW OF PERUVIAN SOCIETY

Peru is a republic located on the west coast (on the Pacific Ocean) of the South American subcontinent. It is bounded to the north by Ecuador and Colombia (also members of the Andean Community), by Brazil to the east (with a notably poor transport network) and by Chile to the south (with which it has recently redefined its maritime borders). According to data provided by the Peruvian National Institute for Statistics and Informatics (Oficina Nacional de Gobierno Electrónico e Informática (Ongei), 2014), the country has a total surface area of 1.28 million km^2

(greater than the sum of the areas of Spain and France) and a population of over 27 million in 2007 (according to the last census), which is estimated at around 30 million today.

Peru has experienced an economic growth of between 4% and 9.75% of its gross domestic product (GDP) since 2002. As noted by Arellano (2011, cited in Figueroa, 2014) in recent decades, profound changes have taken place in the social landscape, including: (i) the rural exodus from the 1970s onwards (which led to growth, especially in Lima); (ii) the incorporation of women into the labour market; (iii) the role of young people in modern society; and (iv) the emergence of sectors with fewer resources. All these changes have given rise to further changes in ways of thinking, acting and consumption among the population. However, as Sen points out (1999, 2000, 2007, cited in Figueroa, 2014), when analysing human development it is important to take into account more than only economic benchmarks.

Administratively, Peru is divided into a total of 25 'Departments' (Fig. 14.1). According to the aforementioned 2007 census (Ongei, 2014), the most populated Department is Lima, the capital, with over 8 million people. The following seven Departments in terms of population are very far behind, with between 1.6 and 1.1 million people: Piura, La Libertad, Cajamarca, Puno, Junin, Cusco and Arequipa. Regarding geography and climate, Peru has 11 ecoregions and 84 of the 117 life zones existing in the world, according to the Commission for the Promotion of Peru for Exportation and Tourism (PromPerú) (PromPerú, Ministerio de Turismo del Perú, 2013). Despite the enormous wealth of natural resources, only three major regions have traditionally been identified: (i) the coast; (ii) the Sierra (the Andes, the dividing zone between the coast and the jungle); and (iii) the jungle (inland, around the Amazon basin). The PromPerú homepage (official website aimed at the end customer) provides a textual presentation of the country, highlighting, in this order, its Incan past, its natural heritage and landscape, its surviving cultural diversity and, in short, the great variety of the country as a whole. The promotional message can be summed up in the phrase 'Experience Peru and discover a variety of worlds with their own landscapes, sounds, colours and flavours; travel through time to ancient civilizations and share with Peruvians its great cultural heritage' (PromPerú, Ministerio de Turismo del Perú, 2013).

THE HISTORICAL DEVELOPMENT AND GOVERNANCE OF TOURISM IN PERU

According to Figueroa's thesis (2014), tourism was included in the portfolio of policies of the Peruvian government after the 1960 campaign of the World Bank and the United Nations Educational, Scientific and Cultural Organization (UNESCO), which was aimed at giving developing countries a way to grow. Thus, in 1964 COTURPERU was created, an organization with the specific purpose of restoring the architectural heritage of the country and developing tourist infrastructure. Only 5 years later, the Special Commission for the Peru UNESCO Cultural tourism plans (Plan Copesco) was created. It was responsible for drafting a tourism development plan of national scope. Its main objective was to improve the living conditions of the areas

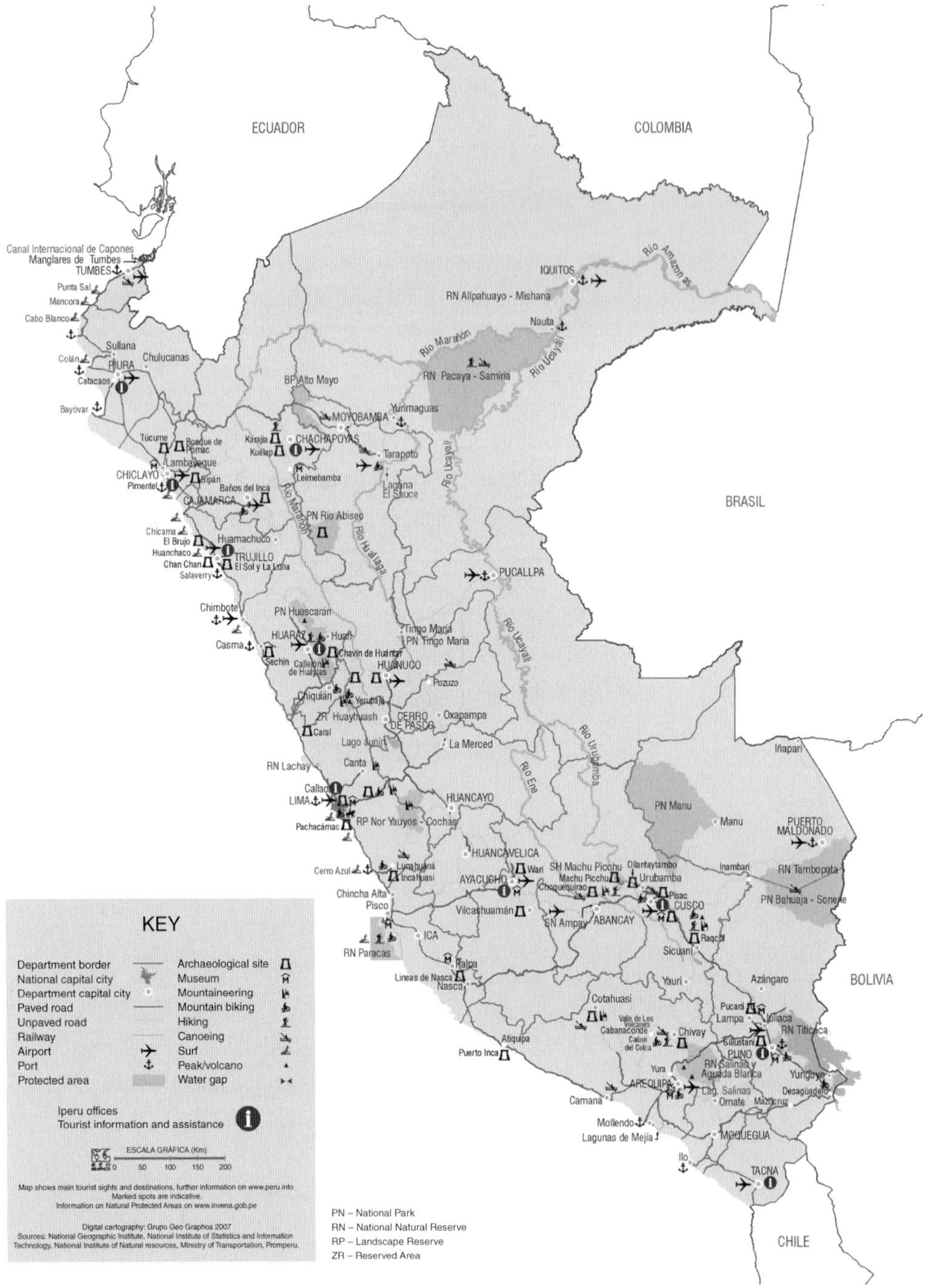

Fig. 14.1. Map of Peru showing the most common touristic attractions. (From PromPerú, Ministerio de Turismo del Perú, 2008.)

with tourism potential and, paradoxically, in situations of extreme poverty. Currently the plan is still in operation, under the direction of the Ministerio de Comercio Exterior y Turismo (MINCETUR) and the regional government of Cusco.

The government's involvement in tourism development started between the 1960s and 1970s. In these years Peru received the support of the American States Organization. Along the years, international organizations, other than the UNESCO and the World Bank, such as the Japan International Cooperation Agency, domestic stakeholders including government ministries of tourism and their regional agencies (today MINCETUR), marketing agencies (PromPerú) and local and regional guild associations (tour guides, hospitality, restaurants and so on) have become involved and they often act as lobbying organizations and are a catalyst for the introduction of laws. Over the following 15 years, however, government policies encouraged private investment in the sector, giving tax advantages and benefits for the purchase of state land, among others. In 1975, Peru joined the UNWTO.

The decade between 1990 and 2000 marked the start of the crisis of the 1990s and the government of Alberto Fujimori. During this period several studies and tourism plans were produced, including: (i) Studio Monitor (1994) on the competitiveness of the Peruvian tourism industry; (ii) Master Plan of National Tourist Development of the Republic of Peru (1999) – the first national sectorial plan, which lacked sufficiently detailed programmes and budgets; and (iii) the Programme of Integrated Tourism Development, in Peru in 1999 – this boosted the structure of the private sector and diversified tourism products. At the turn of the century, the country was placed in a context of economic growth, which still continues today. The Ministry of Foreign Trade and Tourism was established, and public policies were aimed at creating and promoting a country brand, while domestic tourism was promoted and private investment encouraged (particularly accommodation in hotels). The Observatory for Tourism of Peru identified one of the main weaknesses – the fact that promotion of some destinations is made without prior planning, which makes it exceed the carrying capacity, hence damaging tourism resources.

As it will be discussed later, in the case of the Colca Valley, there was some earlier planning for tourism, but it did not stress promotional campaigns or tourism development management. Hence, there was saturation of the market, with all travel agencies selling the same product, which presents obvious content problems for an international audience (i.e. the number of stops at souvenir shops) but at the same time that audience is highly satisfied with the beauty of the natural landscapes.

In 2008, the final document of the National Tourism Strategic Plan, 2008–2018 (PENTUR) was presented. It aimed to promote sustainable and competitive development of tourism in the country. Thus, the strategic objectives are: (i) to increase the safety of visitors; (ii) to create a competitive tourism industry; (iii) to increase demand; and (iv) overall institutional strengthening. After the signing of the Letter of Commitment for the Development of Tourism – the Golden Book promoted by the UNWTO, PENTUR 2012–2021, was produced. Again, this document aimed to achieve a competitive and sustainable tourism industry. Its aims are: (i) to increase foreign exchange; (ii) to increase domestic and international tourists; (iii) to diversify supply; (iv) to attract new market segments; (v) to enhance human resources; and (vi) to promote tourism and tourism culture by services with international quality standards.

This document also included a focus on human resources in the tourism industry. After the boom of international tourism, the demand for skilled workers has increased and has been recognized as a crucial factor in the visitors' experience. Hence, one of the Specific Action Plans to be developed in the coming years has acknowledged that a key factor is to enhance the quality of the final tourist product (i.e. tours and other services) and to enhance the quality and quantity of skilled workers in the industry. This will be achieved by the creation of training programmes and working groups to strengthen staff training.

Table 14.1 summarizes the most important tourism authorities in Peru with regard to the scope of this chapter, indicating the geographic reach and main functions of each organization.

The 2012–2021 National Strategic Tourism Plan (PENTUR) states that tourism accounts for 15% of the Peruvian GDP and identifies six main areas to promote Peru in international markets:

- Modern-day Peru: urban tourism, focusing on the history from colonial times to the present;
- Millennial Peru: historical and archaeological pre-Incan and Incan legacies;
- Living Culture Peru: experience of interacting with communities that preserve old traditions and customs;
- Natural Peru: enjoying nature, observation of flora, fauna and landscapes;
- Adventure Peru: recreational activities that require physical and/or sports ability, with some degree of risk; and
- Fascinating Peru: unique and different experiences (includes gastronomy).

According to the *Travel and Tourism Competitiveness Index* (TTCI), produced by the World Economic Forum (2013), Peru has fallen from 69th position in 2011 to 73rd in 2013. In terms of tourist attractions, it highlights natural heritage (it is the third richest country in the world for fauna) and the several different sites declared as World Heritage Sites (seven cultural, two natural and two mixed), as well as the improved effectiveness of branding and state investment in tourism and travel. Among the negatives, it mentions a lack of safety and poor transport infrastructure, as well as the loss of competitiveness due to taxes levied on the sector. However, the volume of international tourist arrivals has continued to grow. With regard to demand, in 2012 Peru received approximately 3.5 million international tourists (Ongei, 2014).

Meanwhile, the Machu Picchu Archaeological Park recorded 699,680 international visitors and 271,299 domestic visitors in 2012, making a total of 970,979 people. According to the Master Plan that governs this archaeological site, the maximum number of daily visitors is at 2500, having previously reached 3100 in a single day. In the interests of preservation and decongestion (which also affects the visitor experience: Graefe and Vaske, 1987; Navarro Jurado *et al.*, 2013) visits are promoted to the nearby areas of Huaynapicchu, Vilcambamba and Choquequirao. However, despite interest from the Ministry of Culture, in

Table 14.1. Main current tourism stakeholders in Peru. (Compiled by the authors, 2014.)

Organization	Scope	Description
MINCETUR (Ministerio de Comercio Exterior y Turismo)	Country	Analyses tourism data, encourages investment, enhances the quality of services, protects tourists, creates 'tourism awareness' and encourages the diversification of the tourism product
PromPerú	Country	Organization affiliated with MINCETUR. Creates and executes development plans, coordinates the functions of the various stakeholders involved in tourism promotion and creates tourism offices in places where they cannot be supported by local or regional organizations
Cámara Nacional de Turismo del Perú	Country	Contributes to economic and social development of the country, promotes domestic and incoming tourism and organizes professional training and debate
Autocolca	Regional (Caylloma)	Governmental organization that works for the restoration, protection, exploitation and development of natural, archaeological, historical and economic resources associated to the tourist circuit, raising the quality of life of the local population
Agencia Española de Cooperación. Oficina Tècnica de Perú	International (delegació d'abast estatal al Perú)	International cooperation agency. Its objectives include promoting social and economic development, contributing to peace and security, strengthening democratic systems and promoting human rights and links between countries

maintaining the maximum number of people admitted, UNESCO itself has 'understood that it is not right to prevent a tourist who has come so far to see Machu Picchu' (*La República*, 2012).

Puigmartí Cansick (2013) identifies as the most visited tourist sites, Lima, Cusco (as a point of entry to the Sacred Valley and Machu Picchu) and Puno, in this order; the aforementioned receive 75% of international arrivals, while the rest of the country is mostly excluded from the major tourist circuit. The same author and Vásquez Cordano (2010) identify 12 main nodes outside this 'golden triangle': (i) the beaches to the north (on the border with Ecuador); (ii) Iquitos (the National Reserves of Pacaya – Samiria and Allpahuayo – Mishana); (iii) Kuelap – High Utcumbamba; (iv) the Moche Route; (v) Huascarán National Park (Callejón de Huaylas, adventure tourism and new Andean cuisine); (vi) Lima; (vii) the Nazca Lines; (viii) Cusco and the Sacred Valley; (ix) Machu Picchu; (x) Manu National Park (ecotourism); (xi) Arequipa and the Colca Valley; and (xii) Lake Titicaca.

The country's branding strategy, in relation to the National Strategic Tourism Plan, has been key in this respect. In March 2011, PromPerú launched a logo which summarizes the basic symbolic contents: the national colours (white and red) and the monkey tail from the Nazca Lines. Branding has improved the position of Peru in the aforementioned index (the TTCI), from 73rd in 2013 to 40th in 2014 (World Economic Forum, 2013). (Switzerland is number one, followed by Canada, Japan, Sweden and New Zealand. First in Latin America is Costa Rica, ranked 25th overall.) Clearly, the Peru brand stands out above other Latin American countries as a result of three areas: (i) heritage and culture; (ii) history; and (iii) art and culture. The launch of the Peru brand has also had an effect domestically, becoming an element of national pride and identity, associated with other factors such as the economy, sports and the boom in Peruvian cuisine. In short, the new brand must act as an umbrella for all of the country's tourism products.

THE COLCA CANYON: THE LANDSCAPE, PEOPLE AND GEOGRAPHY

The Colca Valley is located within the Department of Arequipa, some 160 km from the capital of the region (Arequipa), and within the province of Caylloma. It is relatively close to Lake Titicaca on the Bolivian and Chilean borders. Arequipa, known by the nickname of 'The White City', was declared a World Heritage Site and is considered perhaps one of the most beautiful cities in Peru, and the second city after Lima. Tourist excursions to the Colca Valley, as well as a regular bus service, depart from the Department capital. The name 'Colca', given to the river and its valley, comes from the words Collaguas and Cabanas, the two pre-Incan ethnic groups that inhabited it. With a depth of 4150 m, it is the second deepest canyon in the world and offers some of the most spectacular natural scenery in Peru. The river is approximately 450 km in length. The upper reaches, where the main valley is located, is about 100 km in length and 40 km wide. This area plays host to very different regions: peaks of up to 4000 m, glaciers and the lower valley, where agriculture is the principal activity.

Administratively, the province of Caylloma is divided into 19 districts (Achoma, Cabanaconde, Callalli, Caylloma, Chivay, Coporaque, Huambo, Huanca, Ichupampa, Lari, Lluta, Maca, Madrigal, San Antonio de Chuca, Sibayo, Tapay, Tisco, Tuti and Yanque); the capital is Chivay, which is also the starting point for almost all of the sightseeing tours that leave from Arequipa. With regard to the economy, a high percentage of the population is still rural, which decreases considerably slower than in the other regions, although it is true that in recent years tourism has been identified as an inductive agent of socio-economic change. According to a study on socio-economic development and heritage exploitation by the Institute for Peruvian Studies (2007), there are numerous different stakeholders in the area with a great variety of strategies regarding the capitalization of heritage (despite, as we will see, standardization of the tourism product). As they see it, the only element that apparently unites all initiatives is the brand 'Colca Valley', which has become a trademark of tourism and culture.

The context of tourism in the Colca Valley

Focusing on the Colca Valley, 1986 saw the creation of the Colca Autonomous Authority (Autoridad Autónoma del Colca, Autocolca). Its mission includes ensuring the restoration, protection, development, use and promotion of natural, archaeological, historical and economic resources on the tourist circuit. Furthermore, it is also responsible for organizing and managing growth and consolidating tourism in the region in order to achieve economic development that will positively affect the quality of life of local people (Autoridad Autónoma del Colca (Autocolca), 2014). Therefore, it is entrusted with awarding value to cultural and natural heritage, as pillars of tourist activity, and obtaining the active involvement of all stakeholders.

However, Poma Bonifaz and Aronés Ochoa (2011) situate the beginnings of tourism promotion mainly in the Colca Valley in 1995, largely due to its attractive landscape and for observing the condor. About 85% of tourists arriving in Arequipa go there, making it the region's second most important tourist destination. It was in this same year that the round table was created within the Development Council of the Province of Caylloma (Poma Bonifaz and Aronés Ochoa, 2011).

For the period 1998–2003, Autocolca detects a very high evolution in tourism terms. The subsidiary of the Association of Hotels, Restaurants and related sectors was formed in 2006, but then it was not until 2009 that partnerships and associations dedicated to tourism began to become established, with the support of MINCETUR. In addition to this, the Bureau for Sustainable Tourism was also formed, which later became the Provincial Tourism Board for the Colca Valley. Meanwhile, MINCETUR created the managing body 'Colca tourist destination, Valley of the Volcaques', which is responsible for defining and implementing both policies and actions.

According to the 2010 Strategic Regional Plan for Tourism in Arequipa, MINCETUR considers both the city of Arequipa and the Colca area to have sufficient tourism facilities to cover the present demand, although it does recognize some weaknesses:

- Accommodation: international quality standards are not met, lack of quality accommodation in some areas with significant tourist attractions.
- Restaurants: most do not meet quality and health standards.
- Complementary services: lack of entertainment establishments adapted to tourism (culture and folklore shows), lack of such services in inland villages with potential in terms of tourism resources.

As stated above, since the inception of public tourism, the desire of agents has been for the tourism sector in the Colca Valley to be structured in such a way that it has a positive effect on the quality of life of the resident population. Tourism in the Colca Valley has led to changes in both employment and the resident population of the city of Arequipa. However, much of the development in tourism and its benefits are reaped by exogenous companies, forcing local councils and organizations to search for alternative formulas, so that local people can also take advantage of the existing tourist circuit by offering new products and services and directing local agricultural production towards the tourist market.

According to data provided by Poma Bonifaz and Aronés Ochoa (2011), in 2005 tourism employed 25% of the population, and only 5 years later, 80%. The jobs, however, were concentrated in the areas of cleaning, gardening, reception, maintenance, support and the like. Conversely, administrative and skilled positions are occupied by foreigners. These authors also state that working conditions in SMEs (small and medium enterprises) tend to be insecure as a result of: (i) salaries of between 250 and 1200 sol a month; (ii) a high staff turnover (due to low standards of service provision and also to avoid payment of social and work benefits); and (iii) the generally informal conditions that reign in these companies. In parallel with this, local investments have resulted in a rise in family businesses, employing only local people. More than three-quarters of these do not comply with current regulations and their clients are 'free tourists', representing around 15% of visitors to the valley. They therefore have relatively little weight in the dominant established tourism industry. Interest in the tourism industry among local families began in 2005. Seeing its sustained growth, they seek jobs as guides or companions and incorporate local handicraft production into the tourism market. Newly created accommodation, restaurant or tourism agency companies are mainly initiatives of young people who have received higher education in tourism.

It is worth noting in their favour that the development of such industries has improved access to education for locals. For example, 2005 saw incorporation of the specialization in Tourism Company Administration at the Higher Institute of Technology in Chivay and the Vocational School set up by the Spanish Cooperation Agency, which aims to provide training and employment for young people of limited resources in trades related to the preservation and sustainable management of cultural heritage, tourism and agriculture (AECID, 2011). With regard to the system used to market the tourist product (i.e. packages for visits to the Colca Valley), it follows the classic system of sales commissions previously agreed between retail and wholesale agents. This system has resulted in disadvantages for local business people, who lacked sufficient experience to negotiate fees that allow them to achieve profitability and continue operations. Links between local business and tourist agencies becomes inevitable, as approximately 85% of the purchase of packages for the Colca Valley are conducted through offices established in Lima and Arequipa.

Returning to the findings of Poma Bonifaz and Aronés Ochoa (2011), the development of an exclusive tourist industry has a number of limitations, which exert very high pressure on the local population:

- insufficient participation of residents in the business and benefits;
- minimum agricultural production geared towards the tourist market;

- weak institutionalization of tourism in the province;
- low quality of local tourist services;
- seasonality;
- insufficient strengthening of local skills in terms of entrepreneurship;
- insufficient diversification of tourism products and services; and
- absence of a management plan for the Colca area.

One of the clearest examples of unconventional tourism products and services offered, and rooted in the local population, is the case of Sibayo, a neighbouring district of Chivay and mostly excluded from conventional tour packages. It offers a rural community-based tourism product based on associative administration (generating less internal competition and increasing the number of people who can benefit from it). The management body is the Sibayo Association of Tour Services, which includes 16 families, and is acknowledged as a successful experience by MINCETUR, although it captures less than 0.09% of the tourists who visit the valley.

The challenge of the supply of tourism resources and products in the Colca Valley

Thus far, we have seen that the main elements to bring tourists to the Colca Valley were its impressive landscapes and observation of the condor. The tourism inventory published by MINCETUR (2009) includes the province of Caylloma in the Valley of the Volcanoes and basically defines it in terms of its landscapes and natural attributes, highlighting, once again, observation of the condor. The various public authorities have stated that awarding value to both cultural and natural heritage for tourism purposes will help to improve the returns and benefits of tourism by local people, achieving 'true' community tourism.

The '2007–2011 Master Plan for Sustainable Tourism Development and Management in the Colca Valley, the National Reserve of Salinas and Aguada Blanca and Valley of the Volcanoes' (Autocolca, 2006), lists an inventory of tourism resources, which includes a total of 232 items. Of these, 67 (29%) are natural (geological or related to fauna and flora), whereas 165 (71%) are cultural (including crafts; myths, folklore and traditions; cuisine; music; dance and similar; and archaeology). According to Autocolca (2006), the main strength of the tourism product in the Colca Valley is its positioning in the European market and the growing volume of foreign and domestic tourists. However, there are a number of weaknesses:

- The tourism product is incomplete: lack of accessibility and basic services, few observation points, trails and rest areas; lack of interpretation sites (museums, interpretation centres, etc.).
- Tour packages are very limited and concentrate on the axis Chivay–Cruz del Condor–Cabanaconde–Tapay, with no other alternatives.
- The tourism product is largely disconnected from the local population and there is a lack of information regarding the development of suitable products.
- There are no 'Site Plans' or criteria for sustainable tourism administration (like those at Machu Picchu, for example).
- There is low competitiveness in international terms.

This last development opportunity is linked with one of the main weaknesses of the area: competition with other domestic and international destinations, deformation of the tourism market by the agency sector (constriction of the tourist product) and lack of available information and action by the authorities. As a result, the population remains expectant and foreign companies develop tourism products without any clear and applicable strategic plan to mark both the framework for action and the discourse to adhere to and around which supply will be structured. With the aim of diversifying supply, Autocolca published in its Strategic Plan (Autocolca, 2006) a summary table of tourism products, with relevant associated resources, related to motivations for demand. A possible application of this matrix, which seems not to have been considered when publishing this text, would be to disseminate it among travel agencies in order that they have a better understanding of which resources are available and are able to design other routes according to the profile of their own segment of the demand. Despite this analysis, the data collected for compilation of the Strategic Plan only evaluate the main elements of tourism intermediaries (guides and travel agencies), providing quantitative data (number of affiliated guide associations, number of people arriving in Arequipa on packages, etc.), but forgetting the content of the message transmitted to visitors, which has a direct impact on the forming of the tourist image on site and consequently behaviour (even civic) and choices relating to demand.

Spanish cooperation with local communities in the Colca

The Spanish Agency for International Development and Cooperation (AECID) was created in 1988 by the Spanish government's Ministry of Foreign Affairs. Its main objectives are summarized as fostering development in economic and social terms, enhancing peace and security, fostering humanitarian aid, strengthening democratic systems, promoting human rights and relations between different countries. The Technical Cooperation Office, integrated within the organizational structure of the AECID, has worked in Peru since before the creation of this body: in 1986 it started work in the region by identifying, evaluating and monitoring cooperation programmes and projects agreed in joint committees with the Peruvian government.

The 2007–2010 Hispanic Peruvian Cooperation Programme contained eight different programmes associated with specific projects: (i) democratic governance (five projects); (ii) basic social services (seven); (iii) rural development and combatting hunger (four); (iv) environmental sustainability, combatting climate change and habitat (three); (v) culture and development (two); (vi) gender in development (one); (vii) regional Andean (no programme during this period); and (viii) economic growth for reducing poverty (two). Cultural tourism was identified as an activity and a potential source for generating sustainable development in the Colca Valley, meaning that three of these projects are directly linked to the tourism industry.

The first of these projects is the Vocational School, located 500 m from the Achoma district, which opened on 14 September 2009 and is currently closed. Its founding has its origins in the institutional tripartite agreement between AECID, Autocolca and the Provincial Municipality of Caylloma. Its purpose is to improve the training and employment of young people with few resources (men and women aged between 16 and 25 living in poverty and extreme poverty) under the criteria of gender equality, recovering endangered methods and trades and improving cultural identity. Four specialized subjects were taught there, which have been modified over time to better suit the demand and idiosyncratic characteristics of the area: (i) labouring and stonemasonry; (ii) restorative carpentry; (iii) hospitality and tourism; and (iv) agricultural production.

In June 2011, the first classes graduated in hospitality and tourism, a total of 20 students (ten men and ten women). The report compiled by Quesada (2012) states that the majority of these (59%) had found work after training at the Vocational School. Of these, 55% were working within the Province of Caylloma and 73% in jobs related to their training, although it does not specify in which cases these are entrepreneurial projects and which are salaried under the conditions described earlier in this chapter. It is important to note that this professional training school, which is closed down at the time this chapter is being written, was a first step in enhancing the benefits of the tourism industry for the local population, before PENTUR 2012–2021 was drafted. However, the new planning framework calls for a revision of this training model and also the existing network of intrinsic-extrinsic companies (endogenous versus exogenous capital) operating in tourism services in the area.

Case Study: The Heritage Resource and the Tourism Product – Colonial Churches in the Colca

The project: 'Development and strengthening community rural tourism', in Peru, affects various parts of the country. According to AECID's own the definition: 'The project presents an alternative path to traditional tourism development strategies, seeking the inclusion of marginalized rural social sectors, with territories with low production potential, in developing tourism in Peru'. This project provides the framework for rural community development in Sibayo (Colca), described in the previous section. An example of a community development project is the Conserving Cultural Heritage in the Colca Valley. This began in 1997 and had a duration of 10 years. The basic aims were to contribute to the integral development of the area through the conservation of its historical heritage and recovering the cultural identity of its people. This includes aspects such as: (i) training local managers, so that local people become involved in this work; (ii) institutional strengthening of tourism development; (iii) conservation of natural and intangible heritage; (iv) territorial planning; and (v) support for artisan production. The National Institute of Culture in Arequipa, the Archbishopric of Arequipa and local municipalities in the Colca Valley all collaborate with the AECID, providing staff and materials.

The specific work that was undertaken was to completely restore ten Baroque churches from the Hispanic era (of the existing 23), which were abandoned during the 20th century and had suffered several robberies. Among the churches are those of Achoma, Lari, Sibayo, Chivay, Huambo, Madrigal, Yanque and Maca. Although the latter is

(Continued)

Case Study. Continued.

the last stop on virtually all sightseeing tours, the church is rarely visited as tourists tend to stay in the craft market. However, the second part of the project has not been undertaken, which focused on awarding value to the churches for tourism purposes and dissemination of this heritage.

The 23 churches from the Hispanic era (17th–18th centuries) in the Colca Valley are unique examples of the blend of Spanish Baroque art and indigenous spirituality which, rather than being eradicated by Christian preachers, entered into syncretism with Christianity (Urbano, 1993). Following the failure of the first attempts at forcible evangelization, missionaries adopted more flexible criteria. Thus the natives appropriated Christian symbols and various of their own deities were identified with Christian saints and symbols in order to continue expressing their own religiosity. The churches of the Colca Valley reflect precisely this mix: a fusion of European art, local indigenous expressions which contributed ornaments from nature and local worldviews and evangelizing zeal on the part of the Church through visual language. With ten of these churches restored (those located on the left bank), there is a need to award value to them so they do not only contribute to the economic development of the region, but also to local self-esteem.

In tourism terms, these are sites neglected by tour packages; stopping in the village, does not guarantee a visit to the church, or even the recommendation of one. An analysis of the packages offered by travel agencies that form part of the Association of Travel and Tourism Agencies of Arequipa, reveals that the most visited town is Chivay, used as a base for most tours. The field observation also reveals that Arequipa agencies often promote stops at more villages and churches than are actually visited, as well as advertising observation points with views of terraces and pre-Incan places, when actually guiding groups to shop in craft markets established elsewhere. Finally, these practices result in a 'poor tourism' product that provides an easily improved experience, since the sale (and therefore expectations generated) rarely correspond exactly with the services provided. Moreover, they are standard packages, that is, identical from one agency to another, so it often does not matter which retail agency tourists contract because the minibuses that leave from Arequipa to Colca mix customers from all of them, and the only substantial difference between what each customer purchased (and, therefore, the price they paid) is the accommodation option.

CONCLUSION

Tourism development in Peru has been the subject of a considerable volume of scientific studies, regarding the region of Latin America. Thus, over the last decade a large number of Peruvian authors have conducted sector-based and global studies on various aspects of the industry, especially with regard to sustainability and benefits to local populations of regions with a higher influx of tourists. Despite this, however, as a region it has gone virtually ignored in the international scientific literature, which tends to focus Latin American research on the Caribbean region, with a longer and more established history of tourism internationally.

Peru is one of the countries with the most notable growth at present, not only in terms of tourism but also other economic sectors. According to research conducted by Figueroa (2014), this economic expansion was initially concentrated in a few hands and had not been extended to the bulk of the population, but this situation is now changing. Several different factors are increasing the self-esteem of the population, including the Peru brand and the gastronomy boom due especially to the figure of Gastón Acurio. Diverse government and international cooperation projects, such as the AECID's Vocational School, are laying the groundwork for a broader sector of the population accessing the benefits of progress, benefits which, as Sen acknowledges (1999, 2000 and 2007, cited in Figueroa, 2014), cannot be considered to be limited only to economic aspects.

For its part, MINCETUR considers the existing tourism resources in Arequipa to be sufficient, although there are shortcomings regarding quality (referring to accommodation). In its assessment, however, it does not mention what most directly affects the tourist experience. The data provided by this body indicate that the most popular sights among international tourists are cultural ones. Contrarily, the different local or regional development plans (developed by both the city and province of Arequipa and the Colca Valley) do not contemplate awarding value to heritage for tourism purposes (referring to spoken discourse and interpretive elements), limiting themselves to their cataloguing, which only constitutes a first step.

It is the existence of these sites in the country (monuments, nature, fauna, etc.) that attracts visitors and it is the experience of their visit that will determine satisfaction, and therefore also their advertising (word of mouth, virtual communities) of the destination once the trip has finished as well as their behaviour at the destination (certain perceptions motivate more civil behaviour than others). Training the local population in the transmission of the values inherent in their local heritage constitutes a step in the sustainable tourism chain in its various dimensions: social, economic and the visitor experience. This social dimension refers to inclusion of local people in the tourism industry in a dignified position with regard to those tourism organizations that dominate the market, exemplified primarily by travel agencies and tour operators. Their participation on equal terms (proposed here as local or sectorial tour guides), contributing to the awarding of value to their own heritage which is attracting the 'gaze' from outside, also constitutes a source of national or local pride. Moreover, those natives who become tourism professionals have the ability to train new generations, making this profession a tool for intergenerational cohesion, and introducing the concept of a need for continuous training.

This dimension is closely linked to the economic aspect. A population that knows and is trained to properly convey the values of its heritage to tourists in an interesting way is also able to establish professional working relationships with dominant companies in the sector, without adopting a position of subservience or leaving the local population occupying only the lowest (and most poorly paid) positions in the entire production process, as regularly tends to happen. Equally, it is also worth bearing in mind that training to improve spoken discourse, especially when it is the first generation training the next (and so on), is a very cheap resource in comparison with other systems of interpretation which, in addition, and especially in the case of new technologies, are tied to an expiry date which has often even been scheduled. Finally, as already pointed out above, the visitor experience will be improved substantially when it is the actual owners and trustees of the heritage assets (whether cultural or natural) who transmit the information and deep sense of heritage to tourists.

QUESTIONS

1. What is the indigenous people's current involvement in the Colca Valley tourism sector?

2. What are the strengths and weaknesses of the tourism development plans and training initiatives that have taken place in the Colca Valley to date?

3. What advantages does the improvement of (spoken) heritage discourse regarding tourist sites offer in comparison with other interpretative tools?

BIBLIOGRAPHY

Agencia Española de Cooperación Internacional para el Desarrollo (AECID) (2011) Programas y Proyectos. Escuela Taller del Valle del Colca. *Agencia Española de Cooperación Internacional para el Desarrollo*. Available at: http://www.programapd.pe/ret/et_colca (accessed 19 April 2014).

Agencia Española de Cooperación Internacional para el Desarrollo (AECID) (2014) Oficina Técnica de Cooperación en Perú. *Agencia Española de Cooperación Internacional*. Available at: http://www.aecid.pe/otc/ (accessed 20 April 2014).

Anon. (2003) *El Turismo rural en las Américas y su contribución a la creación de empleo y a la conservación del patrimonio: Asunción (Paraguay) 12–13 de mayo de 2003 (Rural Tourism in the Americas and Its Contribution to Job Creation and Heritage Conservation: Asun)*. United Nations World Tourism Organization, Madrid. Available at: http://cataleg.udg.edu/record=b1182312~S10*cat (accessed 17 April 2014).

Anon. (2009) *III Conferencia Internacional sobre turismo y artesanía: Lima (Perú) 24–26 de septiembre de 2008 (III International Conference on Tourism and Handicrafts: Lima, Peru 24–26 September 2008)*. Organización Mundial del Turismo, Madrid. Available at: http://cataleg.udg.edu/record=b1276978~S10*cat (accessed 17 April 2014).

Asociación de Arequipa de Agencias de Viaje y Turismo (2014) *Asociación de Arequipa de Agencias de Viaje y Turismo*. Available at: http://www.avit-aqp.org/ (accessed 20 April 2014).

Autoridad Autónoma del Colca (Autocolca) (2006) *Plan Maestro para el Desarrollo y Gestión Sostenible del Turismo en el Valle del Colca, la Reserva Nacional de Salinas y Aguada Blanca y el Valle de los Volcanes 2007–2011*, Arequipa. Available at: http://issuu.com/franklinventuraquico/docs/plan_maestro_de_turismo_2007_-_2011_-_2.0_copia (accessed 20 April 2014).

Autoridad Autónoma del Colca (Autocolca) (2014) Cañon del Colca. *Autoridad Autónoma del Colca*. Available at: http://www.colcaperu.gob.pe/institucional/nosotros-colca.html (accessed 19 April 2014).

Barbaza, Y. (1966) *Le Paysage humain de la Costa Brava*. Armand Colin, Paris. Available at: http://cataleg.udg.edu/record=b1144096~S10*cat (accessed 20 April 2014).

Benavente Gianella, L. (2013) Marca Perú: branding y Desarrollo América Economía – El sitio de los negocios globales de América Latina. *América Economía* Online. Available at: http://www.americaeconomia.com/analisis-opinion/marca-peru-branding-y-desarrollo (accessed 18 April 2014).

Besculides, A., Lee, M.E. and McCormick, P.J. (2002) Residents' perceptions of the cultural benefits of tourism. *Annals of Tourism Research* 29(2), 303–319. Available at: http://www.sciencedirect.com/science/article/pii/S0160738301000664 (accessed 19 April 2014).

Collado, M. (2014) Turisme cultural a la Vall del Colca. *Facultat de Turisme UdG, WordPress*. Available at: http://colcatcultural.wordpress.com (accessed 20 April 2014).

Córdova Aguilar, H. (2003) Tursimo en el Valle del Colca y sus impactos en el desarrollo. *Espacio y Desarrollo* 15, 69–88. Available at: http://revistas.pucp.edu.pe/index.php/espacioydesarrollo/article/view/8120/8412 (accessed 1 April 2014).

De la Cuba Málaga, G. (2006) *Iglesias del Valle del Colca*, 1st edn. Universidad Católica de Santa María, Arequipa, Peru.

Figueroa, J.R. (2014) *Turismo, pobreza y desarrollo sostenible en el Perú. Los casos de Cuzco, Cajamarca y La Libertad*. Universitat de Girona, Girona, Spain. Available at: http://www.tdx.cat/handle/10803/132090 (accessed 20 April 2014).

Graefe, A.R. and Vaske, J.J. (1987) A framework for managing quality in the tourist experience. *Annals of Tourism Research* 14(3), 390–404. Available at: http://www.sciencedirect.com/science/article/pii/0160738387901101 (accessed 18 April 2014).

Hampton, M.P. (2005) Heritage, local communities and economic development. *Annals of Tourism Research* 32(3), 735–759. Available at: http://www.sciencedirect.com/science/article/pii/S0160738305000629 (accessed 13 May 2014).

Institute for Peruvian Studies (2007) Institute for Peruvian Studies, Lima. Available at: http://www.iep.org.pe (accessed 21 November 2014).

Izaguirre Sotomayor, M.H. (2009) PENTUR. Ausencia de direccionabilidad política. *Turismo y patrimonio*, 6, 125–130. Available at: http://cataleg.udg.edu/record=b1294319~S10*cat (accessed 17 April 2014).

La República (2012) En Machupicchu recibieron por día a 3.100 turistas. *La República* Online. Available at: http://www.larepublica.pe/07-08-2012/machupicchu-recibio-3100-turistas-por-dia (accessed 18 April 2014).

Ministerio de Comercio Exterior y Turismo (MINCETUR) (2009) Inventario Turístico de Perú. Valle de los Volcanes. *Ministerio de Comercio Exterior y Turismo*. Available at: http://www.mincetur.gob.pe/TURISMO/OTROS/inventarioturistico/Ficha.asp?cod_Ficha=609 (accessed 20 April 2014).

Ministerio de Comercio Exterior y Turismo (MINCETUR) (2010) *Plan Estratégico Regional del Turismo de Arequipa*. Ministerio de Comercio Exterior y Turismo, Arequipa, Spain. Available at: http://www.concytec.gob.pe/portalsinacyt/images/stories/corcytecs/arequipa/pertur_arequipa_2011-2015.pdf (accessed 14 April 2014).

Ministerio de Comercio Exterior y Turismo (MINCETUR) (2012) *Plan Estratégico Nacional de Turismo 2012–2021*. Ministerio de Comercio Exterior y Turismo, Lima, Peru. Available at: http://www.mincetur.gob.pe/newweb/Portals/0/documentos/turismo/PENTUR_2013-Final.pdf (accessed 14 April 2014).

Ministerio de Comercio Exterior y Turismo (MINCETUR) (2013) *Turismo*. Ministerio de Comercio Exterior y Turismo, Lima, Peru. Available at: http://www.mincetur.gob.pe/newweb/Default.aspx?tabid=137 (accessed 19 April 2014).

Navarro Jurado, E., Damian, I.M. and Fernández-Morales, A. (2013) Carrying capacity model applied in coastal destinations. *Annals of Tourism Research* 43, 1–19. Available at: http://www.sciencedirect.com/science/article/pii/S0160738313000595 (accessed 18 April 2014).

Oficina Nacional de Gobierno Electrónico e Informática (Ongei) (2014) *Portal del Estado Peruano*. Available at: http://www.peru.gob.pe/ (accessed 14 April 2014).

Padró Werner, J. (2010) Turismo cultural y desarrollo local. In: Ayuntamiento de Úbeda, A.E.de C. de la C. (ed.) *El turismo como recurso econòmico. La cerámica como recurso turístico*. Asociación Española de Ciudades de la Cerámica, Úbeda, Spain, pp. 1–23. Available at: http://www.ciudades-ceramica.es/documentos/Congreso_Ubeda_2010/01-Ponencia_J_Padro.pdf (accessed 14 April 2014).

Poma Bonifaz, D. and Aronés Ochoa, C. (2011) *Impacto del turismo en el Valle del Colca*. Centro de Estudios y Promoción de Desarrollo, Lima, Peru, pp. 236–250. Available at: http://www.descosur.org.pe/lecturas (accessed 25 March 2015).

Poria, Y., Butler, R. and Airey, D. (2003) The core of heritage tourism. *Annals of Tourism Research* 30(1), 238–254. Available at: http://www.sciencedirect.com/science/article/pii/S0160738302000646 (accessed 13 May 2014).
PromPerú, Ministerio de Turismo del Perú (2008) Mapa turismo Perú, Lima: Wikimmedia Commons. Available at: http://commons.wikimedia.org/wiki/File:Mapa_turismo_peru.jpg?uselang=es (accessed 1 August 2004).
PromPerú, Ministerio de Turismo del Perú (2013) *Perú Travel*. Available at: http://www.peru.travel/es-pe/sobre-peru.aspx (accessed 14 April 2014).
Puigmartí Cansick, S. (2013) *Historia del Turismo y Turismo Receptivo 2013*. SlideShare. Available at: http://www.slideshare.net/susanapuigmarti/per-historia-del-turismo-y-turismo-receptivo-2013 (accessed 18 April 2014).
Quesada, A. (2012) *Pobreza, desarrollo y migraciones: el caso del Valle del Colca*. Available at: http://portfolioquesada.weebly.com/uploads/9/4/8/5/9485235/rapport_perou.pdf (accessed 18 April 2014).
Ruoss, E., Alfarè, L. and Ruzin, I. (2014) Cultural tourism as an opportunity for sustainable regional development: the case of Bitola. *Regions Magazine* 293(1), 14–17. Available at: http://rsa.tandfonline.com/doi/abs/10.1080/13673882.2014.10806802?journalCode=resn20 (accessed 18 April 2014).
United Nations Educational, Scientific and Cultural Organization (UNESCO) (2014) UNESCO World Heritage Centre, Peru. Available at: http://whc.unesco.org/en/statesparties/PE/ (accessed 18 April 2014).
Urbano, H. (1993) Sincretismo y sentimiento religioso en Los Andesapuntes sobre sus orígenes y desarrollo. In: Facultad de Teología San Vicente Ferrer de los Padres Dominicos (ed.) *Etica y teología ante el Nuevo Mundo: Valencia y América: actas del VII Simposio de Teología Histórica*. Facultad de Teología San Vicente Ferrer de los Padres Dominicos, Valencia, Spain, pp. 252–292.
Vásquez Cordano, A. (2010) *Historia de la hoteleria en el Perú*. SlideShare. Available at: http://www.slideshare.net/lidey/historia-de-la-hoteleria-en-el-per (accessed 18 April 2014).
World Economic Forum (2013) *Travel and Tourism Competitiveness Index*. The Travel and Tourism Competitiveness Report 2013. World Economic Forum, Geneva, Switzerland. Available at: http://www3.weforum.org/docs/WEF_TT_Competitiveness_Report_2013.pdf (accessed 18 April 2014).
Zanotti, M. (2011) Antes y después, la nueva 'marca país' de Perú. Blog de diseño gráfico y comunicación visual. *Soy Diseñador Gráfico*. Available at: http://www.soydg.com/blog/2011/04/05/antes-y-despues-la-nueva-marca-pais-de-peru/ (accessed 1 April 2014).

chapter 15

Conclusions: Diverse Paths to Sustainability

Nigel D. Morpeth* and Hongliang Yan

Leeds Beckett University, Leeds, UK

This chapter picks out the main themes which have emerged from authors throughout the book and highlights insights from the themes of 'diverse approaches to sustainability', 'governance for sustainability' and 'pedagogy and praxis' and then offers philosophical reflections to inform future planning processes for sustainability. The intention in inviting a range of authors from diverse parts of the world was to provide both practitioner and academic perspectives on how diverse destinations articulate commitments to sustainability, and how this might be manifested in the activities which might be construed as being sustainable forms of tourism. The range of chapters written from different cultural contexts offer diverse 'paths' to sustainability which are reflected in different approaches to planning for tourism.

DIVERSE APPROACHES TO SUSTAINABILITY

While there might be similarities and commonalities in narratives, there are diverse approaches to sustainability and the connections with planning for tourism that emerged within the different chapters.

At a local level, within the city of York, the narrative of sustainability and sustainable tourism largely focused on initiatives for sustainable transport which has implications for both residents and tourists. In particular, Tempest's chapter (Chapter 5) provided in-depth evidence of the priority of planning for the motor car, in UK cities, in the 1960s and 1970s, with the emphasis on creating ease of access for the motor car to reach the

*n.morpeth@leedsmet.ac.uk

heart of the city, something which Tempest argues would be unthinkable now. However, he pointed out that the motor car was seen then as a symbol of both personal freedom and modernization.

The consideration of an inner ring road, which did not achieve fruition, was viewed as being tangible evidence that, in addition to losing a significant amount of buildings in its construction, the car would impact negatively on schools and neighbourhood areas in the city. Tempest demonstrates the challenges that an historic destination has had to respond to over the last five decades. In his view, there has been a transformation of the physical environment of the city through the strategies to encourage sustainable forms of transport. He argues that traffic management in the city, in 2014, still raises serious concerns and the experimentation of additional traffic restrictions in closing a bridge (Lendal Bridge) to all motor vehicles has proved difficult through opposition to the initiative. The case study focus in Chapter 4 demonstrated that Canterbury, another historic city within the UK, faces similar problems to York in managing the flow of motor vehicles in the city. Both York and Canterbury have tried to introduce modifications to the motor car entering the heart of the city but have faced strong opposition from the car lobby. However, what has been successful in both destinations is tourist arrival by train and coach, in conjunction with the success of park-and-ride schemes significantly curbing the use of motor vehicles in the city. Substantial modifications to create 'footstreets', areas of pedestrianization and the development of a bicycle infrastructure have promoted sustainable transport options for residents and tourists in both cities. Both chapters have revealed mechanisms to involve local communities in contributing to a Local Plan and engagement in future visions for the two cities. In the case of York, Tempest highlights how the Residents' Festival initiative has been a successful approach in engaging with local people to enable them to access facilities largely used by tourists.

At a national level in Latvia, Atstāja, Brīvers and Līviņa (Chapter 12) emphasize how planning for the sustainability of tourism is done in parallel with other sectors in an integrated way, with the *Sustainable Development Strategy of Latvia 2030* intended to strategically coordinate planning to create commonalities between different sectors. This entails combining issues to do with the development of cultural space and an eco-efficient economy, to ensure the conservation of nature as a future capital. In a set of policy indicators that are intended to measure these themes of sustainability, 'measurement' of sustainable tourism is partly through monitoring the length of stay of tourists. Within the planning development and policy documents the Latvian Tourism Development Agency has extended the measurement of sustainability to other economic indicators. The development and promotion of the national tourism brand 'Latvia. Best Enjoyed Slowly' positions niche tourism products as exemplars of sustainable tourism. However, while preserving natural resources in nature-based tourism and cultural resources through dark tourism preservation, the overarching sustainability goals appear to be economic and cultural. However, as we will see shortly in the next section, there are governance arrangements focusing on the environmental dimension of sustainability and a coordinated approach to coastal (beach management).

Prayag (Chapter 13) in his analysis of Mauritius, while emphasizing an embryonic attempt at governance arrangements which involve community participation in tourism planning processes, emphasizes that the creation of a Ministry of Environment and Sustainable Development signalled a commitment to trying to plan for sustainable development on the island. The promotion of the product-led tourism through enclave tourism has put a strain on local ecosystems of the island. He argues that ostensibly enclave tourism has served to widen the 'social and cultural gap' and there are limited community-based tourism opportunities. He also argues that the private hospitality sector has taken the lead over the public sector in introducing corporate social responsibility initiatives working with disadvantaged children (he speculated whether the motivation was altruistic or an act of compliance). In presenting a case study of the Grand Baie development he emphasized how a local fishing village has been transformed into a 'concrete tourism', with a reduction in access to public spaces such as beaches. Overall, the move towards sustainability is contradictory and it is difficult to locate a consistency in progress.

In Chapter 9, Eurico and Oliveira emphasize that both the preservation and the regeneration of the cultural heritage of the rural schist villages, in Portugal, have been established partly by the creation of a network of shops. Within this initiative communities determine what might be sold as representative of the views of local people and which reflects local cultural identity. They emphasize that they feel that 'theoretical concepts and methodologies related to the concept of sustainable tourism can be well applied' (Eurico and Oliveira, Chapter 9, this volume) in this respect and in the next section the governance arrangements for this project emphasize a strong community involvement as an example of participatory sustainable development.

In Chapter 10, Farra Haddad's analysis of sustainable tourism development highlights the ongoing challenge for Lebanese society to be able to sustain political initiatives for sustainability in the context of regional insecurity. The country has had to absorb both the destructive nature of warfare and the influx in 2014 of nearly 1.5 million refugees, largely from neighbouring Syria. This has prevented a consistent flow of tourist visitation and the creation of suitable tourism infrastructure. The main initiatives for sustainable tourism are based on forms of niche tourism, such as nature-based tourism, religious tourism, health tourism and an ecotourism industry, which seeks a legal framework to develop the industry sustainably. Farra Haddad suggests that niche tourism has the potential to promote national identity and cohesion in society at a time of uncertainty and insecurity. The Beyond Beirut forum, in 2011, staged by a range of international organizations, interpreted sustainable tourism initiatives in the Lebanon as community-based rural tourism initiatives with aspirations to strengthen the role of women in the tourism industry.

Vidal-Casellas and Crous-Costa, in Chapter 14, when referring to the regeneration of the Colca churches in Peru, focus on niche heritage tourism. In doing so they emphasize the importance of the social dimension of sustainability, exemplified by the inclusion of local people (in a 'dignified position') working as communicators of their indigenous culture to tourists. The creation of the 2010 Strategic Regional Plan for Tourism in Arequipa (MINCETUR) sought to influence greater access for local people to educational and employment opportunities.

The opening of a vocational school, in 2009, in the Provincial Municipality of Caylloma, in the Colca region, aimed to improve training and employment for young people, with tourism and hospitality featuring within this educational programme. In the case study on the heritage tourism opportunities of colonial churches, Vidal-Casellas and Crous-Costa demonstrated how the social, economic and cultural dimensions of sustainability are exemplified through the involvement of the local community and their interaction with visitors in this community-based project. Importantly, they also emphasize that with the emergence of locals trained as tourism professionals, there is the scope to create 'intergenerational cohesion', with locals training new generations of tourism professionals. They also promoted this as an example of social equity and as empowering local people in their engagement with tourism businesses.

In the chapter on Dubai (Chapter 8), Kotsi and Michael highlight the aspirational aspects of sustainability in a destination which has developed rapidly in a relatively short period of time with tourism a major element of its diversification process. The rapid development of the city for tourism, commerce and industrialized processes has raised concerns of both the environmental and the social aspects of sustainability. The Department of Tourism and Commerce Marketing's current 'Vision for Tourism' anticipates a doubling of tourist arrivals annually, from 10 million international arrivals in 2010, to 20 million in 2020. 'The Mall of the World' development emphasizes the environmental sustainability challenge of such a venture (which will occupy 4.46 million m^2). While they identify that the concept of sustainable tourism is a fledgling industry in Dubai, with a current focus (as with the chapters on Latvia and the Lebanon) on developing forms of niche tourism as an exemplar of sustainability, Kotsi and Michael highlight wider sectoral initiatives for sustainability through the United Arab Emirate's commitment to renewable and 'Green Energy' and a 'Green Lifestyle' initiative. These initiatives are at the forefront of the development of neighbouring Masdar City, an exemplar of utilizing green technologies and an example of a sustainable city, such as harnessing solar energy. They acknowledge, however, that creating a balance between a booming economy and environmental and socio-cultural sustainability is challenging. In particular there are environmental concerns about the scale of development and Kotsi and Michael highlight this dilemma in their Expo 2020 case study, which is intended to promote the theme of sustainability contained within an accompanying large-scale development programme.

In Chapter 7, Beardmore argues that an integrated approach to planning and management of visitation is linked closely to understanding the dynamic of the interaction and impacts of both leisure and tourism on host communities within the Yorkshire Dales National Park. She identifies that greater understanding of these impacts and interactions will lead to sustainable tourism appropriate to national park environments. This holistic approach applied to an integrated approach to planning is concerned with understanding the mutual dependence of leisure, tourism and the everyday activities of residents.

GOVERNANCE FOR SUSTAINABILITY

Within a range of chapters the concept of governance emerged as a key consideration of the way in which destinations enable a range of community representatives, organizations and agencies to have an input on the creation of policies and initiatives informed by the principles of sustainability.

In Chapter 4, the work of Boddy and Parkinson (2004) helpfully presented the theoretical possibilities of what 'healthy' governance arrangements can achieve, where the views of many actors within a locality can, in theory, create inclusive and collaborative approaches to sustainability. The planner who was interviewed in the chapter revealed how working across multi-sectoral boundaries has its advantages, in being able to draw a multitude of perspectives into policy and planning processes. However, these challenges require the planner to have special communication skills, in assimilating potentially dissonant voices and perspectives as to what constitutes the 'public interest' and the 'public good', in realizing the collective agreement and action on visions for planning informed by the principles of sustainability.

Within Chapter 5, Tempest highlights the success of local governance arrangements for tourism, with the private sector taking a key lead, in conjunction with the local authority, arguing that this partnership working has led to the transformation of the landscape of the city (e.g. 'footstreets', park and ride, pavement cafés, transformations to the city's events programme and VIC service). He cites organizations such as the York Tourism Bureau and Visit York, who, historically, as well as having a key role in the promotion of the city, stimulated partnership work with the Council, Yorkshire Tourist Board (now Welcome to Yorkshire) and a myriad of businesses, all committed to develop tourism in the city.

In Chapter 7, Beardmore highlights how the strategic vision for sustainable tourism is embodied within the objectives of the National Park Management Plan, together with separate strategies for tourism and recreation. The operationalization of these strategic considerations emerges in the everyday partnership working, for example with the Dales Tourism Partnership with the Nidderdale Area of Outstanding Natural Beauty. Partnership work with other organizations and tourism businesses is central to the management of the National Park. In this respect a key consideration of planning and partnership working for sustainable tourism is through EUROPARC and the accreditation of the Yorkshire Dales National Park, in line with European Sustainable Tourism Charter principles.

In Chapter 9, Eurico and Oliveira emphasize the tangible example of how the communities of the schist villages in Portugal, since 2000, have been revitalized by the joint partnership work of local government, regional government bodies and, crucially, community residents and local businesses. They emphasize that the process of sustainable tourism development emerged through residents being part of decision making, on creating a balance between quality of life and residential considerations, combined with the promotion of entrepreneurship of a network of 'Schist Villages Shops'. It was emphasized by the authors that the ongoing development of the network should be predicated on the maxim that all future actions should be on the basis of the guarantee of residents' involvement at each stage of development.

In Chapter 13, Prayag, in contrast to the 'Schist Village' model, identifies how the governance arrangements for enclave tourism in

Mauritius are an example of non-participatory tourism planning, top-down tourism policy implementation with central government not having an holistic view of sustainability. Furthermore, he argues that the public sector feels reliant on these external organizations and agencies to ensure 'successful tourism development'. Prayag suggests that there is a lack of more inclusive governance structures to enable community tourism decision making. This he suggests might potentially increase community ownership of tourism as a viable option for economic development. He argues that this could be done through a community monitoring programme that would measure economic, social and environmental indicators of change. However, while outlining the virtues of community-based tourism in Mauritius, Prayag emphasizes that the reality of organization of tourism means a dominant role for external investors, the national government and consultants. Nevertheless, while not the norm, he points out that there is evidence emerging that there is the start of a community-based consultation process, with community workshops used to develop the Le Morne Cultural Landscape planning process in 2010, which included a statement of a community vision. However, he emphasizes that even if the community is involved in decision making for tourism, these decisions are superseded by past decisions on things like real estate development.

In Chapter 11 on the Balaton, Raffay and Clarke emphasize the complex organizational structure of governance for tourism in Hungary and not least in the Balaton, which reflects the remnants of a centralized system of planning, top-down in emphasis. Governance, based on fostering community trust (informal) and bringing together public and private organizations (formal) coordinated by the Balaton Development Agency, focuses on developing spa and health tourism and additional recreational activities at Lake Balaton. With 20 Destination Management Organization (DMO) areas, which are distinctive localities, effectively in competition with each other, the organization and agreement of sustainable tourism priorities, according to Raffay and Clarke, for the future remains problematical.

Farra Haddad in Chapter 10 highlights the complexity of the country of Lebanon trying to plan for the future governance of the tourism industry, in a period of uncertainty and insecurity, stating that:

> Stakeholders in the tourism field know very well that the country will not be successful in attracting tourists back or investment without achieving sustainable peace among its various secular factions and with its regional neighbours. But they know also that even with the political turbulence in the region and, at times, inside the country, tourism has always exhibited remarkable resilience and there are always crisis management solutions.
>
> (Farra Haddad, Chapter 10, this volume)

She highlights the lack of a current tourism strategy and a coherent planning vision for the future and one of the problems of diversifying tourism in rural areas is that there is a concentration of tourism resources in Beirut.

The structure and processes of governance of tourism in Latvia (Chapter 12) communicated by Atstāja, Brīvers and Līviņa highlight the cohesive structural arrangements from the national to the local level in the planning and organization of tourism. An example of linkages in governance structures related to sustainability is the co-ordination of the *Coastal Spatial Development Strategy (2011–2017)* which in conjunction with the *Sustainable Development Strategy of Latvia 2030* has introduced a long-term strategic approach to the Baltic coast. However, they note that in advance of this strategy local actions have restricted accessibility to coastal resources.

PEDAGOGY AND PRAXIS

There are two aspects to pedagogy that emerged within Chapter 2: (i) the dissemination of ideas and initiatives for sustainability within educational institutions; and (ii) policy learning within policy communities, particularly within public sector organizations that might be engaged in planning and policy processes for sustainability. In policy terms, the lexicon of the concept of sustainability can have 'insider' meanings, to the exclusion of external audiences – this is ironic as it is for the intended benefit of these audiences that these concepts are crafted into policy formulation for cross-sector delivery, at all levels of governance. What appears will remain a challenge for the future is the durability of policy initiatives which inform planning processes for sustainability. In the case study on policy initiatives for sustainability in Chapter 2, Morpeth and Yan highlight that initiatives such as Norwich 21 have stalled in their development because the UK national government decided that Local Agenda 21 initiatives would be disbanded and sustainability would become part of corporatized processes, rather than through an articulation of community-based grass-roots initiatives. Sustainability as a policy concern would appear to go through issue cycles, when it struggles for visibility in the public realm and in a UK context sustainability at a national political and policy level is now intrinsically linked to budget deficit reduction.

The other aspect of the important role of pedagogy in the dissemination of learning on sustainability, it was argued by Morpeth and Yan (Chapter 2), is the role of educational institutions in influencing, informing and critically stimulating learning towards sustainability. The idea of critical pedagogy and praxis is perhaps best exemplified through the work of Grimes in Chapter 6, as an example of stimulating levels of both critical and social consciousness and encouraging students to view everyday experience through a different critical lens.

It is important to note that in both Chapter 3 and Chapter 6 it was highlighted that had there been more scope in the book, it would have been used to provide more in-depth analysis of women, sexualized and racialized spaces, and planning to consider the wider issues of social equity. In profiling disability and planning, Grimes's chapter (Chapter 6) was not intended to speak for other sectional issues in society but to provide insights into how historically disabled people were absent from tourism literature. The emergence of the concepts of accessible tourism and universal design are key innovations that can potentially transform inaccessible spaces for both disabled people and the able bodied.

Grimes presented a compelling argument for transferring theory into praxis and demonstrating that social agency can inform change. In writing cogently about background theory to disability, universal design and accessible tourism, he offered insights into his own experiences and initiatives which have changed attitudes (and behaviour) to disabled people. Also he has demonstrated that the introduction of universal design is not about offering disabled people greater independence but greater *interdependence* with all sections of society. He concluded that:

> Therefore, if governments and planners truly want to create *destinations for all* they must consult with local disabled people and their representative organizations and continuously update information to ensure that it is detailed and reliable to give everyone the confidence to visit.
>
> (Grimes, Chapter 6, this volume)

This is a maxim for interactions with different groups in society that should be at the heart of planning processes.

PHILOSOPHICAL REFLECTIONS

This chapter has provided a summary and synthesis of key issues which have emerged from the preceding chapters within this book. It offers thoughts on the challenges of working towards sustainability, which remains the key challenge for humankind to respond to for the remainder of the 21st century and beyond.

Applying existentialist thoughts to discourses of sustainability is comforting, only in so far that it guards against always wanting to have a completeness in our cognizance, perception and interaction of the material and social world around us (and within us). This is borne out in the work of Macquarrie (1972) who suggests that:

> In the existentialist view there are always loose ends. Our experience and our knowledge are always incomplete and fragmentary: only a divine Mind, if there is one, could know the world as a whole – and perhaps even for such a Mind there would be gaps and discontinuities.
>
> (Macquarrie, 1972, p. 13)

Macquarrie's words put into perspective the relentless search for answers to always be able to draw 'loose ends' smoothly together, to establish epistemological insights into initially understanding and then operationalizing the complex concept of sustainability – initially, in everyday life, but not least in planning processes that seek to help organize those everyday spaces for people.

In Chapter 2 we cited the work of Morse and Bell (2008) who argued, perhaps consistently with Macquarrie's views, that we should adopt positions of honesty, humility and reflection in how far loose ends of data and positivistic assessments of the 'real world' can accurately measure progress towards sustainability. In a socially constructed world, the challenge of the grand narratives of sustainability will remain constant, for ours and future generations, to respond in our everyday lives with both actions or in applying the philosophy of 'wu wei' inaction. In the words of Tempest (Chapter 5, this volume) 'planning is as much about influencing behavioural change as it is about technical solutions' and for Wiertsema (1995) the paths to sustainability are diverse.

BIBLIOGRAPHY

Boddy, M. and Parkinson, M. (eds) (2004) *City Matters: Competitiveness, Cohesion and Urban Governance*. The University of Chicago Press, Chicago, Illinois.
Macquarrie, J. (1972) *Existentialism: An Introduction, Guide and Assessment*. Penguin Books, London.
Morse, S. and Bell, S. (2008) *Sustainability Indicators: Measuring the Immeasurable?*, 2nd edn. Earthscan Publications, London.
Wiertsema, W. (1995) Paths to sustainability. In: Goldsmith, E., Khor, M., Norberg-Hodge, H. and Shiva, V. (eds) *The Future of Progress: Reflections on Environment and Development*. Green Books Ltd, Dartington, UK, pp. 172–178.

Index